Sustainability, Circular Economy and Waste Recycling: Advances in Materials Research

Sustainability, Circular Economy and Waste Recycling: Advances in Materials Research

Editor

Daniela Fico

Basel • Beijing • Wuhan • Barcelona • Belgrade • Novi Sad • Cluj • Manchester

Editor
Daniela Fico
National Research
Council—Institute of
Heritage Science (CNR-ISPC)
Lecce
Italy

Editorial Office
MDPI AG
Grosspeteranlage 5
4052 Basel, Switzerland

This is a reprint of articles from the Special Issue published online in the open access journal *Materials* (ISSN 1996-1944) (available at: https://www.mdpi.com/journal/materials/special_issues/89Y7749MJ7).

For citation purposes, cite each article independently as indicated on the article page online and as indicated below:

Lastname, A.A.; Lastname, B.B. Article Title. *Journal Name* **Year**, *Volume Number*, Page Range.

ISBN 978-3-7258-1749-8 (Hbk)
ISBN 978-3-7258-1750-4 (PDF)
doi.org/10.3390/books978-3-7258-1750-4

Cover image courtesy of Daniela Fico

Contents

About the Editor

Daniela Fico

Daniela Fico obtained her PhD in Analytical Chemistry in 2015 at the Department of Cultural Heritage of the University of Salento (Lecce, Italy). From 2016 to 2020, she worked at the Department of Cultural Heritage (Lecce, Italy) dealing with the morphological and chemical characterization of materials, the development of innovative non-toxic and environmentally friendly coatings, and the development of innovative analytical protocols for the analysis of organic compounds using spectroscopic and chromatographic techniques, provenance studies, and statistical data processing. From 2021 to 2022, she worked at the Department of Engineering for Innovation of the University of Salento (Lecce, Italy), carrying out research activities related to material processing up to complete characterization (morphological–structural, chemical, physical–mechanical, rheological analysis), by means of a multidisciplinary approach (differential scanning calorimetry, mechanical analysis, permeability, colorimetry, modelling and 3D printing, etc.), and to the development of innovative materials (polymers, composites, biomaterials), with particular attention given to topics such as Sustainability and Circular Economy. Daniela Fico is currently working at the National Research Council-Institute of Heritage Science (CNR-ISPC, Lecce, Italy), and she is mainly involved in diagnostic and chemical analyses of Cultural Heritage, the study of manufacturing processes, alterations, conservation, and preservation, through the use of state-of-the-art instrumentation for advanced diagnostics and archaeometry (mapping and imaging spectroscopies, high-resolution mass spectrometry). She has participated in national and international projects, has supervised several theses, and is the author of several scientific articles in international Scopus-indexed journals. She currently serves as a Guest Editor for MDPI publishing house and as a reviewer for several international journals.

Preface

Today, ecological problems have led to the limited production of plastics and non-biodegradable materials and their replacement by materials with a low environmental impact. Biodegradable macromolecules and their composites are desirable candidates for a wide range of applications to overcome the difficulties of waste disposal. Moreover, one of the most crucial trends in current research on the development of new materials is associated with the use of raw waste materials or industrial by-products. This approach takes into account not only ecological issues, but also economic ones, since recycled and waste materials are significantly cheaper than virgin raw materials and their use produces low-cost end products. The substitution in industries, of raw materials that must be disposed of, by-products and waste with renewable and recyclable raw materials, constitutes an important transition to sustainable development and the circular economy: topics that many countries have already introduced into their environmental agenda through the creation of specific legislation.

This Reprint aims to highlight advanced research on the development of new eco-friendly materials and new technologies for sustainability and the circular economy.

Topics discussed by several experts in the field include, but are not limited to, the following:
• Biopolymers and biocomposites from natural raw materials;
• Innovative materials from recycled waste and industrial by-products;
• Synthesis, preparation and processing, and applications;
• Characterization, properties and potential of new biodegradable and eco-friendly materials;
• Studies on durability and biodegradability under different conditions and environments;
• Life cycle assessment of new materials;
• The circular economy, sustainability, and innovative and green materials.

Daniela Fico
Editor

Article

Bio-Composite Filaments Based on Poly(Lactic Acid) and Cocoa Bean Shell Waste for Fused Filament Fabrication (FFF): Production, Characterization and 3D Printing

Daniela Fico [1,2], Daniela Rizzo [3], Valentina De Carolis [1] and Carola Esposito Corcione [1]

1   Department of Engineering for Innovation, University of Salento, Edificio P, Campus Ecotekne, s.p. 6 Lecce-Monteroni, 73100 Lecce, Italy; daniela.fico@unisalento.it (D.F.); valentina.decarolis@unisalento.it (V.D.C.); carola.corcione@unisalento.it (C.E.C.)
2   Italian National Council of Research-Institute of Heritage Sciences (CNR-ISPC), Campus Ecotekne, 73100 Lecce, Italy
3   Department of Cultural Heritage, University of Salento, Via D. Birago 64, 73100 Lecce, Italy
*   Correspondence: daniela.rizzo@unisalento.it

Abstract: In this study, novel biocomposite filaments incorporating cocoa bean shell waste (CBSW) and poly(lactic acid) (PLA) were formulated for application in Fused Filament Fabrication (FFF) technology. CBSW, obtained from discarded chocolate processing remnants, was blended with PLA at concentrations of 5 and 10 wt.% to address the challenge of waste material disposal while offering eco-friendly composite biofilaments for FFF, thereby promoting resource conservation and supporting circular economy initiatives. A comprehensive analysis encompassing structural, morphological, thermal, and mechanical assessments of both raw materials and resultant products (filaments and 3D printed bars) was conducted. The findings reveal the presence of filler aggregates only in high concentrations of CBSW. However, no significant morphological or thermal changes were observed at either CBSW concentration (5 wt.% and 10 wt.%) and satisfactory printability was achieved. In addition, tensile tests on the 3D printed objects showed improved stiffness and load resistance in these samples at the highest CBSW concentrations. In addition, to demonstrate their practical application, several 3D prototypes (chocolate-shaped objects) were printed for presentation in the company's shop window as a chocolate alternative; while retaining the sensory properties of the original cocoa, the mechanical properties were improved compared to the base raw material. Future research will focus on evaluating indicators relevant to the preservation of the biocomposite's sensory properties and longevity.

Keywords: waste; cocoa shell; 3D printing; fused filament fabrication; PLA; circular economy

Citation: Fico, D.; Rizzo, D.; De Carolis, V.; Esposito Corcione, C. Bio-Composite Filaments Based on Poly(Lactic Acid) and Cocoa Bean Shell Waste for Fused Filament Fabrication (FFF): Production, Characterization and 3D Printing. *Materials* **2024**, *17*, 1260. https://doi.org/10.3390/ma17061260

Academic Editor: Young-Hag Koh

Received: 31 January 2024
Revised: 28 February 2024
Accepted: 6 March 2024
Published: 8 March 2024

1. Introduction

Respect for the environment and sustainability are issues of great concern today due to the growing problem of the disposal of industrial production waste. The circular economy (CE), a sustainable system that aims to reduce the consumption of raw materials and the production of waste by reintegrating them into a closed production cycle, underpins this research effort [1,2]. Disposal poses a significant challenge in the food industry, exacerbated by factors such as population growth. In particular, the escalating global demand for chocolate, a popular foodstuff in confectionery around the world, has led to a dramatic increase in the production of cocoa bean shell waste (CBSW) [3]. CBSW (100% cocoa bean shell) is typically repurposed as organic fertilizer, animal feed, or fuel, albeit with adverse environmental consequences [4]. In very few studies, CBSW is used as a renewable resource and as an additive to polymer matrix such as poly(lactic acid), PLA, which is known for its versatility and suitability in 3D printing with FFF technology [5,6]. Among the various methods for recycling waste materials, additive manufacturing (AM), or 3D printing, which facilitates the production of complex prototypes from computer-aided design (CAD) models while minimizing material waste [7,8], may be one of the most promising techniques. Its versatility extends to various industries, including advanced medicine and food processing, aligning with the transition from a linear to a circular economy model [9–11]. Previous research has demonstrated AM's potential to reduce waste production, enhance product value, and promote sustainability [12–14]. Studies by Sanyang et al. [15], Papadopoulou et al. [16], and Tran et al. [17] have explored the use of cocoa waste in composite films and biocomposites, showcasing its potential as a natural filler with ecological benefits. Additionally, Leonard and Berrio [18] investigated PLA-based filaments incorporating cocoa husk for biomedical applications, noting improvements in compressive strength without altering thermal properties [19,20]. Andres J. Garcia-Brand et al. [21] pioneered a method for utilizing CBSW as reinforcement to enhance the bioactivity of synthetic polymeric materials in biomedical research. TGA and DSC thermograms unveiled the feasibility of employing various processing techniques, such as compression molding, extrusion, and potentially 3D printing. However, they did not explore the utilization of any 3D printing technology for fabricating 3D models using composite materials. In this study, biofilaments comprising PLA and CBSW at varying concentrations were developed and used to produce 3D models by FFF, offering a novel approach to waste recycling and reuse through FFF printing. The findings highlight the feasibility of integrating waste materials into the production cycle while emphasizing the importance of sustainable practices in modern manufacturing [22–25]. Based on this scenario, the aim of this study was to develop and characterize biocomposite filaments made from poly(lactic acid) and 5–10 wt.% of cocoa bean shell waste for FFF printing technology, from a CE perspective. The developed green filaments were subjected to morphological–structural, thermal and mechanical analyses. For the first time, the authors tested the feasibility of the FFF printing process to produce new 3D objects from waste materials, with the aim of permitting their integration into the same material production cycle (company-to-company). The work highlights the importance of sustainable practices in modern manufacturing while using solvent-free, inexpensive and easy-to-use methods.

2. Materials and Methods

The technical specifications of the materials utilized for the creation of composite biofilaments, manufacturing methodologies, and analytical techniques for structural, morphological, thermal, and mechanical characterization are outlined as follows.

2.1. Materials

The Ingeo 4043D poly(lactic acid) (NatureWorks LLC, Blair, NE, USA) employed as the polymer matrix for fabricating the composite filaments was obtained in pellet form. It possesses a density of 1.24 g cm^{-3} and a melt flow index (MFI) of 6 g/10 min at a temperature of 210 °C. PLA was utilized both independently to produce a virgin filament (100PLA) and as a polymer matrix for creating two biofilaments incorporating CBSW as fillers at 5 wt.% and 10 wt.%, respectively. Initially, the PLA pellets, with a diameter of approximately 5 mm (manufacturer's data sheet), were stored in an oven at 60 °C for 24 h before being reduced into particles using the Retsch ZM 100 Ultracentrifugal mill (Retsch GmbH, Haan, Germany) equipped with a sieve with a mesh diameter of 0.75 mm.

The cocoa bean shell waste (CBSW) utilized in the production of the new composite filaments originates from the manufacturing residues of the company "Maglio Arte Dolciaria S.R.L." (Maglie, Lecce, Apulia, Italy). Supplied by the company in the form of coarse-grained and heterogeneous powder, it exhibits the characteristic brown color of cocoa and a robust chocolate-like odor. Prior to utilization, the CBSW was subjected to storage in an oven at 60 °C for 24 h and subsequently reduced into particles using the Retsch ZM 100 Ultracentrifugal mill (Retsch GmbH, Haan, Germany) equipped with a sieve with a mesh diameter of 0.25 mm.

2.2. Production of Biofilaments for FFF

The PLA and CBSW powders were initially combined manually at room temperature before being introduced into the extrusion chamber of the 3Devo Composer 450 Filament Maker single-screw extruder (Utrecht, The Netherlands,), as reported in Figure 1.

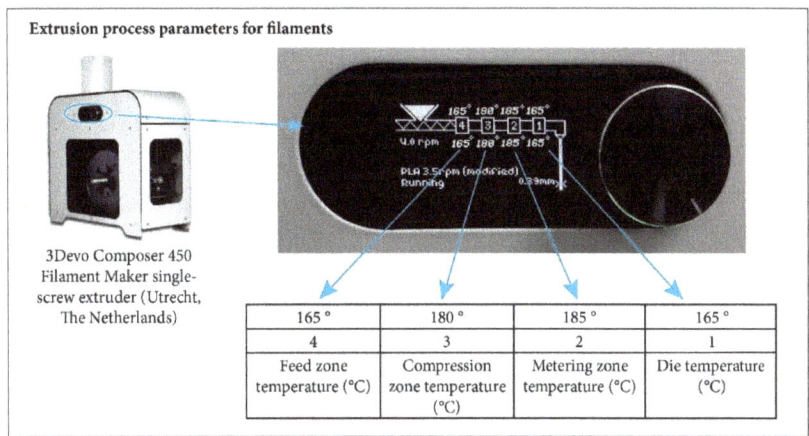

Figure 1. 3Devo Composer 450 Filament Maker single-screw extruder with indication of different temperature zones.

The parameters used to produce the filaments by the extrusion process are outlined in Table 1, while the labels and compositions of the materials used and developed are listed in Table 2.

Table 1. Extrusion process parameters for filaments.

Extrusion Process Parameters	100PLA	95PLA/5CBSW	90PLA/10CBSW
Screw speed (rpm)	3.5	4	4
Fan cooling speed (%)	0	30	30
Feed zone temperature (°C)	170	165	165
Compression zone temperature (°C)	185	180	180
Metering zone temperature (°C)	190	185	185
Die temperature (°C)	200	165	165

Table 2. Raw Materials, biofilament and 3D printed sample labels and weight composition (wt.%).

Label	Type	Composition (wt.%)
PLA	Raw material	100 PLA Ingeo 4043D pellet
CBSW	Raw material	100 Cocoa bean shell waste powder
100PLA	Filament	100 PLA Ingeo 4043D
95PLA/5CBSW	Filament	95 PLA Ingeo 4043D and 5 Cocoa bean shell waste
90PLA/10CBSW	Filament	90 PLA Ingeo 4043D and 10 Cocoa bean shell waste
100PLA_3D	3D sample	100 PLA Ingeo 4043D
95PLA/5CBSW_3D	3D sample	95 PLA Ingeo 4043D and 5 Cocoa bean shell waste
90PLA/10CBSW_3D	3D sample	90 PLA Ingeo 4043D and 10 Cocoa bean shell waste

2.3. 3D Printing

Three-dimensional samples were printed on the Creality CP-01 printer (Creality, London, UK) in accordance with the European standard used for the tensile tests ISO 527-2(2012) [26]. The following printing parameters were used: extrusion temperature 200 °C, plate temperature 50 °C, print speed 50 mm/s, fill 100% and fan speed 100%.

The CAD model was developed using Fusion 360 software (Autodesk, San Rafael, CA, USA) and converted to a G-code file using Cura software, Ultimaker Cura version 5.1.1. (Ul-timaker B.V., Utrecht, The Netherlands). Finally, after studying the structural, thermal and mechanical properties of the biocomposites, some design objects (some traditional chocolate shapes) were printed for display in the window of the company that supplied the cocoa waste. The CAD models were developed using Fusion 360 software (Autodesk, San Rafael, CA, USA) and converted to Gcode files using Cura software, (Ultimaker B.V., Utrecht, The Netherlands). Prints were made with the Creality CP-01 printer (Creality, London, UK) with the following operating parameters: extrusion temperature 190 °C, plate temperature 50 °C, print speed 60 mm/s and infill 20%.

Three-dimensional objects (chocolate shaped objects) were also printed to demonstrate the printability of complex shapes with the recycled filaments. For this purpose, the BQ Hephestos 2 printer (BQ, Barcelona, España) was used with the following printing parameters: extrusion temperature 200 °C, plate temperature 0 °C, printing speed 60 mm/s and infill 20%.

A comprehensive list of all materials employed in this study is provided in Table 2.

2.4. Methods

The morphological characterization of the raw materials and the biofilaments produced was carried out using a scanning electron microscope (SEM), model Zeiss E Evo 40 (Oberkochen, Germany).

Fourier transform infrared spectroscopy (FTIR) analyses were performed on the raw materials (PLA and CBSW powders) to perform a preliminary chemical structural characterization. FTIR spectra were obtained on KBr pellets using a JASCO FT/IR 6300 spectrometer (Easton, MD) with a resolution of 4 cm^{-1}, setting up 64 scans in the region between 4000 and 600 cm^{-1}. Five spectra were considered for each replicate sample.

XRD measurements of the raw materials and biofilaments were performed with a Rigaku Ultima + diffractometer (Tokyo, Japan) using CuKα radiation (λ = 1.5418 Å) in the step scan mode, recorded in the 2θ range from 2–60°, with a step size of 0.02° and a step duration of 0.5 s. Five spectra were considered for each replicate sample.

DSC analysis (Mettler Toledo DSC1 StareSystem) was performed on raw materials, filaments and 3D printed rods to investigate thermal properties by measuring the glass transition temperature (T_g), crystallization temperature (T_c) and melting temperature (T_m). The analyses were performed over a temperature range of 25 °C to 200 °C at a heating rate of 10 °C/min.

According to the ISO 527-2(2012) standard [26], tensile tests were performed on the 3D printed sample using a Lloyd LR5K dynamometer (Lloyd Instruments Ltd., Bognor Regis, UK) at a strain rate of 1 mm/min. The grips used in these tests were wedge-type grips, the distance between grips was 58 mm, and the load cell was 5 kN. The dimensions of the specimen refer to the 1BA type, according to the standard. Five replicates were performed for each specimen.

Magnified images of the sections of the 3D printed tensile samples were obtained using the Dino Lite digital microscope instrument (AnMo Electronics Corporation, New Taipei City, Taiwan) to understand the morphology and adhesion of the different deposited layers.

3. Results and Discussion

3.1. Characterization of Raw Materials

The morphological–structural, chemical and thermal features of all raw materials used to produce biofilaments were investigated. The results of the SEM, FTIR, XRD and DSC analyses of the raw materials are shown below.

After the milling process, SEM images (Figure 2A,B) at high magnifications (100× and 1000×) of the neat PLA particles show an irregular, square shape and a diameter up to about 500 μm. The cocoa bean shell waste (CBSW) was supplied in the form of a coarse-grained, inconstant powder, after grinding the cocoa bean shell. Furthermore, as visible from the SEM image (Figure 2C,D), the CBSW particle size varies from about 5 μm to about 200 μm.

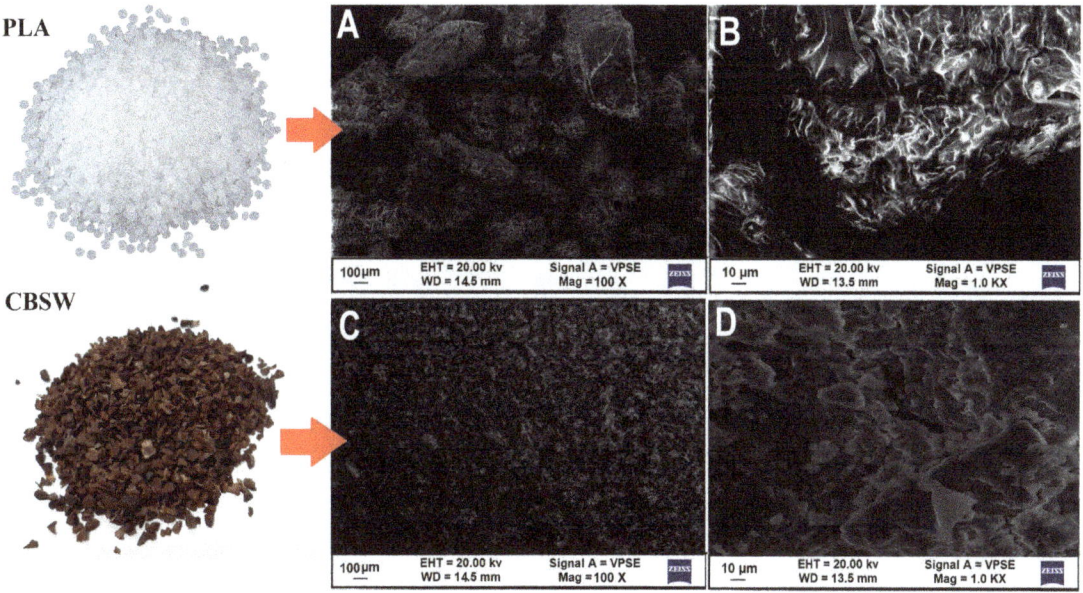

Figure 2. SEM images of raw materials: (**A**) PLA pellet magnification 100× and (**B**) magnification 1K×; (**C**) cocoa bean shell waste magnification 100× and (**D**) magnification 1K×.

The chemical characterization of the raw materials was carried out using FTIR spectroscopy. The infrared spectra of PLA powder and CBSW are shown in Figure 3.

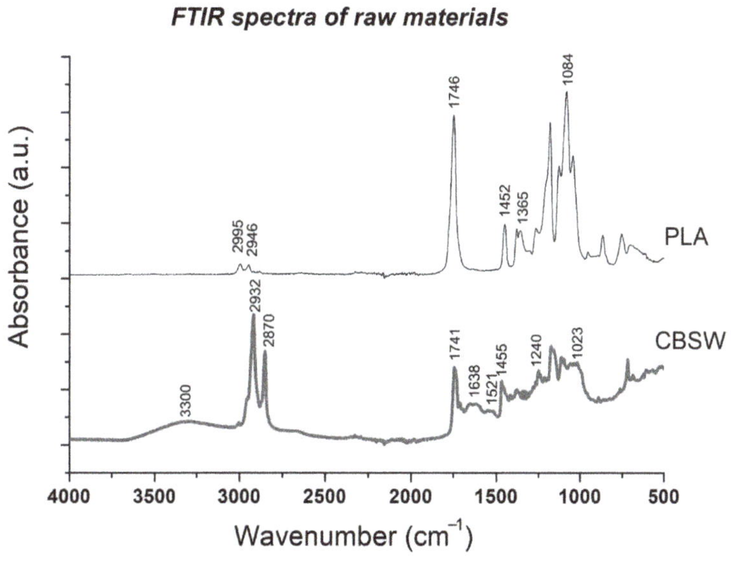

Figure 3. FTIR spectra of PLA pellet and cocoa bean shell waste and main infrared peaks.

The infrared spectrum of the PLA shows asymmetrical and symmetrical $-CH_3$ and $-CH_3$ stretching frequencies at 2995 cm^{-1} and 2946 cm^{-1}, respectively, and a characteristic peak of C=O stretching at about 1746 cm^{-1} [27]. The peaks at 1452 cm^{-1} and 1365 cm^{-1} are associated with the asymmetric and symmetric bending vibrations of the $-CH_3$ groups, respectively, while the band at about 1084 cm^{-1} corresponds to the C–O bond [27].

In the FTIR spectrum of the CBSW, the band present at around 3300 cm^{-1} is associated with OH groups, while the peaks at 2932 cm^{-1} and 2870 cm^{-1} correspond to the asymmetric and symmetric stretching vibrations of the C–H groups of cellulose and hemicellulose. The infrared peak at 1741 cm^{-1} corresponds to the C=O stretching of saturated esters, while the three infrared bands at approximately 1638 cm^{-1}, 1521 cm^{-1} and 1401 cm^{-1} correspond to amide I, II and III, respectively. These bands can be associated with proteins, lignin and polysaccharides present in the biomass. Finally, the peaks at 1240 cm^{-1} and 1023 cm^{-1} are associated with holocellulose biomolecules [28].

In the XRD diffractogram of the PLA (Figure 4A), more defined diffraction peaks emerge at $2\theta = 16.70°$ and at $2\theta = 19.10°$ over a broad amorphous band, assigned to the crystalline plane (110) and (203), respectively [14]. The diffractogram of the CBSW sample (Figure 4A) shows peaks at $2\theta = 3.09°$, 19.5°, 21.20°, 22.02° (002), 23.11° and 24.30°, some of which are characteristic of crystalline polymorphs of cellulose [29,30]. The cocoa bean from which the powder was obtained is composed of about 60 wt.% water and holocellulose (cellulose, hemicellulose, pectin) and then lignin [29] and its XRD diffractogram is very similar to cocoa butter [31].

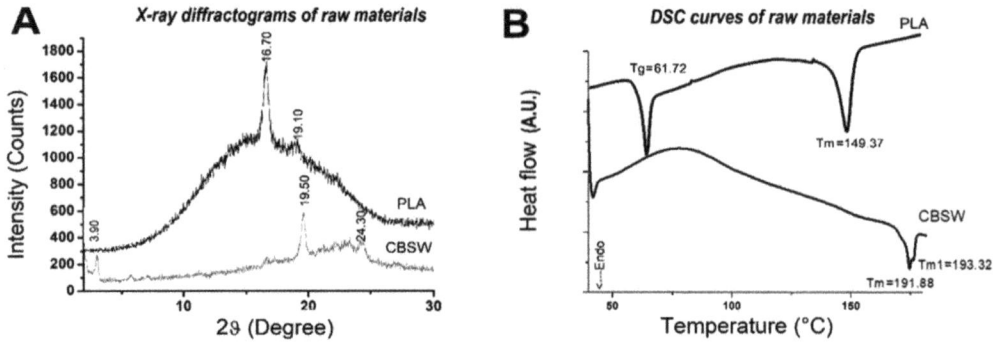

Figure 4. (**A**) XRD diffractograms and (**B**) DSC curve of raw materials.

A DSC analysis was performed on raw materials to investigate thermal properties (Figure 4B). In the DSC curve of poly(lactic acid), the glass transition temperature (T_g) was measured using the inflection point method with the STARe System version 11 software (Mettler Toledo, Milan, Italy) [12] and corresponds for PLA to 61.72 °C. In Figure 4B, it is also observed that the T_g is accompanied by an endothermic peak (relaxation enthalpy) due to the transformation of the polymer from the glassy to the liquid-viscous or rubbery state [32]. The DSC curve also shows a melting peak at temperature T_m of 149.37 and melting enthalpy ΔH_m of 19.90 J/g. In the present investigation, the DSC curves of CBSW samples were also examined. As with PLA, the average values obtained from the five measurements are shown in Figure 4B. The DSC analysis was carried out in the maximum temperature range commonly used in FFF printing with PLA filaments, i.e., between 25 and 200 °C. CBSW is a complex biomass consisting of several components, such as proteins, lignin and polysaccharides, as previously reported. Moreover, similar to woody biomass, the main polysaccharide components, consisting of hemicellulose, cellulose and lignin, manifest maximum endothermal peaks in different ranges according to the scientific literature [33,34]. Therefore, it can be concluded that the double endothermic peak present in the CBSW-DSC curve only refers to the thermal decomposition of hemicellulose and partially of cellulose.

3.2. Characterization of Biofilaments

SEM analyses were performed on the external surfaces of the composite filaments to investigate surface roughness and on the cross-section of filaments to verify homogeneity (Figure 5). The surface of the composite filament becomes slightly irregular after the addition of CBSW (Figure 5D–F), and some discontinuity areas are observed (Sample 95PLA/5CBSW). This phenomenon becomes more pronounced with the increase in the amount of filler added to the polymer matrix (Sample 90PLA/10CBSW), where the surface appears completely rough (Figure 5G–I). Furthermore, as shown by the cross-sectional images of the two produced FFF biofilaments, the cocoa particles aggregate in some places and these aggregates are even more visible in the filament containing 10 wt.% CBSW. The phenomenon appears to be attributable not only to a different diameter of the particles in the mixture (PLA approximately 500 μm measured by SEM and CBSW 5 μm to 1 mm diameter measured by SEM), but also to a different chemical nature that creates poor interfacial adhesion. This occurrence is already known in the literature for various PLA-based composites and waste fillers with different polarities and may indicate the possibility in the future of using plasticizer agents that improve the compatibility and adhesiveness of the different substances [35].

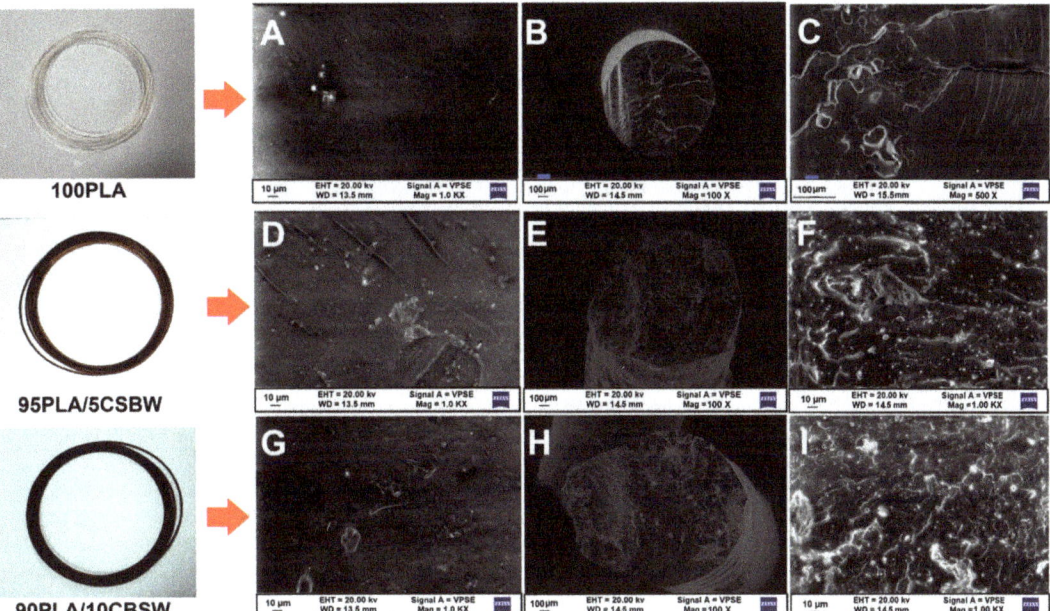

Figure 5. SEM images of filaments: (**A**) 100PLA external filament surface magnification 1.0K×, (**B**) section magnification 100× (**C**) section magnification 500×; (**D**) 95PLA/5CSBW external filament surface magnification 1.0K×, (**E**) section magnification 100×, (**F**) section magnification 1.0K×; (**G**) 90PLA/10CSBW external filament surface magnification 1.0K×, (**H**) section magnification 100×, (**I**) section magnification 1.0K×.

XRD spectra of filaments show only amorphous bands and a remarkable decrease in the intensity of diffraction peaks relative to the raw materials (Figure 6A). However, no major differences are shown between the 100PLA filament and the 95PLA/5CBSW and 90PLA/10CBSW filaments, indicating that the reticular parameters of the PLA crystal structure remained unchanged after coextrusion of the polymer with CBSW. Similar observations were also reported by Tran et al. 2017 for poly(ε-caprolactone) matrix biocomposites [17]. Therefore, to obtain information on the variation in the thermal properties of composite filaments compared to pure materials, a DSC analysis was performed (Figure 6B). The DSC thermograms of the PLA-based biofilaments or PLA and CBSW-based filaments are shown in Figure 5B. The glass temperatures (T_g) of all materials were measured using the inflection point: the T_g coincides with the point at which the second derivative is equal to zero. The T_g measured for the 100PLA sample was 59.19 °C, a value similar to that reported in the literature [1] and slightly lower than that of the starting pellet.

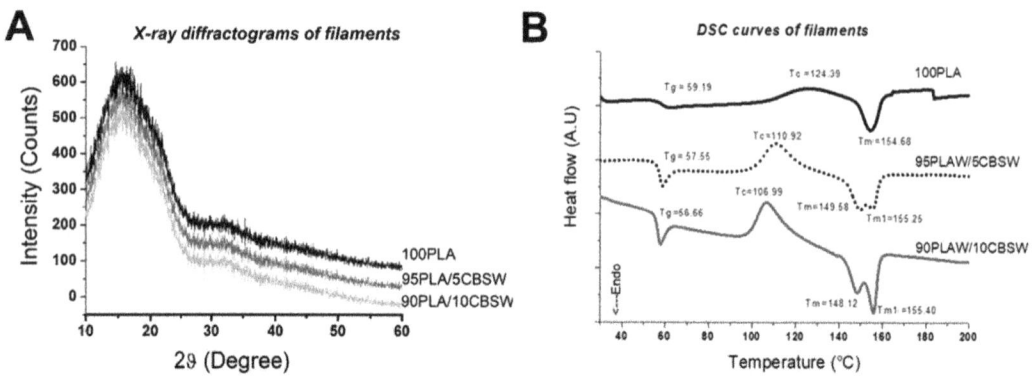

Figure 6. (**A**) XRD diffractograms and (**B**) DSC curves of filaments.

Composite filaments show a lower T_g than the PLA filament. This suggests that the cocoa particles interacted with the PLA on a molecular scale. However, the main differences due to this interaction process between the polymer chains and the filler are found in the crystallization (T_c) and melting (T_m) temperatures. The crystallization temperature T_c in the 95PLA/5CBSW and 90PLA/10CBSW filaments compared to the 100PLA filament having a T_c of 124.39 °C decreases to 110.92 °C and 106.99 °C, respectively. Furthermore, the melting process seems to be particularly influenced by the presence of the CBSW filler, which leads to the development of a double melt peak (T_m and T_{m1}) in composite biofilaments, which is absent in 100PLA (Figure 6B). Similar behaviour and the presence of the double melting peak were observed by the authors in biocomposites based on polylactic acid and olive wood waste, indicating a different behaviour during the phase transition between the polymer and the organic filler [35].

Leonard and Berrio, on the other hand, highlighting the same characteristic in the PLA-based filaments and cocoa bean shells they have developed and which have not undergone any chemical treatment, attribute the existence of two peaks during melting to the different degradation temperatures of hemicellulose and cellulose present in the filler [18]. This last interpretation seems to be in agreement with the DSC curve of CBSW obtained in our study and shown in Figure 4B.

3.3. Characterization of 3D Samples

Three-dimensional specimens for tensile testing were produced by FFF printing in accordance with ISO 527-2(2012) [26]. The overall average results are shown in Figure 7. The results of the pure PLA samples are consistent with those reported in the literature [1].

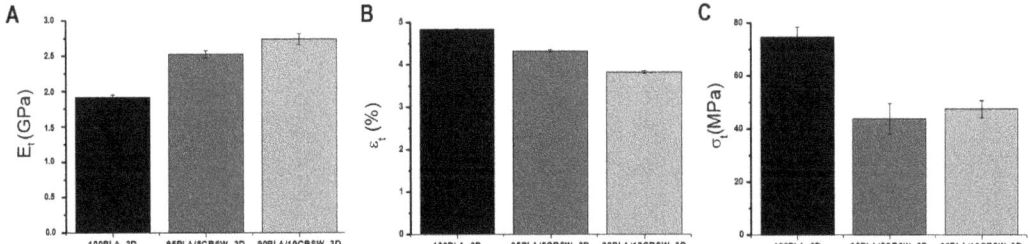

Figure 7. Mechanical properties from tensile tests of 3D samples: (**A**) Young's modulus E_t (GPa), (**B**) elongation at break ε_t (%), (**C**) maximum strength (MPa).

The Young's modulus (Et) increased in the CBSW-based samples. Indeed, an inevitable consequence of the addition of filler to the polymer matrix is an increase in the viscosity of the polymer melt, which usually depends on the volume fraction of filler and the size and shape of the particles (Figure 7A) [36]. Overall, the presence of filler appears to have hindered the mobility of the polymer chain of the PLA matrix, thereby increasing the stiffness of the 3D sample composite. Furthermore, this increase appears to be linear, as observed by the Et value increasing with filler wt.%. It has been reported in the scientific literature that the addition of the filler actually modifies the polymer phase and the polymer interacts with the filler surface to form an adsorbed polymer interphase [36]. The thickness of this interphase (as well as the value of the Young's modulus) depends on the bond between the polymer and the filler, which in turn depends on the surface area (i.e., the shape and size of the filler particles). Based on these considerations, the different behaviours of the composites could probably be due to a stronger interaction between the polymer and the filler in the 90PLA/10CBSW_3D sample, caused by the larger specific active surface area of the CBSW particles, as well as the higher volume fraction of the filler [36].

The elongation at break εt (%) always decreases with the addition of filler to pure PLA (Figure 7B), in agreement with the literature [17,30], leading to an increase in the fragility of the composites compared to the polymer. Similarly, the value of the maximum strength σt (MPa) decreases in both the 95PLA/5CBSW_3D and 90PLA/10CBSW_3D samples (Figure 7C) with a non-linear trend. This is probably due to a different orientation of the filler particles dispersed in the polymer matrix. Overall, the tensile test results show a lower ductility and a change in the mechanical properties of CBSW based composites; the non-uniform distribution of CBSW particles in the composite structure is the main cause.

The DSC analysis performed on the FFF printed 3D samples is shown in Figure 8. A decrease in glass transition temperature (T_g) is observed for all 3D samples compared to the pure PLA 3D sample. The crystallization (T_c) and melting temperatures (T_m and T_{m1}) of the CBSW-based samples also decrease but remain higher than those measured for the corresponding composite filaments, indicating a greater interaction between the polymer chains and the filler and a higher degree of crystallinity after the printing process. This also explains the increase in Young's modulus of the CBSW-based 3D samples described above. There is also always a double melting peak in 3D samples due to the presence of the filler in the polymer matrix, i.e., different phases.

To better understand the printability of the developed biofilaments, a microscopic analysis was performed along the side surface of the 3D samples. The images obtained by light microscopy are shown in Figure 9 and overall show good adhesion between the 3D layers in both the 100PLA_3D sample (Figure 9A) and in the 95PLA/5CBSW_3D sample (Figure 9B), in contrast to the 90PLA/10CBSW_3D sample (Figure 9C). The addition of CBSW filler to the polymer matrix, which causes the formation of particle aggregates in the filament, causes adhesion problems between the layers in the 90PLA/10CBSW_3D; as in the other 3D samples, the layers are perfectly aligned with each other and do not show swelling, deformation, or detachment; however, the presence of some air voids is observed, and this porosity probably inhibits the mechanical performance of the composite material.

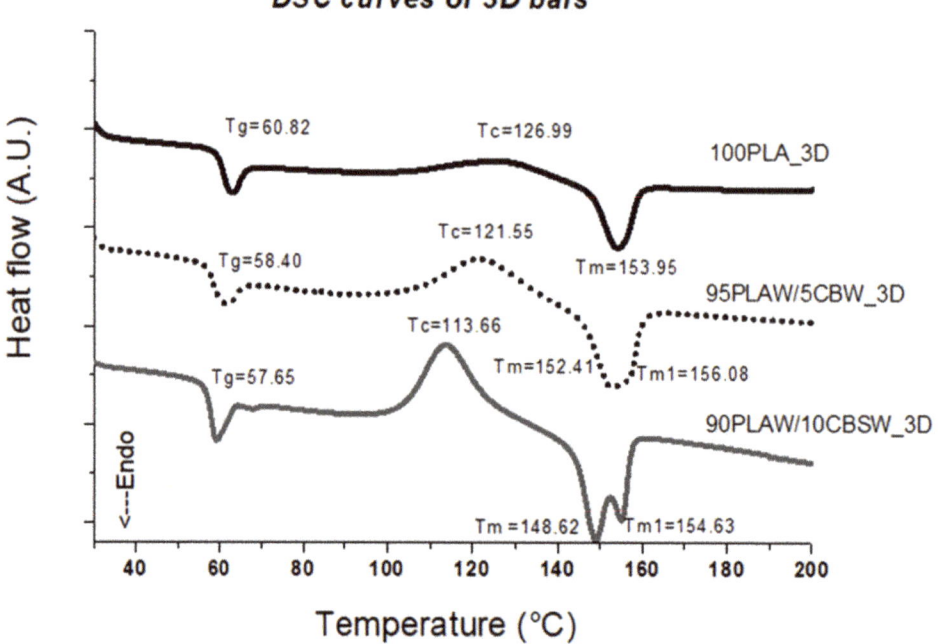

Figure 8. DSC curves of 3D samples.

3.4. From Agro-Industrial Waste to New Objects

In this work, the waste materials, i.e., cocoa bean shell waste CBSW, were supplied as a residue from the chocolate manufacturing process in the form of a coarse-grained, non-homogeneous powder. These are residues from chocolate processing that the company must necessarily discard, with a significant time and economic effort. In this section, the authors want to illustrate an example of one of the many sustainable applications that can be imagined, i.e., reusing recycled materials to create a new object that can be redeployed by the same company that produced the waste. Based on the tested performance of the developed cocoa and poly(lactic acid) biofilaments, the 90PLA/10CBSW filament was selected for 3D printing the chocolate-shaped objects (prototypes) using Fused Filament Fabrication 3D printing technology (Figure 10). Specifically, the authors chose to produce 3D printed objects in the shape of traditional chocolates or other shapes, so that they could be used by the supplier company for display in the shop window or inside the shop as design and furnishing objects, having greater durability and resistance to environmental conditions than real chocolate. The final object after printing preserves the same environmental resistance characteristics as the 3D bars tested in the previous paragraph. Most interestingly is that already the 5 wt.% of cocoa powder (CSBW) product allows the organoleptic characteristics of pure chocolate to be preserved in the 3D object, giving customers the feeling that it is a real product exhibited in a store window.

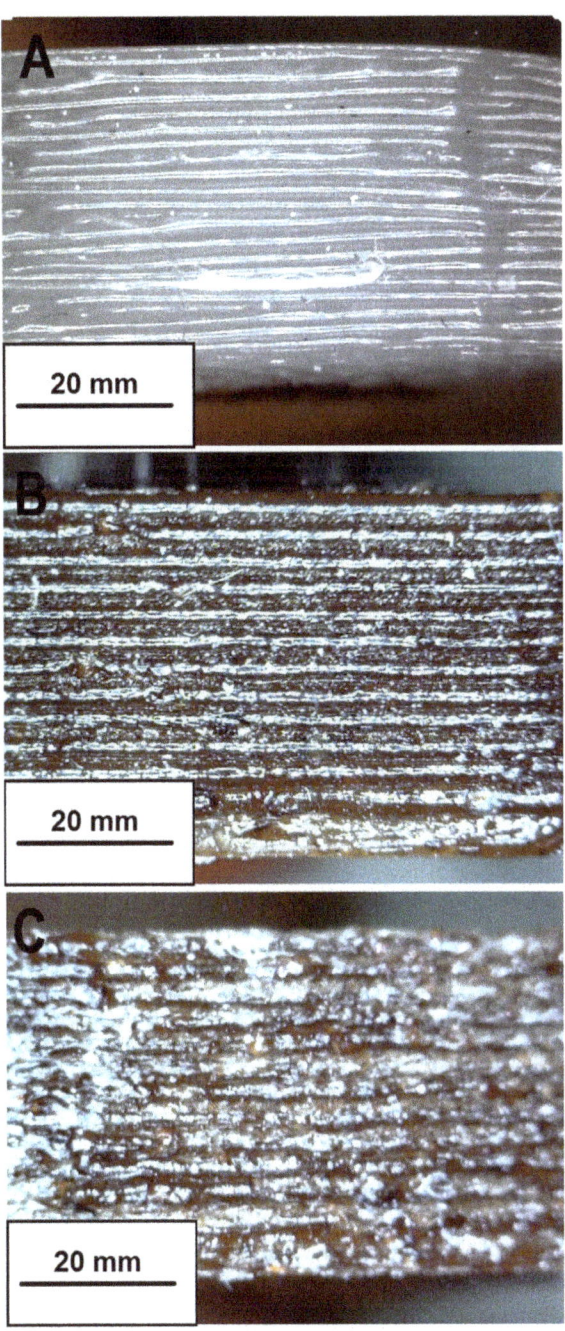

Figure 9. Magnified images (50 X) of the 3D bars: (**A**) printing surface of the 100PLA_3D specimen; (**B**) printing surface of the 95PLA/5CBSW_3D specimen; (**C**) printing surface of the 90PLA/10CBSW_3D specimen.

Figure 10. Manufactured filament (**A**), CAD model created with Rhinoceros version 7 software (Robert McNeel & Associates, Seattle, DC, USA) (**B**) and representative printed samples (**C**).

4. Conclusions

In this study, novel green composite filaments for FFF were developed and characterized, utilizing cocoa bean shell waste and polylactic acid. Structural–morphological, thermal, and mechanical characterization of both raw materials and products (extruded filaments and 3D printed specimens) was comprehensively conducted. Overall, the findings revealed significant filler aggregates at 10 wt.% CBSW concentration in the PLA matrix. However, at both CBSW concentrations (5 wt.% and 10 wt.%), no notable morphological or thermal alterations were observed, and satisfactory printability was achieved. Moreover, tensile tests on the 3D printed objects demonstrated enhanced rigidity in these specimens and better load resistance at the highest CBSW concentrations. Finally, the biofilament exhibiting superior properties (containing the 10 wt.% of CBSW) was utilized to fabricate some simple 3D printed chocolate models, showcasing the efficacy of the proposed approach in successfully recycling agro-industrial waste within the same production company, leading to economic and ecological benefits. Future research endeavors aim to identify biomarkers confirming the longevity of cocoa organoleptic properties post-printing process and investigate the durability of biocomposites under controlled environmental conditions.

Author Contributions: Conceptualization, C.E.C.; software, V.D.C. and D.R.; validation, C.E.C. and D.F.; formal analysis, D.F. and D.R.; data curation, C.E.C., D.F. and D.R.; writing—original draft preparation, C.E.C., D.F. and D.R.; writing—review and editing, C.E.C., D.F. and D.R.; visualization, C.E.C., D.F. and D.R.; supervision, C.E.C., D.F. and D.R. All authors have read and agreed to the published version of the manuscript.

Funding: This research was funded by Ministero dell'Istruzione dell'Università e della Ricerca (project PON R&I 2014–2020, D.M. 1061 del 10/8/2021-37° ciclo-Azione IV.5, Dottorati su Tematiche Green, 2021).

Institutional Review Board Statement: Not applicable.

Informed Consent Statement: Not applicable.

Data Availability Statement: Data are contained within the article.

Acknowledgments: The authors thank the company "Maglio Arte Dolciaria S.R.L." (Maglie, Lecce, Italy) for providing the recycled raw material.

Conflicts of Interest: The authors declare no conflicts of interest.

References

1. Fico, D.; Rizzo, D.; Casciaro, R.; Corcione, C.E. A Review of Polymer-Based Materials for Fused Filament Fabrication (FFF): Focus on Sustainability and Recycled Materials. *Polymers* **2022**, *14*, 465. [CrossRef]
2. Zhang, P.; Wang, Z.; Li, J.; Li, X.; Cheng, L. From materials to devices using fused deposition modeling: A state-of-art review. *Nanotechnol. Rev.* **2020**, *9*, 1594–1609. [CrossRef]
3. Verter, N. Cocoa export performance in the world's largest producer. *Bulg. J. Agric. Sci.* **2016**, *22*, 713–721.
4. Mendoza-Meneses, C.J.; Feregrino-Pe'rez, A.A.; Gutie´rrez-Antonio, C. Potential Use of Industrial Cocoa Waste in Biofuel Production. *J. Chem.* **2021**, *2021*, 3388067. [CrossRef]

5. Morettini, G.; Palmieri, M.; Capponi, L.; Landi, L. Comprehensive characterization of mechanical and physical properties of PLA structures printed by FFF-3D-printing process in different directions. *Progr. Addit. Manuf.* **2022**, *7*, 1111–1122. [CrossRef]
6. Pang, X.; Zhuang, X.; Tang, Z.; Chen, X. Polylactic acid (PLA): Research, development and industrialization. *Biotechnol. J.* **2010**, *5*, 1125–1136. [CrossRef]
7. Ivanova, O.; Williams, C.; Campbell, T. Additive manufacturing (AM) and nanotechnology: Promises and challenges. *Rapid Prototyp. J.* **2013**, *19*, 353–364. [CrossRef]
8. González-Henríquez, C.M.; Sarabia-Vallejos, M.A.; Rodriguez-Hernandez, J. Polymers for additive manufacturing and 4Dprinting: Materials, methodologies and biomedical applications. *Prog. Polym. Sci.* **2019**, *94*, 57–116. [CrossRef]
9. Wang, X.; Jiang, M.; Zhou, Z.; Gou, J.; Hui, D. 3D printing of polymer matrix composites: A review and prospective. *Compos. Part B Eng.* **2017**, *110*, 442–458. [CrossRef]
10. Daminabo, S.C.; Goel, S.; Grammatikos, S.A.; Nezhad, H.Y.; Thakur, V.K. Fused deposition modeling-based additive manufacturing (3D printing): Techniques for polymer material systems. *Mater. Today Chem.* **2020**, *16*, 100248. [CrossRef]
11. Wu, H.; Fahy, W.P.; Kim, S.; Kim, H.; Zhao, N.; Pilato, L.; Kafi, A.; Bateman, S.; Koo, J.H. Recent developments in polymers/polymer nanocomposites for additive manufacturing. *Prog. Mater. Sci.* **2020**, *111*, 100638. [CrossRef]
12. Fico, D.; Rizzo, D.; De Carolis, V.; Montagna, F.; Esposito Corcione, C. Sustainable polymer composites manufacturing through 3D printing technologies by using recycled polymer and filler. *Polymers* **2022**, *14*, 3756. [CrossRef]
13. Khalid, M.Y.; Arif, Z.U. Novel biopolymer-based sustainable composites for food packaging applications: A narrative review. *Food Packag. Shelf Life* **2022**, *33*, 100892. [CrossRef]
14. Fico, D.; Rizzo, D.; De Carolis, V.; Montagna, F.; Palumbo, E.; Esposito Corcione, C. Development and characterization of sustainable PLA/Olive wood waste composites for rehabilitation applications using Fused Filament Fabrication (FFF). *J. Build. Eng.* **2022**, *56*, 104673. [CrossRef]
15. Sanyang, M.L.; Sapuan, S.M.; Haron, M. Effect of cocoa pod husk filler loading on tensile properties of cocoa pod husk/polylactic acid green biocomposite films. *AIP Conf. Proc.* **2017**, *1891*, 1.
16. Papadopoulou, E.L.; Paul, U.C.; Tran, T.N.; Suarato, G.; Ceseracciu, L.; Marras, S.; d'Arcy, R.; Athanassiou, A. Sustainable Active Food Packaging from Poly(lactic acid) and Cocoa Bean Shells. *ACS Appl. Mater. Interfaces* **2019**, *11*, 34. [CrossRef]
17. Tran, T.N.; Bayer, I.S.; Heredia-Guerrero, J.A.; Frugone, M.; Lagomarsino, M.; Maggio, F.; Athanassiou, A. Cocoa Shell Waste Biofilaments for 3D Printing Applications. *Macromol. Mater. Eng.* **2017**, *302*, 1700219. [CrossRef]
18. Leonard, J.; Berrio, V. *Desarrollo y Caracterización de un Material Compuesto con Fibra de Cacao Para la Producción de Filamento Para Impresión 3D*; Departamento de Ingeniería Química y de Alimentos, Universidad de los Andes: Bogotá, Colombia, 2021.
19. Shanmugam, V.; Pavan, M.V.; Babu, K.; Karnan, B. Fused deposition modeling based polymeric materials and their performance: A review. *Polym. Compos.* **2021**, *42*, 5656–5677. [CrossRef]
20. Gardan, G. *Additive Manufacturing Technologies in Additive Manufacturing Handbook*; CRC Press: Boca Raton, FL, USA, 2017; p. 20.
21. Garcia-Brand, A.J.; Morales, M.A.; Hozman, A.S.; Ramirez, A.C.; Cruz, L.J.; Maranon, A.; Muñoz-Camargo, C.; Cruz, J.C.; Porras, A. Bioactive Poly(lactic acid)–Cocoa Bean Shell Composites for Biomaterial Formulation: Preparation and Preliminary In Vitro Characterization. *Polymers* **2021**, *13*, 3707. [CrossRef]
22. Yu, W.; Shi, J.; Qiu, R.; Lei, W. Degradation Behavior of 3D-Printed Residue of Astragalus Particle/Poly(Lactic Acid) Biocomposites under Soil Conditions. *Polymers* **2023**, *15*, 1477. [CrossRef]
23. De Kergariou, C.; Saidani-Scott, H.; Perriman, A.; Scarpa, F.; Le Duigou, A. The influence of the humidity on the mechanical properties of 3D printed continuous flax fibre reinforced poly(lactic acid) composites. *Compos. Part A Appl. Sci. Manuf.* **2022**, *155*, 106805. [CrossRef]
24. Depuydt, D.; Balthazar, M.; Hendrickx, K.; Six, W.; Ferraris, E.; Desplentere, F.; Ivens, J.; Van Vuure, A.W. Production and characterization of bamboo and flax fiber reinforced polylactic acid filaments for fused deposition modeling (FDM). *Polym. Compos.* **2019**, *40*, 5. [CrossRef]
25. Garcia-Brand, A.J.; Maranon, A.; Muñoz-Camargo, C.; Cruz, J.C. Potential Bone Fillers Based on Composites of Cocoa Bean Shells and Poly(Lactic Acid): Compression Molding Manufacturing. In Proceedings of the IEEE 2nd International Congress of Biomedical Engineering and Bioengineering (CI-IB&BI), Bogota D.C., Colombia, 13–15 October 2021.
26. *UNI EN ISO 527-2:2012*; Materie Plastiche—Determinazione Delle Proprietà a Trazione—Parte 2: Condizioni di Prova per Materie Plastiche per Stampaggio ed Estrusione. ISO: Geneva, Switzerland, 2012.
27. Chieng, B.W.; Ibrahim, N.A.; Yunus, W.M.Z.W.; Hussein, M.Z. Poly (lactic acid)/Poly (ethylene glycol) Polymer Nanocomposites: Effects of Graphene Nanoplatelets. *Polymers* **2014**, *6*, 93–104. [CrossRef]
28. Hoyos, C.G.; Márquez, P.M.; Vélez, L.P.; Guerra, A.S.; Eceiza, A.; Urbina, L.; Vélasquez-Cock, J.; Rojo, P.G.; Acosta, L.V.; Zuluaga, R. Cocoa shell: An industrial by-product for the preparation of suspensions of holocellulose nanofibers and fat. *Cellulose* **2020**, *27*, 10873–10884. [CrossRef]
29. Lionetto, F.; Del Sole, R.; Cannoletta, D.; Vasapollo, G.; Maffezzoli, A. Monitoring Wood Degradation during Weathering by Cellulose Crystallinity. *Materials* **2012**, *5*, 10. [CrossRef]
30. Puglia, D.; Dominici, F.; Badalotti, M.; Santulli, C.; Kenny, J.M. Tensile, Thermal and Morphological Characterization of Cocoa Bean Shells (CBS)/Polycaprolactone-Based Composites. *J. Renew. Mater.* **2016**, *4*, 3.
31. Le Révérend, B.J.D.; Fryer, P.J.; Coles, S.; Bakalis, S. A Method to Qualify and Quantify the Crystalline State of Cocoa Butter in Industrial Chocolate. *J. Am. Oil Chem. Soc.* **2010**, *87*, 3. [CrossRef]

32. Reignier, J.; Tatibouët, J.; Gendron, R. Effect of Dissolved Carbon Dioxide on the Glass Transition and Crystallization of Poly (lactic acid) as Probed by Ultrasonic Measurements. *J. Appl. Polym. Sci.* **2009**, *112*, 1345–1355. [CrossRef]
33. Bryś, A.; Bryś, J.; Ostrowska-Lige̜za, E.; Kaleta, A.; Górnicki, K.; Głowacki, S.; Koczoń, P. Wood biomass characterization by DSC or FT-IR spectroscopy. *J. Therm. Anal. Calorim.* **2016**, *126*, 27–35. [CrossRef]
34. Tsujiyama, S.; Miyamori, A. Assignment of DSC thermograms of wood and its components. *Thermochim. Acta* **2000**, *351*, 177–181. [CrossRef]
35. Carichino, S.; Scanferla, D.; Fico, D.; Rizzo, D.; Ferrari, F.; Jordá-Reolid, M.; Martínez-García, A.; Esposito Corcione, C. Poly-Lactic Acid-Bagasse Based Bio-Composite for Additive Manufacturing. *Polymers* **2023**, *15*, 4323. [CrossRef] [PubMed]
36. De Armitt, C.; Hancock, M. Filled Thermoplastics. In *Particulate-Filled Polymer Composites*, 2nd ed.; Rothon, R.N., Ed.; Rapra Technology Limited: Shrewsbury, UK, 2003.

Article

Synthesis and Surface Strengthening Modification of Silica Aerogel from Fly Ash

Lei Zhang [1,2,*], Qi Wang [1], Haocheng Zhao [1], Ruikang Song [1], Ya Chen [1], Chunjiang Liu [1] and Zhikun Han [1]

[1] School of Geology and Environment, Xi'an University of Science and Technology, Xi'an 710054, China; 18391871932@163.com (Q.W.); haocheng202302@163.com (H.Z.); srk781388595@163.com (R.S.); c1431332271@163.com (Y.C.); leocj2000@163.com (C.L.); han619332950@163.com (Z.H.)

[2] Key Laboratory of Coal Resources Exploration and Comprehensive Utilization, Ministry of Natural Resources, Xi'an 710021, China

* Correspondence: leizh1981@sohu.com

Abstract: This study focuses on using activated fly ash to prepare silica aerogel by the acid solution–alkali leaching method and ambient pressure drying. Additionally, to improve the performance of silica aerogel, $C_6H_{16}O_3Si$ (KH-570) and $CH_3Si(CH_3O)_3$ (MTMS) modifiers were used. Finally, this paper investigated the factors affecting the desilication rate of fly ash and analyzed the structure and performance of silica aerogel. The experimental results show that: (1) The factors affecting the desilication rate are ranked as follows: hydrochloric acid concentration > solid–liquid ratio > reaction temperature > reaction time. (2) KH-570 showed the best performance, and when the volume ratio of the silica solution to it was 10:1, the density of silica aerogel reached a minimum of 183 mg/cm^3. (3) The optimal process conditions are a hydrochloric acid concentration of 20 wt%, a solid–liquid ratio of 1:4, a reaction time of two hours, and a reaction temperature of 100 °C. (4) The optimal performance parameters of silica aerogel were the thermal conductivity, specific surface area, pore volume, average pore size, and contact angle values, with $0.0421 \text{ W·(m·K)}^{-1}$, $487.9 \text{ m}^2 \cdot \text{g}^{-1}$, $1.107 \text{ cm}^3 \cdot \text{g}^{-1}$, 9.075 nm, and 123°, respectively. This study not only achieves the high-value utilization of fly ash, but also facilitates the effective recovery and utilization of industrial waste.

Keywords: fly ash; silica aerogel; acid dissolution–alkali leaching; ambient pressure drying; reinforcing modification

Citation: Zhang, L.; Wang, Q.; Zhao, H.; Song, R.; Chen, Y.; Liu, C.; Han, Z. Synthesis and Surface Strengthening Modification of Silica Aerogel from Fly Ash. *Materials* **2024**, *17*, 1614. https://doi.org/10.3390/ma17071614

Academic Editors: Haejin Hwang and Daniela Fico

Received: 27 January 2024
Revised: 15 March 2024
Accepted: 21 March 2024
Published: 1 April 2024

1. Introduction

Fly ash is an industrial waste that is difficult to effectively utilize and is often stored in piles. This not only takes up a large amount of land, but also poses harm to humans and the environment when harmful substances migrate to the soil, atmosphere, and water [1,2]. But, the fly ash can still be utilized in various ways. Presently, its primary bulk application lies in the manufacturing of construction materials [3–7]. Nevertheless, fly ash is also rich in chemical elements such as Si, Al, Fe, and C, which constitutes a significant rationale for exploring high-value utilization pathways for fly ash [8–12]. Extracting and utilizing these chemical elements from fly ash can truly achieve the high-value utilization of fly ash [8,13]. Therefore, the comprehensive utilization of fly ash has important ecological and socio-economic benefits.

Aerogel is a porous solid material composed of nanoparticles with a three-dimensional network structure [14]. It commonly includes materials such as silica aerogel and Al_2O_3 aerogel [15–18]. Silica aerogel, in particular, has gained attention due to its high porosity, high specific surface area, low density, and low thermal conductivity, and these characteristics make it suitable for applications in industrial insulation materials, electrical fields, acoustic fields, and environmental fields [19–24].

Currently, the preparation methods for silica aerogel typically involve the sol–gel method, aging method, solvent replacement method, and drying process [25–28]. However, the use of expensive organic alkoxides as silicon sources and the high cost, danger, and complexity associated with the supercritical drying method limit its large-scale industrial application [17,29–32]. Therefore, it is significant to explore low-cost, efficient, and environmentally friendly silicon sources and safe, cost-effective drying methods for the preparation of silica aerogel.

Considering the high-silica (SiO_2) content in fly ash, it can be used as a raw material for the preparation of silica aerogel. The utilization of fly ash through methods such as acid dissolution, alkali leaching, sol–gel processing, surface modification, and atmospheric drying can effectively reduce the production cost of preparing silica aerogels from fly ash, while also offering new avenues for the utilization of fly ash [15,33–36]. Consequently, the utilization of fly ash and the preparation of silica aerogel using fly ash as a raw material have the potential to reduce costs, increase value-added utilization, and have significant ecological and socio-economic benefits. However, further research is required to optimize the preparation methods and achieve aerogel with desired properties.

In the first step of this study, the sodium silicate solution was prepared from activated fly ash by the acid dissolution and alkali leaching method. During this process, the single-factor effects of different experimental conditions (the HCl mass fraction, reaction temperature, reaction time, and solid–liquid ratio) on the desilication rate of fly ash were investigated. Subsequently, the interaction effects of four experimental conditions on the desilication rate of fly ash were studied through an orthogonal test, and the optimal experimental conditions for preparing a sodium silicate solution from fly ash were determined. In the second step, silica aerogels were prepared using the sodium silicate solution prepared under the optimal experimental conditions through processes such as the sol–gel process and ambient pressure drying. Additionally, to enhance the performance of silica aerogels, a surface enhancement modification process was introduced during the preparation, and the modification mechanisms of different modifiers (KH-570 and MTMS) were investigated. Eventually, surface-enhanced modifiers with excellent performance were selected based on the experimental results.

Finally, to analyze the microstructure, functional groups, crystalline phases, thermal conductivity, hydrophobicity, and specific surface area of the silica aerogels, the characterizations of the composition and properties of the silica aerogels were conducted using X-ray Diffraction (XRD), a Fourier Transform Infrared Spectrometer (FT-IR), Brunauer–Emmett–Teller (BET), contact angle measurement, and a thermal conductivity meter. Meanwhile, the changes in the composition and properties of the silica aerogels during the preparation process were studied to explain the reaction mechanism of preparing silica aerogels by ambient pressure drying. The experimental principle flow chart is shown in Figure 1.

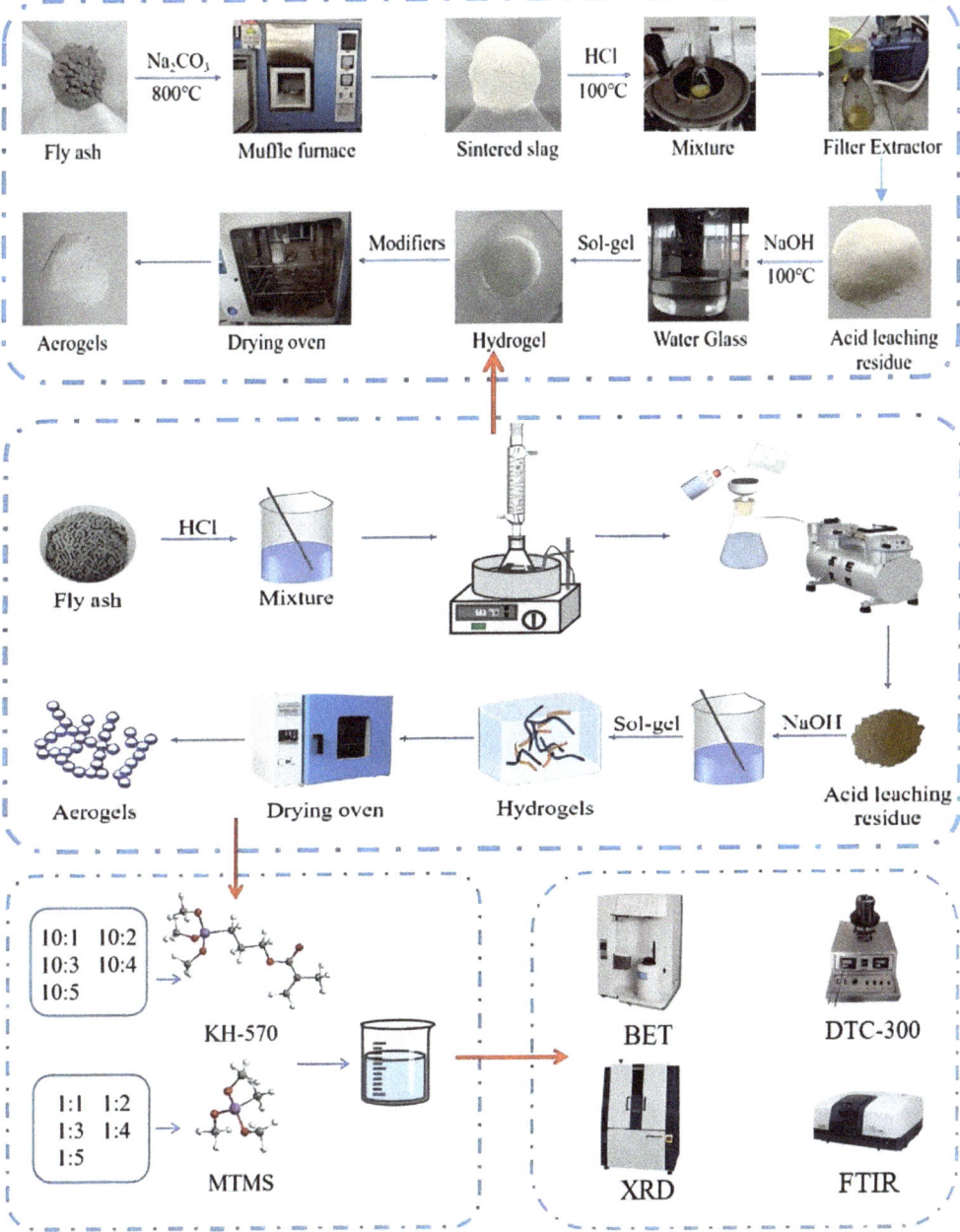

Figure 1. Experimental schematic diagram of preparation and modification of silica aerogel.

2. Experiment

2.1. Materials

The fly ash utilized in the experiment is derived from a coal-fired furnace. Table 1 displays the chemical composition of the fly ash employed in the study. As observed, the

SiO$_2$ and Al$_2$O$_3$ content in this particular fly ash surpasses 80%, with SiO$_2$ constituting over 50% of the composition.

Table 1. Chemical composition of fly ash.

Component (wt%)	SiO$_2$	Al$_2$O$_3$	Fe$_2$O$_3$	CaO	MgO	Na$_2$O	K$_2$O	SO$_3$	MnO	TiO$_2$	Loss
Fly ash	50.04	33.52	6.60	4.16	0.33	0.02	1.33	0.94	0.075	1.48	1.52

2.2. Characterization Methods

The main experimental equipment as shown Supplementary Table S2.

2.2.1. XRD

XRD analysis was carried out using an XD-3X (Beijing Puxi General Instrument Co., Ltd., Beijing, China). Prior to testing, SiO$_2$ aerogel is ground into powder. The initial angle is 5°, the final angle is 90°, the step width is 0.02, and the wavelength is 1.54, utilizing Cu target Kα radiation, with an operating voltage of 36 kV, and an operating current of 20 mA.

2.2.2. BET

BET analysis was carried out using an ASAP2020 (Mike, Detroit, MI USA). Prior to measurement, samples are degassed under vacuum conditions at 200 °C for 6 h. Subsequently, high-purity nitrogen is used as the adsorbate, and adsorption–desorption measurements are conducted at 77 K.

2.2.3. FTIR

FTIR analysis was carried out using a VERTEX70 (Bruker, Karlsruhe, Germany) after grinding SiO$_2$ aerogel. Samples are prepared by mixing with KBr and pressing into pellets for testing and analysis. The measurement range is 400~4000 cm^{-1}, with 28 scans, and a resolution of 0.4 cm^{-1}.

2.2.4. Contact Angle

Contact angle analysis was carried out using a CA100B (Shanghai Yingnuo Precision Instrument Co., Ltd., Shanghai, China), using the sessile drop method at room temperature.

2.2.5. Thermal Conductivity

Thermal conductivity analysis was carried out using a TC-3000E (Xi'an Xiaxi Electronic Technology Co., Ltd., Xi'an, China), pressing the SiO$_2$ aerogel into blocks (Thickness $\geq$ 0.1 mm, Length $\geq$ 25 mm) at room temperature.

2.3. Preparation of Silica Aerogel Modified Silica Aerogel

2.3.1. Extraction of SiO$_2$ through Acid Dissolution and Alkali Leaching from Fly Ash

In this processing stage, Si-Al bonds in fly ash are initially disrupted using an acid dissolution method, followed by alkali leaching to extract SiO$_2$ and prepare a sodium silicate solution. The experimental reagents as shown Supplementary Table S1.

1. Acid Dissolution: Pre-treated fly ash is mixed with hydrochloric acid solutions of varying mass fractions (10 wt%, 15 wt%, 20 wt%, 25 wt%, and 30 wt%) in specific ratios (1:3, 1:4, 1:5, 1:6, and 1:7). The mixture is stirred in a magnetic stirrer at a designated temperature (80 °C, 90 °C, 100 °C, 110 °C, and 120 °C) for a specified duration (1 h, 2 h, 3 h, 4 h, and 5 h). After completion, the reaction mixture is filtered and repeatedly washed with distilled water until neutral. The residue is then dried at 105 °C for 3–5 h in a vacuum drying oven.
2. Alkali Leaching: The filtered residue is mixed with sodium hydroxide solutions of various mass fractions (10 wt%, 15 wt%, and 20 wt%) in a 1:5 ratio. The mixture is

stirred in a magnetic stirrer at a certain temperature (80 °C, 90 °C, and 100 °C) for a specific duration (1 h, 1.5 h, and 2 h). After completion, the mixture is filtered and distilled water is added for washing until neutral. The filtrate is collected and stored in a beaker, representing the NaSiO$_3$ solution.

2.3.2. Preparation of Strengthened and Modified Silica Aerogel

In this processing stage, modified silica aerogels are prepared through the processes of wet gel formation, surface enhancement modification, aging, and ambient pressure drying.

1. Preparation of wet gel and surface enhancement modification: The sodium silicate solution obtained from fly ash is poured into an ion exchange column containing strongly acidic cation exchange resin, resulting in a silicon acid solution with a pH of 2–3. Modified liquid is prepared by mixing MTMS and KH-570 with anhydrous ethanol in a 1:4 ratio. The silicon acid solution is mixed with the modified liquid in a certain proportion and stirred for 1 h. After stirring, the pH of the mixed solution is adjusted to 5–6 using 1 mol/L ammonia solution. The sealed gel is left to age, observed by tilting the beaker at a 45° angle to check for gel formation.
2. Aging: deionized water is added to the beaker containing the wet gel, sealed, and placed in a 50 °C water bath for aging for 48 h.
3. Solvent replacement: the aged wet gel is soaked in anhydrous ethanol and placed in a 50 °C water bath for solvent replacement for 24 h.
4. Post-treatment modification: the wet gel, after solvent replacement, is soaked in a mixture of n-hexane, HMDSO, and ethanol (with HMDSO to wet gel volume ratio of 1:1) at 50 °C for surface modification for 24 h.
5. Solvent replacement: the modified gel is soaked in n-hexane and placed in a 50 °C water bath for solvent replacement for 12 h.
6. Ambient pressure drying: the gel, after solvent replacement, is placed in a vacuum drying oven and dried at 60 °C, 80 °C, 120 °C, and 180 °C for 2 h each, resulting in strengthened and modified SiO$_2$ aerogels.

2.4. The Influence of Acid Leaching Conditions on the Desilication Rate of Fly Ash

Mix the fly ash thoroughly with hydrochloric acid and allow it to react. Once the reaction is complete, filter the mixture while it is still hot and then proceed to wash the filter residue. Subsequently, add 20% sodium hydroxide solution to the dried acid leaching residue at a ratio of 1:5. Place the mixture in a magnetic stirrer and allow it to react for 2 h at a temperature of 100 °C. After the reaction, filter the mixture using hot water and wash it until it reaches a neutral state. The resulting filtrate is a sodium metasilicate solution. Finally, select appropriate reaction conditions for the orthogonal test using an L9 (3^4) orthogonal design table. The experimental conditions include the mass fraction of hydrochloric acid (15%, 20%, and 25%), solid–liquid ratio (1:3, 1:4, and 1:5), reaction temperature (90 °C, 100 °C, and 110 °C), and reaction time (2 h, 3 h, and 4 h). Through these experiments, we investigated the impact of these four factors on the desilication rate.

3. Results and Discussion

3.1. Effect of Acid Leaching Conditions on the Desilication Rate of Fly Ash

After acid leaching, the fly ash undergoes a certain level of activation, leading to the formation of a significant amount of amorphous SiO$_2$ as a result of mullite structure disruption. The reaction equation is as follows:

$$NaAlSiO_4 + 4HCl \rightarrow NaCl + AlCl_3 + SiO_2 + 2H_2O \tag{1}$$

3.1.1. Effect of Hydrochloric Acid Concentration on the Desilication Rate of Fly Ash

By maintaining a constant solid–liquid ratio (1:4), reaction temperature (100 °C), and reaction time (2 h), we can investigate the impact of varying hydrochloric acid concentrations on the desilication rate of fly ash. Figure 2 illustrates the obtained results.

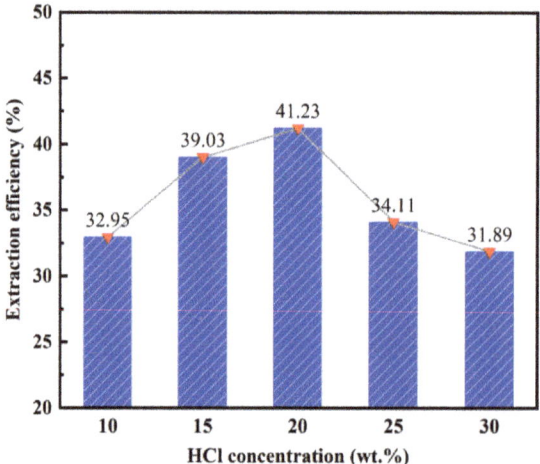

Figure 2. Effect of HCl concentration on the desilication rate of coal fly ash.

According to the findings presented in Figure 2, the desilication rate of fly ash exhibits an initial increase followed by a decrease with the rise in the hydrochloric acid concentration. At a concentration of 10 wt%, the desilication rate reaches its lowest point at 31.89%. This can be attributed to the insufficient reaction between hydrochloric acid and SiO_2 in fly ash caused by the low acid concentration. Conversely, at a concentration of 20 wt%, the desilication rate peaks at 41.23%. However, as the concentration of the hydrochloric acid exceeds 20–30 wt%, the desilication rate gradually declines. This decline may be attributed to side reactions occurring between the hydrochloric acid and other substances present in fly ash due to the high acid concentration. These side reactions increase the overall amount of reactants and consequently reduce the desilication rate.

3.1.2. Effect of Solid–Liquid Ratio on the Desilication Rate of Fly Ash

By maintaining a constant hydrochloric acid concentration (20 wt%), reaction temperature (100 °C), and reaction time (2 h), we can explore the impact of varying solid–liquid ratios on the desilication rate of fly ash. The obtained results are illustrated in Figure 3.

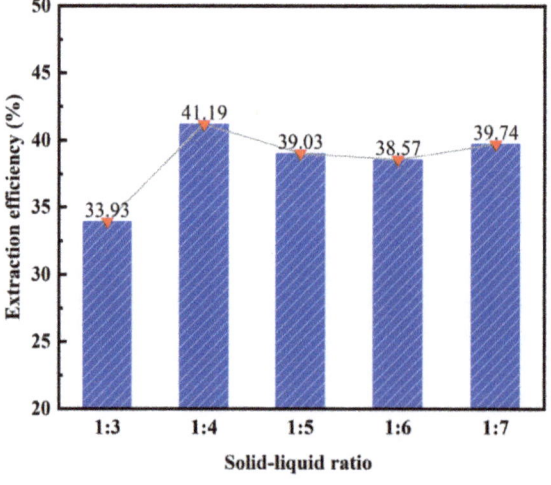

Figure 3. Effect of solid–liquid ratio on the desilication rate of coal fly ash.

According to the data presented in Figure 3, the desilication rate of fly ash initially increases and then stabilizes as the solid–liquid ratio decreases. At a solid–liquid ratio of 1:3, the desilication rate reaches its lowest point at 33.93%. This can be attributed to the insufficient addition of hydrochloric acid, resulting in an inadequate reaction between the fly ash and hydrochloric acid. Conversely, at a solid–liquid ratio of 1:4, the highest desilication rate is observed at 41.19%. At this ratio, the reaction between the fly ash and hydrochloric acid is deemed complete. With further increases in the solid–liquid ratio, specifically reaching 1:7, there is no significant change in the desilication rate of fly ash. Consequently, a solid–liquid ratio of 1:4 was chosen as the subsequent reaction condition.

3.1.3. Effect of Reaction Temperature on the Desilication Rate of Fly Ash

By maintaining a constant hydrochloric acid concentration (20 wt%), solid–liquid ratio (1:4), and reaction time (2 h), we can investigate the impact of varying reaction temperatures on the desilication rate of fly ash. The obtained results are illustrated in Figure 4.

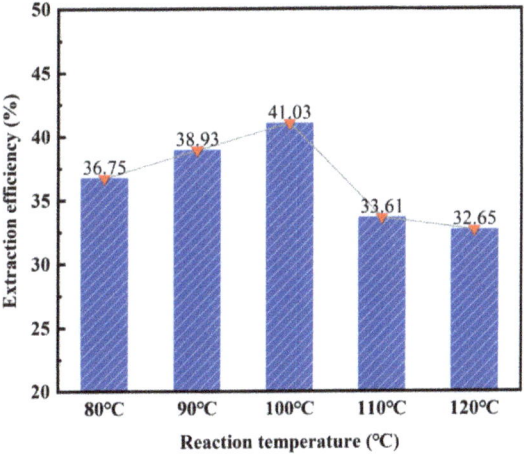

Figure 4. Effect of reaction temperature on the desilication rate of fly ash.

According to the findings presented in Figure 4, the desilication rate of fly ash initially increases and then decreases as the reaction temperature increases. A turning point is observed at 100 °C, with the highest desilication rate reaching 41.03%. However, the desilication rate is found to be the lowest at 120 °C, measuring 32.65%. The primary reason for this is that the boiling point of 20 wt% hydrochloric acid is 110 °C. When the temperature exceeds 110 °C, the solution reaches its boiling point and undergoes rapid evaporation, thereby affecting the acid-leaching effect of fly ash. Consequently, a reaction temperature of 100 °C was chosen as the subsequent reaction condition.

3.1.4. Effect of Reaction Time on the Desilication Rate of Fly Ash

By maintaining a constant hydrochloric acid concentration (20 wt%), solid–liquid ratio (1:4), and reaction temperature (100 °C), we can explore the impact of varying reaction times on the desilication rate of fly ash. The obtained results are illustrated in Figure 5.

According to the data presented in Figure 5, the desilication rate of fly ash initially increases and then stabilizes as the reaction time increases. At a reaction time of 1 h, the desilication rate reaches its lowest point at 33.06%. This can be attributed to the short reaction time, which hinders the complete reaction between the fly ash and hydrochloric acid. Conversely, at a reaction time of 2 h, the highest desilication rate is observed at 41.19%, indicating that the acid leaching reaction is completed at this point. When the reaction time is extended to 5 h, there is no significant change in the desilication rate. Consequently, a reaction time of 2 h was chosen as the subsequent reaction condition.

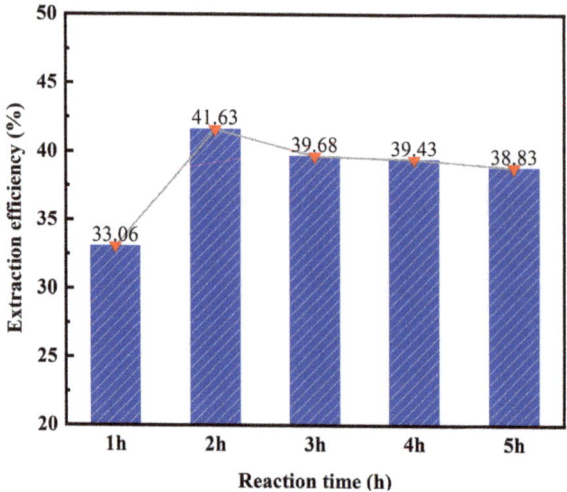

Figure 5. Effect of reaction time on the desilication rate of fly ash.

3.1.5. Interactive Effects of Different Reaction Conditions on the Desilication Rate of Fly Ash

To investigate the combined effect of reaction conditions on the desilication rate of fly ash while keeping the alkali-leaching reaction conditions constant, four factors were chosen for the test, as shown in Table 2. Each factor was tested at three different levels. The details of the orthogonal test results and the corresponding analysis can be found in Tables 3 and 4. The orthogonal test results analysis table as shown Supplementary Table S3.

Table 2. Orthogonal experimental design for the influence of different reaction conditions on the desilication rate of fly ash.

Level Number of Group	HCl/(wt%)	Solid–Liquid Ratio	Reaction Time/h	Reaction Temperature/°C
1	15%	1:3	2 h	90 °C
2	20%	1:4	3 h	100 °C
3	25%	1:5	4 h	110 °C

Table 3. Analysis table of orthogonal test.

Test Group Number	Factor				Desilication Rate/%
	A	B	C	D	
1	15%	1:3	2 h	90 °C	34.99
2	15%	1:4	3 h	100 °C	41.19
3	15%	1:5	4 h	110 °C	41.5
4	20%	1:3	3 h	110 °C	35.7
5	20%	1:4	4 h	90 °C	39.88
6	20%	1:5	2 h	100 °C	42.67
7	25%	1:3	4 h	100 °C	32.58
8	25%	1:4	2 h	110 °C	37.28
9	25%	1:5	3 h	90 °C	33.19

Note: A represents the factor of HCl concentration, B represents the factor of Solid–liquid ratio, C represents the factor of reaction time, and D represents the factor of reaction temperature.

Table 4. Orthogonal test results analysis table.

Level	A	B	C	D
1	15%	1:3	2 h	90 °C
2	20%	1:4	3 h	100 °C
3	25%	1:5	4 h	110 °C
Extreme deviation	5.07	5.03	1.62	2.79

Note: A represents the factor of HCl concentration, B represents the factor of Solid–liquid ratio, C represents the factor of reaction time, and D represents the factor of reaction temperature.

According to the data provided in Table 4, the hydrochloric acid concentration exhibits the largest range of 5.07, indicating that it has the most significant impact on the desilication rate of fly ash. In contrast, the reaction time shows the smallest range of 1.62, suggesting that its effect on the desilication rate is relatively minor. The solid–liquid ratio demonstrates a range of 5.03, indicating that both the hydrochloric acid concentration and solid–liquid ratio have a substantial influence on the desilication rate. Thus, the order of importance for the four influencing factors is as follows: the hydrochloric acid concentration > solid–liquid ratio > reaction temperature > reaction time. Based on the k-value analysis, the optimal conditions for the desilication of fly ash are as follows: a hydrochloric acid concentration of 20 wt%, a solid–liquid ratio of 1:4, a reaction temperature of 100 °C, and a reaction time of 2 h.

3.1.6. Optimization of Acid Leaching Conditions

The desilication rate of fly ash is limited to approximately 42% under the optimal acid solution alkali leaching condition. Consequently, to enhance the activity of fly ash, a method involving the activation of fly ash through baking with sodium carbonate is employed to generate a sodium metasilicate solution, which ultimately leads to a higher desilication rate. The reaction equation is as follows:

$$3Al_2O_3 \cdot 2SiO_2 + 4SiO_2 + 3Na_2CO_3 \rightarrow 6NaAlSiO_4 + 3CO_2 \tag{2}$$

$$NaAlSiO_4 + 4HCl \rightarrow NaCl + AlCl_3 + SiO_2 + 2H_2O \tag{3}$$

$$SiO_2 + 2NaOH = Na_2SiO_3 + H_2O \tag{4}$$

During the acid leaching experiment, the reaction unexpectedly ceased as a result of fly ash agglomeration. Further investigation, as indicated by a previous study, revealed that the incomplete reaction was due to an excessively low solid–liquid ratio of fly ash to hydrochloric acid [33]. To address this issue, the experimental conditions were maintained at a constant (Ceteris paribus), and the influence of different solid–liquid ratios on the desilication rate of fly ash was investigated. The results of this study can be observed in Figure 6.

As depicted in Figure 6, the desilication rate of fly ash exhibits an initial increase followed by stabilization as the solid–liquid ratio decreases. Notably, when the solid–liquid ratio is 1:5, the desilication rate reaches its lowest point at 65.66%. This is primarily attributed to the inadequacy of hydrochloric acid, which hinders the complete reaction between fly ash and hydrochloric acid. Conversely, when the solid–liquid ratio is 1:7, the highest desilication rate of 81.72% is achieved. At this point, the reaction between fly ash and hydrochloric acid is completed. The further reduction of the solid–liquid ratio to 1:9 does not yield any significant change in the desilication rate. Consequently, after careful consideration, a solid–liquid ratio of 1:7 was selected as the subsequent reaction condition.

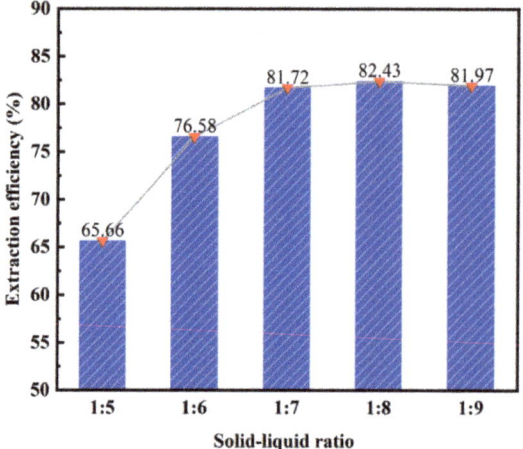

Figure 6. Effect of solid–liquid ratio on the desilication rate of activated fly ash.

3.2. Surface Strengthening Modification of Silica Aerogel

Effect of Modifier on the Density of Silica Aerogel

This section focuses on analyzing the process of producing modified silica aerogel from calcined, activated fly ash. The objective is to investigate the impact of different modifiers on the density of the aerogel. The ratio of the modified solution can be found in Table 5, while the corresponding experimental results are illustrated in Figures 7 and 8.

Table 5. Surface-enhanced modified silica aerogel solution proportioning.

| Serial Number | Silicon Source | Modified Liquid 1 | | Modified Liquid 2 | |
	Silicic Acid Solution/mL	KH-570/mL	EtOH/mL	MTMS/mL	EtOH/mL
1	10	5	20	10	40
2	10	4	16	20	80
3	10	3	12	30	120
4	10	2	8	40	160
5	10	1	4	50	200

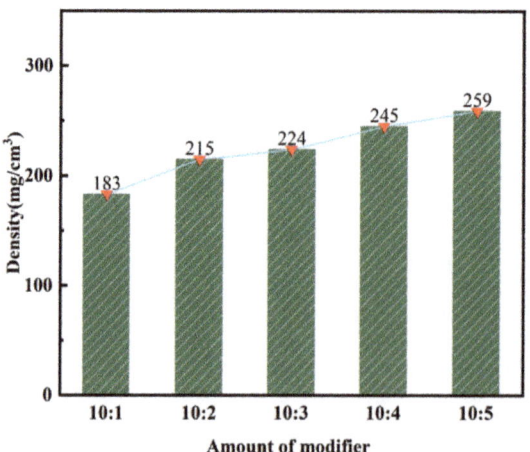

Figure 7. Effect of KH-570 dosage on the density of silica aerogel.

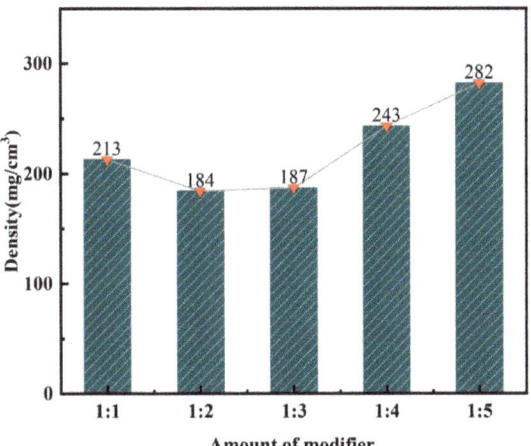

Figure 8. Effect of MTMS dosage on the density of silica aerogel.

As depicted in Figure 7, the density of silica aerogel shows an increasing trend with the addition of KH-570. Specifically, at a volume ratio of 10:1 for the silicic acid solution to KH-570, the minimum density is recorded at 183 mg/cm^3. Conversely, at a volume ratio of 10:5, the maximum density reaches 259 mg/cm^3. This can be attributed to the connection of the silane coupling agent KH-570, which contains a C=C double bond, to the surface of SiO$_2$ particles. Subsequently, during radical polymerization, a protective film forms on the particle surface, resulting in a higher solid content of silica aerogel and, consequently, an increased density. For subsequent experiments, a volume ratio of 10:1 for silicic acid solution to KH-570 was chosen for performance analysis.

As illustrated in Figure 8, the density of silica aerogel exhibits an initial decrease followed by an increase as the dosage of MTMS is increased. At a volume ratio of 1:2 for the silicic acid solution to MTMS, the minimum density is recorded as 184 mg/cm^3. Conversely, at a volume ratio of 1:5, the maximum density reaches 282 mg/cm^3. This can be attributed to the hydrolysis of MTMS, which leads to the formation of -OH and -CH$_3$ groups. During the polycondensation reaction, hydrophobic methyl groups infiltrate the gel matrix, enhancing the structure of silica aerogel and resulting in a decrease in density. However, when the MTMS content becomes excessive, some sol ions fail to effectively cross-link with -OH, causing an increase in the density of silica aerogel. For subsequent experiments, a volume ratio of 1:2 between silica solution and MTMS was selected for performance analysis.

3.3. Analysis of Surface Enhancement and Modification Mechanism

3.3.1. KH-570 Modification Mechanism

The molecule of 3-methacryloxypropyl Trimethoxy silane (KH-570) contains the -OCH$_3$ group which, after hydrolysis, can undergo a single displacement reaction with the -OH group on the gel surface. This reaction contributes to an increase in the structural strength of silica aerogel. Additionally, apart from the -OCH$_3$ group, the C=C in the KH-570 molecule can undergo radical polymerization when a radical initiator is introduced. The introduction of species reintroduction polymers further enhances the performance of aerogel through KH-570 modification. During the modification process, KH-570 is typically mixed with anhydrous ethanol first and then combined with a silicic acid solution. This gradual mixing helps slow down the reaction rate and ensures a more uniform cross-linking of sol particles, thereby reducing the capillary force of silica aerogel [37,38].

KH-570 is utilized to reinforce and modify silica aerogel, and the polycondensation process involves two scenarios: with water and anhydrous conditions. In the presence of water, KH-570 undergoes hydrolysis initially, followed by a condensation reaction with the hydroxyl group (Si-OH) on the surface of the wet gel. Simultaneously, the hydrolyzed KH-570 also undergoes a condensation reaction itself. The chemical reaction equations involved in this process are denoted as 5~7 [39].

$$(5)$$

$$(6)$$

$$(7)$$

In the absence of water, KH-570 undergoes a condensation reaction directly with Si-OH. During the experiment, anhydrous ethanol is employed to facilitate the removal of water from the wet gel matrix. Consequently, KH-570 primarily reacts with the silicon hydroxyl groups on the surface of the wet gel through a condensation reaction in the absence of water.

$$(8)$$

3.3.2. MTMS Modification Mechanism

Three -OCH_3 groups in the molecules of alkyl tri alkoxysilane (MTMS) undergo a reaction with -OH groups after hydrolysis. After the hydrolysis of water glass, each silicon atom carries four hydroxyl groups, while after MTMS hydrolysis, it carries an inert -CH_3 group, which can condense under the influence of an alkaline catalyst. Due to the presence of the methyl group in MTMS after hydrolysis, the polycondensation reaction is slowed down, leading to the preferential formation of primary particles of a specific size through the polycondensation of $Si(OH)_4$. As the reaction progresses, the hydrolysis product CH_3-$Si(OH)_3$ of MTMS also participates in the polycondensation reaction, resulting in a significant amount of unreacted Si-OH on the surface of the primary particles. Through the reaction with the primary particles, the -OH group is replaced, and the -CH_3 group is successfully grafted onto the gel skeleton. As the condensation reaction continues, the primary particles grow into secondary particles, and the -OH groups on the surface are continuously replaced by -CH_3 groups [40] (show in Figure 9).

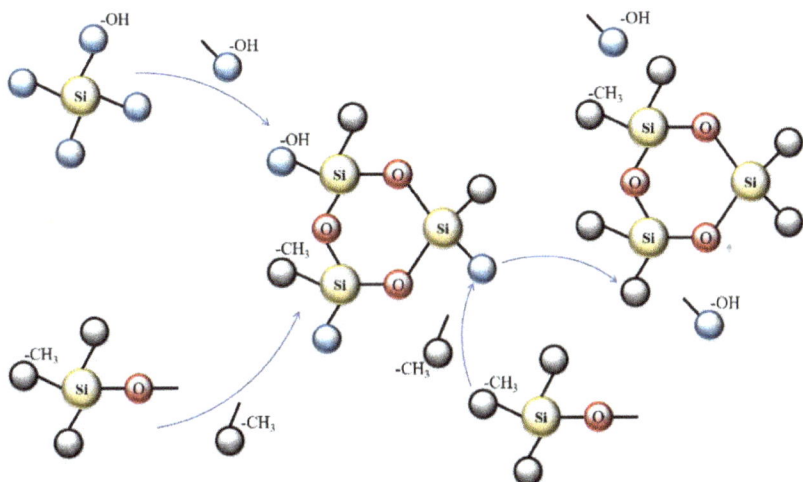

Figure 9. MTMS-enhanced modified silica aerogel mechanism diagram [40].

3.4. Effects of Surface Enhancement Modification on the Structure and Properties of Silica Aerogel

3.4.1. Phase Analysis

In order to investigate the impact of various strengthening modifiers on the crystal structure of silica aerogel, X-ray diffraction (XRD) analysis was performed on the KH-570, MTMS-modified, and unmodified silica aerogel. The results of this analysis are depicted in Figure 10.

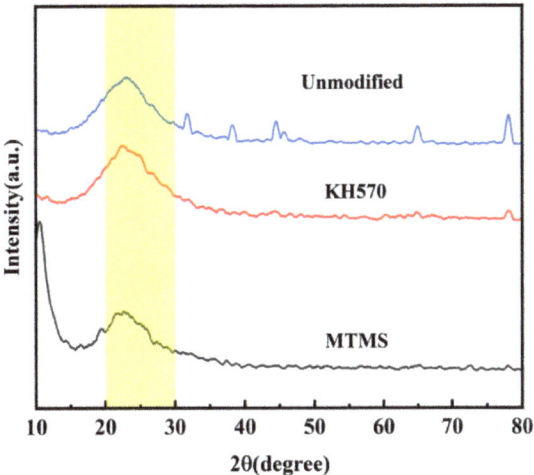

Figure 10. XRD patterns of polymer-reinforced modified silica aerogel.

In Figure 10, it can be observed that the XRD patterns of the aerogel samples under the three different modification conditions exhibit prominent peak broadening in the range of 20~30°, indicating a predominantly amorphous structure of silicon in the samples. The unmodified samples display diffraction peaks corresponding to LiH and Fe_2O_3, which can be attributed to the presence of high levels of impurities in the water glass. Despite the use of a strong acidic cation exchange resin during pre-treatment, the complete removal of metal ions could not be achieved. On the other hand, no such peaks are observed after

KH-570 and MTMS enhancement, suggesting that the modifiers might have undergone a coordination reaction with the metal ions in the water glass, forming compounds like metal silicates. These compounds typically have low solubility and tend to precipitate, resulting in a significant reduction in the metal ion content after modification.

3.4.2. Physical Property Analysis

In order to validate the modification mechanism of various strengthening modifiers, the thermal conductivity of KH-570- and MTMS-modified, as well as unmodified silica aerogel was measured, and its thermal insulation performance was analyzed. The results of this analysis are presented in Figure 11.

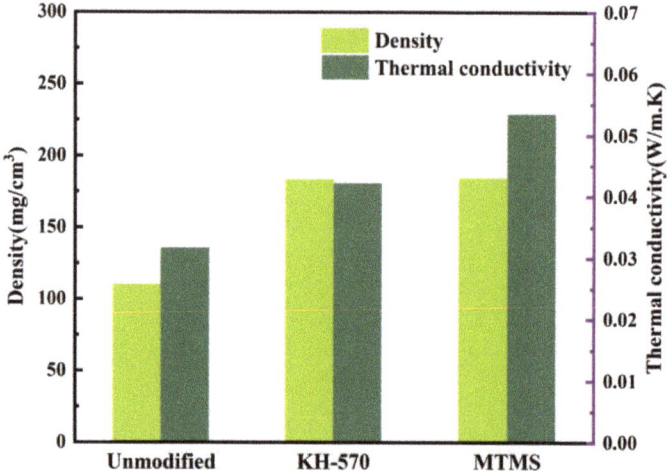

Figure 11. Effect of polymer-reinforced modification on density and thermal conductivity of SiO_2.

As depicted in Figure 11, both the density and thermal conductivity of the unmodified silica aerogel are low, measuring 110 mg·cm^{-3} and 0.0316 W·(m·K)$^{-1}$, respectively. However, when enhanced with KH-570, the density and thermal conductivity of the silica aerogel increase significantly to 183 mg·cm^{-3} and 0.0421 W·(m·K)$^{-1}$, respectively. While the density of the MTMS-enhanced silica aerogel is not considerably different from the KH-570-enhanced version, there is an improvement in the thermal conductivity. The density of the silica aerogel modified with MTMS is measured at 184 mg·cm^{-3}, while its thermal conductivity is 0.0534 W·(m·K)$^{-1}$.

Figure 12 presents the visual representation of the reinforced and modified silica aerogel after 30 min of grinding. Based on the density and thermal conductivity of the silica aerogel, it can be observed that the order is as follows: MTMS > KH-570 > unmodified. However, in terms of appearance, all samples appear as powder, with the unmodified samples lacking any discernible particle sensation, having smaller particle sizes and exhibiting poorer mechanical properties during extrusion. Conversely, the samples modified with KH-570 and MTMS display a pronounced particle sensation, larger particle sizes, and higher hardness during extrusion. Although there is a slight increase in density and thermal conductivity for the reinforced and modified silica aerogel, it still falls within the low density and low thermal conductivity range. However, a significant improvement in mechanical properties is achieved. Therefore, strengthening and modification effectively enhance the mechanical properties of silica aerogel.

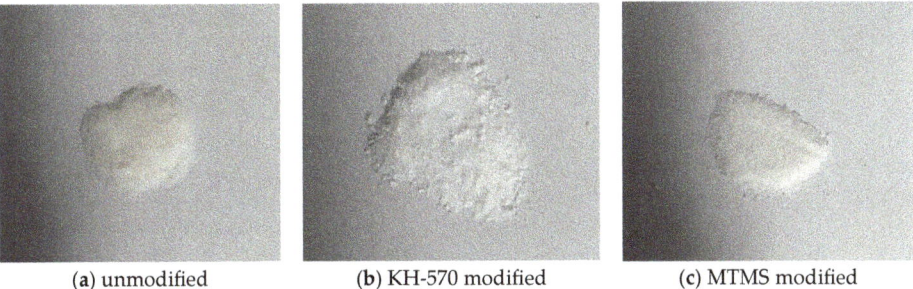

(a) unmodified   (b) KH-570 modified   (c) MTMS modified

Figure 12. Diagram of different reinforced modified silica aerogel.

3.4.3. Chemical Property Analysis

In order to validate the modification mechanism of various strengthening modifiers, the infrared spectra of KH-570- and MTMS-modified, as well as unmodified silica aerogel were measured. The functional groups present in the samples were then analyzed. The results of this analysis are depicted in Figure 13.

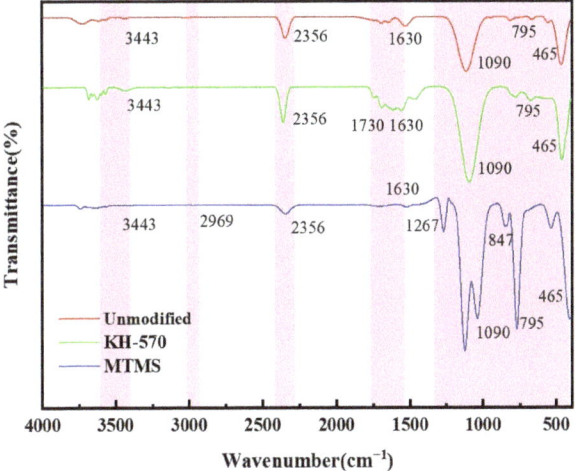

Figure 13. Infrared spectra of polymer-reinforced modified silica aerogel.

As illustrated in Figure 13, all samples exhibit distinct bands at approximately 465 cm^{-1}, 795 cm^{-1}, and 1090 cm^{-1}. These bands correspond to the bending vibration, symmetric stretching vibration, and antisymmetric stretching vibration, respectively, of the Si-O-Si bond. These bonds form the skeleton structure of aerogel. Additionally, all samples display an absorption peak at 2356 cm^{-1}, which is attributed to the stretching vibration of the C=O bond in CO_2 present in the air. This peak may be influenced by the testing environment.

The spectra also reveal bands at 2969 cm^{-1}, 1267 cm^{-1}, and 847 cm^{-1}, representing the antisymmetric stretching vibration peak, antisymmetric bending vibration peak, and symmetric bending vibration peak, respectively, of the -CH$_3$ group.

The bands near 3443 cm^{-1} and 1630 cm^{-1} indicate the presence of -OH and H-O-H infrared bands. However, the peak intensity is weak due to the replacement of Si-OH on the surface of the modified aerogel with Si-CH$_3$, thus enhancing the hydrophobicity of the aerogel.

In the spectrum of the aerogel modified with KH-570, bands are observed at 1630 cm^{-1} and 1730 cm^{-1}. The bands at 1630 cm^{-1} may be attributed to the presence of adsorbed water or the bending vibration of the C=C bond. The absorption peak at 1730 cm^{-1} corresponds to the stretching vibration absorption peak of the C=O bond, indicating the successful bonding of KH-570 to the surface of silica aerogel [38,41].

3.4.4. Analysis of Pore Structure and Specific Surface Area

In order to investigate the impact of different reinforcement phases on the specific surface area and pore structure of silica aerogel, the unmodified and KH-570- and MTMS-modified silica aerogels were analyzed using a specific surface area analyzer (BET). The porosity results are presented in Table 6, while the N$_2$ adsorption–desorption contour line, temperature, and related parameters, as well as the pore size distribution curve, are depicted in Figure 14.

Table 6. Effect of different enhancement modifications on the pore properties of silica aerogel.

Sample	Specific Surface Area/(m^2/g)	Pore Volume/(cm^3/g)	Average Pore Diameter/(nm)
Unmodified	538.7	0.991	7.358
KH-570	487.9	1.107	9.075
MTMS	146.8	0.425	11.58

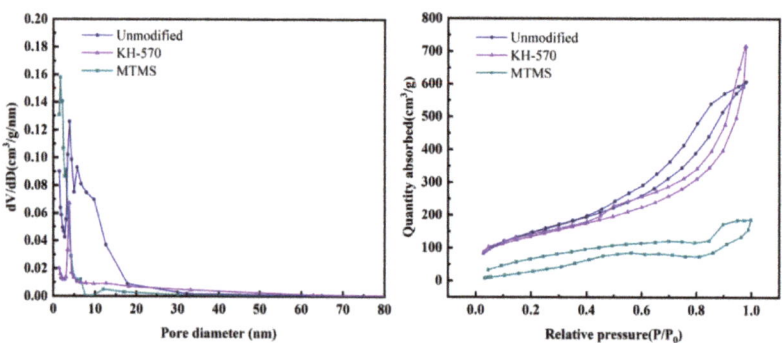

Figure 14. N$_2$ adsorption–desorption isotherms and pore size distribution curves of different enhanced modified silica aerogels.

As depicted in Figure 14, the adsorption–desorption isotherm curves of the unmodified, MTMS-modified, and KH-570-modified silica aerogel samples all exhibit Type IV behavior, indicating that the prepared silica aerogel is a mesoporous material. The unmodified and KH-570-modified silica aerogels show an H3-type hysteresis loop in the middle- and high-pressure range, suggesting the presence of a slit-like mesoporous structure. On the other hand, the MTMS-modified silica aerogel displays an H2 (b)-type hysteresis loop in the middle- and high-pressure range, indicating a relatively wider pore size distribution and potential channel blockage [37].

Under the three modification conditions, the N$_2$ adsorption capacity of the aerogel is 715 cm^3·g^{-1} after KH-570 modification, 607 cm^3·g^{-1} for the unmodified sample, and 187 cm^3·g^{-1} after MTMS modification. From the pore size distribution curve, the pore size ranges for the unmodified, KH-570-modified, and MTMS-modified aerogel samples are 2.7~20 nm, 2.5~15 nm, and 1.5~20 nm, respectively. The most probable diameters are 3.865 nm, 3.82 nm, and 1.661 nm.

Table 6 reveals that the maximum specific surface area of the unmodified silica aerogel is 538.7 m^2·g^{-1}. The specific surface area and pore volume of the KH-570-modified and MTMS-modified aerogel are lower than the unmodified sample, measuring 487.9 m^2·g^{-1} and 146.8 m^2·g^{-1}, respectively. The maximum pore volume for the KH-570-modified aerogel is 1.107 cm^3·g^{-1}, while the minimum pore volume for the MTMS-modified silica aerogel is 0.425 cm^3·g^{-1}.

In summary, the performance of the KH-570-modified silica aerogel is superior to that of the MTMS-modified silica aerogel, and the modified aerogel exhibits a slightly lower performance compared to the unmodified aerogel. However, considering previous research, reinforcement modification enhances the skeleton structure and improves the mechanical properties. Therefore, selecting KH-570 as the reinforcement modifier can lead to a better specific surface area, pore volume, and stronger mechanical properties.

3.4.5. Hydrophobicity Analysis

In order to examine the impact of different reinforcement phases on the hydrophobicity of silica aerogel, contact angle analysis was conducted on the unmodified, KH-570-modified, and MTMS-modified samples using a contact angle measurement. The experimental results are presented in Figure 15.

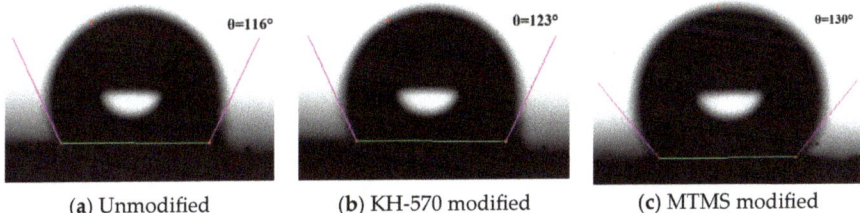

(**a**) Unmodified   (**b**) KH-570 modified   (**c**) MTMS modified

Figure 15. Contact angle of different reinforced modified silica aerogel.

As depicted in Figure 15, the contact angles measured for the three samples indicate hydrophobic silica aerogel properties, with values of 116°, 123°, and 130°, respectively. The presence of residual Si-OH hydrophilic groups in the unmodified silica aerogel may explain the slight hydrophilicity observed, suggesting that the complete conversion of Si-OH to Si-CH$_3$ was not achieved with HMDSO. However, the enhancement and modification with KH-570 and MTMS further replaced the unmodified Si-OH groups with Si-CH$_3$, leading to improved hydrophobic properties. The hydrophobicity of the three samples can be ranked in descending order as follows: MTMS > KH-570 > Unmodified.

4. Conclusions

1. The impact of different acid leaching conditions on the desilication rate of fly ash can be ranked as follows: hydrochloric acid concentration > solid–liquid ratio > reaction temperature > reaction time. The hydrochloric acid concentration and solid–liquid ratio were found to have the most significant effects on the desilication rate.

2. The optimal conditions for acid leaching were determined as follows: a hydrochloric acid concentration of 20 wt%, a solid–liquid ratio of 1:4, a reaction time of 2 h, and a reaction temperature of 100 °C.

3. After the calcination and activation of fly ash using sodium carbonate, the highest desilication rate of 81.72% was achieved with an acid-leaching solid-to-liquid ratio of 1:7.

4. Silica aerogel was prepared by ambient pressure drying using a sodium silicate solution obtained from fly ash as a silicon source. KH-570 and MTMS were used as strengthening modifiers. The lowest density of silica aerogel, 183 mg·cm^{-3}, was obtained with a volume ratio of silica solution to KH-570 of 10:1. Likewise, the lowest density of silica aerogel, 184 mg·cm^{-3}, was achieved with a volume ratio of silicic acid solution to MTMS of 1:2. And after characterizing by XRD, FTIR, and BET, the KH-570-enhanced aerogel exhibited the best overall performance.

5. The optimal performance parameters of silica aerogel were characterized by XRD, FTIR, and BET, resulting in thermal conductivity, a specific surface area, pore vol-

ume, average pore size, and contact angle values of 0.0421 $W \cdot (m \cdot K)^{-1}$, 487.9 $m^2 \cdot g^{-1}$, 1.107 $cm^3 \cdot g^{-1}$, 9.075 nm, and 123° respectively.

Supplementary Materials: The following supporting information can be downloaded at: https://www.mdpi.com/article/10.3390/ma17071614/s1, Table S1: Experimental reagents. Table S2: Main experimental equipment. Table S3: Orthogonal test results analysis table.

Author Contributions: L.Z.: Conceptualization; Q.W.: Methodology; C.L. and Y.C.: Software; Q.W. and H.Z.: Validation; Z.H.: Data curation; R.S.: Investigation; Q.W.: Writing—original Draft; Q.W.: Writing—review and editing. All authors have read and agreed to the published version of the manuscript.

Funding: This research was funded by the Technology Innovation Leading Program of Shaanxi (S2022QFY06-04), the Technology Innovation Leading Program of Shaanxi (2023-YD-CGZH-27), the Key Research and Development Plan of the Ningxia Hui Autonomous Region (2023-BEG02058), and the National Natural Science Foundation of China (Key Program, 42330808).

Institutional Review Board Statement: Not applicable.

Informed Consent Statement: Not applicable.

Data Availability Statement: Data are contained within the article and Supplementary Materials.

Conflicts of Interest: The authors declare no conflict of interest.

References

1. Ahmaruzzaman, M. A review on the utilization of fly ash. *Prog. Energy Combust. Sci.* **2010**, *36*, 327–363. [CrossRef]
2. Hamidi, A.; Nazari, P.; Shakibania, S.; Rashchi, F. Microwave irradiation for the recovery enhancement of fly ash components: Thermodynamic and kinetic aspects. *Chem. Eng. Process. Process. Intensif.* **2023**, *191*, 109472. [CrossRef]
3. Alterary, S.S.; Marei, N.H. Fly ash properties, characterization, and applications: A review. *J. King Saud Univ. Sci.* **2021**, *33*, 101536. [CrossRef]
4. Goswami, K.P.; Pakshirajan, K.; Pugazhenthi, G. Process intensification through waste fly ash conversion and application as ceramic membranes: A review. *Sci. Total. Environ.* **2021**, *808*, 151968. [CrossRef] [PubMed]
5. Öztürk, Z.B.; Çam, T. Performance of eco-friendly fly ash-based geopolymer mortars with stone-cutting waste. *Mater. Chem. Phys.* **2023**, *307*, 128112. [CrossRef]
6. Ferrante, F.; Bertini, M.; Ferlito, C.; Lisuzzo, L.; Lazzara, G.; Duca, D. A computational and experimental investigation of halloysite silicic surface modifications after alkaline treatment. *Appl. Clay Sci.* **2023**, *232*, 106813. [CrossRef]
7. Amran, M.; Debbarma, S.; Ozbakkaloglu, T. Fly ash-based eco-friendly geopolymer concrete: A critical review of the long-term durability properties. *Constr. Build. Mater.* **2021**, *270*, 121857. [CrossRef]
8. Dindi, A.; Quang, D.V.; Vega, L.F.; Nashef, E.; Abu-Zahra, M.R.M. Applications of fly ash for CO_2 capture, utilization, and storage. *J. CO2 Util.* **2019**, *29*, 82–102. [CrossRef]
9. Fauzi, A.; Nuruddin, M.F.; Malkawi, A.B.; Abdullah, M.M.A.B. Study of Fly Ash Characterization as a Cementitious Material. *Procedia Eng.* **2016**, *148*, 487–493. [CrossRef]
10. Ding, J.; Ma, S.; Shen, S.; Xie, Z.; Zheng, S.; Zhang, Y. Research and industrialization progress of recovering alumina from fly ash: A concise review. *Waste Manag.* **2017**, *60*, 375–387. [CrossRef] [PubMed]
11. Mathapati, M.; Amate, K.; Prasad, C.D.; Jayavardhana, M.; Raju, T.H. A review on fly ash utilization. *Mater. Today Proc.* **2022**, *50*, 1535–1540. [CrossRef]
12. Hait, P.; Dhara, D.; Ghanta, I.; Biswas, C.; Basu, P. Simultaneous Utilization of Fly Ash and Waste Plastics for Making Bricks and Paver Blocks. *J. Inst. Eng. India Ser. D* **2024**, 1–8. [CrossRef]
13. Das, D.; Rout, P.K. A Review of Coal Fly Ash Utilization to Save the Environment. *Water, Air, Soil Pollut.* **2023**, *234*, 1–23. [CrossRef]
14. Xu, L.; Jiang, Y.; Feng, J.; Feng, J.; Yue, C. Infrared-opacified Al_2O_3–SiO_2 aerogel composites reinforced by SiC-coated mullite fibers for thermal insulations. *Ceram. Int.* **2015**, *41*, 437–442. [CrossRef]
15. Bin Rashid, A.; Shishir, S.I.; Mahfuz, A.; Hossain, T.; Hoque, E. Silica Aerogel: Synthesis, Characterization, Applications, and Recent Advancements. *Part. Part. Syst. Charact.* **2023**, *40*, 202200186. [CrossRef]
16. Lamy-Mendes, A.; Pontinha, A.D.R.; Santos, P.; Durães, L. Aerogel Composites Produced from Silica and Recycled Rubber Sols for Thermal Insulation. *Materials* **2022**, *15*, 7897. [CrossRef]
17. Shen, M.; Jiang, X.; Zhang, M.; Guo, M. Synthesis of SiO_2–Al_2O_3 composite aerogel from fly ash: A low-cost and facile approach. *J. Sol-Gel Sci. Technol.* **2019**, *93*, 281–290. [CrossRef]
18. Mazrouei-Sebdani, Z.; Begum, H.; Schoenwald, S.; Horoshenkov, K.V.; Malfait, W.J. A review on silica aerogel-based materials for acoustic applications. *J. Non-Crystalline Solids* **2021**, *562*, 120770. [CrossRef]

19. Smirnova, I.; Gurikov, P. Aerogel production: Current status, research directions, and future opportunities. *J. Supercrit. Fluids* **2018**, *134*, 228–233. [CrossRef]

20. Woignier, T.; Primera, J.; Alaoui, A.; Etienne, P.; Despestis, F.; Calas-Etienne, S. Mechanical Properties and Brittle Behavior of Silica Aerogels. *Gels* **2015**, *1*, 256–275. [CrossRef] [PubMed]

21. Lamy-Mendes, A.; Pontinha, A.D.R.; Alves, P.; Santos, P.; Durães, L. Progress in silica aerogel-containing materials for buildings' thermal insulation. *Constr. Build. Mater.* **2021**, *286*, 122815. [CrossRef]

22. Hasan, M.A.; Sangashetty, R.; Esther, A.C.M.; Patil, S.B.; Sherikar, B.N.; Dey, A. Prospect of Thermal Insulation by Silica Aerogel: A Brief Review. *J. Inst. Eng. India Ser. D* **2017**, *98*, 297–304. [CrossRef]

23. Randall, J.P.; Meador, M.A.B.; Jana, S.C. Tailoring Mechanical Properties of Aerogels for Aerospace Applications. *ACS Appl. Mater. Interfaces* **2011**, *3*, 613–626. [CrossRef] [PubMed]

24. Zu, G.; Kanamori, K.; Shimizu, T.; Zhu, Y.; Maeno, A.; Kaji, H.; Nakanishi, K.; Shen, J. Versatile Double-Cross-Linking Approach to Transparent, Machinable, Supercompressible, Highly Bendable Aerogel Thermal Superinsulators. *Chem. Mater.* **2018**, *30*, 2759–2770. [CrossRef]

25. Xue, J.; Han, R.; Li, Y.; Zhang, J.; Liu, J.; Yang, Y. Advances in multiple reinforcement strategies and applications for silica aerogel. *J. Mater. Sci.* **2023**, *58*, 14255–14283. [CrossRef]

26. Wang, S.; Zhu, Z.; Zhong, Y.; Gao, J.; Jing, F.; Cui, S.; Shen, X. Comparative studies on the physicochemical properties of in-situ hydrophobic silica aerogels by ambient pressure drying method. *J. Porous Mater.* **2023**, *30*, 2043–2055. [CrossRef]

27. Koyuncu, D.D.E.; Okur, M. Investigation of dye removal ability and reusability of green and sustainable silica and carbon-silica hybrid aerogels prepared from paddy waste ash. *Colloids Surf. A Physicochem. Eng. Asp.* **2021**, *628*, 127370. [CrossRef]

28. Dorcheh, A.S.; Abbasi, M. Silica aerogel; synthesis, properties and characterization. *J. Am. Acad. Dermatol.* **2008**, *199*, 10–26. [CrossRef]

29. Liu, Z.; Zang, C.; Zhang, S.; Zhang, Y.; Yuan, Z.; Li, H.; Jiang, J. Atmospheric drying preparation and microstructure characterization of fly ash aerogel thermal insulation material with superhydrophobic. *Constr. Build. Mater.* **2021**, *303*, 124425. [CrossRef]

30. Akhter, F.; Soomro, S.A.; Inglezakis, V.J. Silica aerogels; a review of synthesis, applications and fabrication of hybrid composites. *J. Porous Mater.* **2021**, *28*, 1387–1400. [CrossRef]

31. Yue, S.; Li, X.; Yu, H.; Tong, Z.; Liu, Z. Preparation of high-strength silica aerogels by two-step surface modification via ambient pressure drying. *J. Porous Mater.* **2021**, *28*, 651–659. [CrossRef]

32. Çok, S.S.; Gizli, N. Microstructural properties and heat transfer characteristics of in-situ modified silica aerogels prepared with different organosilanes. *Int. J. Heat Mass Transf.* **2022**, *188*, 122618. [CrossRef]

33. Shi, F.; Liu, J.-X.; Song, K.; Wang, Z.-Y. Cost-effective synthesis of silica aerogels from fly ash via ambient pressure drying. *J. Non-Crystalline Solids* **2010**, *356*, 2241–2246. [CrossRef]

34. Wu, X.; Fan, M.; Mclaughlin, J.F.; Shen, X.; Tan, G. A novel low-cost method of silica aerogel fabrication using fly ash and trona ore with ambient pressure drying technique. *Powder Technol.* **2018**, *323*, 310–322. [CrossRef]

35. Nguyen, T.H.; Mai, N.T.; Reddy, V.R.M.; Jung, J.H.; Truong, N.T.N. Synthesis of silica aerogel particles and its application to thermal insulation paint. *Korean J. Chem. Eng.* **2020**, *37*, 1803–1809. [CrossRef]

36. Shao, Z.; He, X.; Niu, Z.; Huang, T.; Cheng, X.; Zhang, Y. Ambient pressure dried shape-controllable sodium silicate based composite silica aerogel monoliths. *Mater. Chem. Phys.* **2015**, *162*, 346–353. [CrossRef]

37. Torres, R.B.; Vareda, J.P.; Lamy-Mendes, A.; Durães, L. Effect of different silylation agents on the properties of ambient pressure dried and supercritically dried vinyl-modified silica aerogels. *J. Supercrit. Fluids* **2019**, *147*, 81–89. [CrossRef]

38. Mahadik, S.; Pedraza, F.; Parale, V.; Park, H.-H. Organically modified silica aerogel with different functional silylating agents and effect on their physico-chemical properties. *J. Non-Crystalline Solids* **2016**, *453*, 164–171. [CrossRef]

39. Pan, Y.; He, S.; Gong, L.; Cheng, X.; Li, C.; Li, Z.; Liu, Z.; Zhang, H. Low thermal-conductivity and high thermal stable silica aerogel based on MTMS/Water-glass co-precursor prepared by freeze drying. *Mater. Des.* **2017**, *113*, 246–253. [CrossRef]

40. Chen, K.; Feng, Q.; Ma, D.; Huang, X. Hydroxyl modification of silica aerogel: An effective adsorbent for cationic and anionic dyes. *Colloids Surf. A Physicochem. Eng. Asp.* **2021**, *616*, 126331. [CrossRef]

41. Wang, Y.; Han, J.; Zhai, J.; Yang, D. The effect of different alkaline catalysts on the formation of silica aerogels prepared by the sol–gel approach. *J. Ceram. Soc. Jpn.* **2020**, *128*, 395–403. [CrossRef]

Article

The Recovery of Vanadium Pentoxide (V_2O_5) from Spent Catalyst Utilized in a Sulfuric Acid Production Plant in Jordan

Hiba H. Al Amayreh [1], Aya Khalaf [2], Majd I. Hawwari [3], Mohammed K. Hourani [4] and Abeer Al Bawab [4,5,*]

[1] Department of Scientific Basic Sciences, Faculty of Engineering Technology, Al-Balqa Applied University, Amman 11134, Jordan; hiba.alamyreh@bau.edu.jo

[2] Department of Basic Sciences, Faculty of Arts and Sciences, Al-Ahliyya Amman University, Amman 19328, Jordan; a.khaled@ammanu.edu.jo

[3] Jordan Atomic Energy Commission, Amman 11934, Jordan; majd.hawari@gmail.com

[4] Chemistry Department, School of Science, The University of Jordan, Amman 11942, Jordan; mhourani@ju.edu.jo

[5] Hamdi Mango Center for Scientific Research, The University of Jordan, Amman 11942, Jordan

* Correspondence: drabeer@ju.edu.jo

Abstract: Vanadium is a significant metal, and its derivatives are widely employed in industry. One of the essential vanadium compounds is vanadium pentoxide (V_2O_5), which is mostly recovered from titanomagnetite, uranium–vanadium deposits, phosphate rocks, and spent catalysts. A smart method for the characterization and recovery of vanadium pentoxide (V_2O_5) was investigated and implemented as a small-scale benchtop model. Several nondestructive analytical techniques, such as particle size analysis, X-ray fluorescence (XRF), inductively coupled plasma (ICP), and X-ray diffraction (XRD) were used to determine the physical and chemical properties, such as the particle size and composition, of the samples before and after the recovery process of vanadium pentoxide (V_2O_5). After sample preparation, several acid and alkali leaching techniques were investigated. A noncorrosive, environmentally friendly extraction method based on the use of less harmful acids was applied in batch and column experiments for the extraction of V_2O_5 as vanadium ions from a spent vanadium catalyst. In batching experiments, different acids and bases were examined as leaching solution agents; oxalic acid showed the best percent recovery for vanadium ions compared with the other acids used. The effects of the contact time, acid concentration, solid-to-liquid ratio, stirring rate, and temperature were studied to optimize the leaching conditions. Oxalic acid with a 6% (w/w) to a 1/10 solid-to-liquid ratio at 300 rpm and 50 °C was the optimal condition for extraction (67.43% recovery). On the other hand, the column experiment with a 150 cm long and 5 cm i.d. and 144 h contact time using the same leaching reagent, 6% oxalic acid, showed a 94.42% recovery. The results of the present work indicate the possibility of the recovery of vanadium pentoxide from the spent vanadium catalyst used in the sulfuric acid industry in Jordan.

Keywords: vanadium pentoxide; leaching process; recovery of V_2O_5; recycling; extraction

Citation: Al Amayreh, H.H.; Khalaf, A.; Hawwari, M.I.; Hourani, M.K.; Al Bawab, A. The Recovery of Vanadium Pentoxide (V_2O_5) from Spent Catalyst Utilized in a Sulfuric Acid Production Plant in Jordan. *Materials* **2023**, *16*, 6503. https://doi.org/10.3390/ma16196503

Academic Editor: Daniela Fico

Received: 25 August 2023
Revised: 21 September 2023
Accepted: 26 September 2023
Published: 30 September 2023

1. Introduction

Vanadium, the 22nd most prevalent element, is a metal of great value that is found extensively in the Earth's crust [1]. Vanadium is mostly used in ferrovanadium or as a steel-enhancing agent and accounts for 85% of all vanadium output [2]. Among the numerous vanadium compounds, the oxide of pentavalent vanadium (V_2O_5) is frequently utilized as a catalyst in diverse chemical transformations and represents an important industrial usage of vanadium in addition to that in steel [3].

The recovery of vanadium pentoxide (V_2O_5) from a spent catalyst is of prime importance from industrial, economic, and environmental standpoints [4]. Vanadium pentoxide (V_2O_5) represents a material of prime importance that has important technological applications due to its spectacular electronic, magnetic, and catalytic properties [5]. The major use of vanadium pentoxide (V_2O_5) is as a catalyst in the manufacturing of sulphuric acid, which can be used in oil refining, treatment of steel, production of phosphoric acid, the later can be used in production of detergents, soap, and fertilizers, in addition to many other applications [6]. Vanadium pentoxide (V_2O_5) supported on a silica substrate is used as a catalyst in the contact method in the manufacturing of sulfuric acid where SO_2 is converted into SO_3 [7]. Huge quantities of spent catalysts, however, are accumulated in the environment as industrial waste [8]. Accumulation of the spent vanadium catalyst, however, poses a real environmental problem [9]. The air above the accumulated waste is loaded with vanadium pentoxide dust and transferred to the neighboring areas by wind. Vanadium pentoxide is known to have adverse effects on humans, animals, and plants [10]. Vanadium pentoxide causes nausea, vomiting, salivation, lacrimation, and the absence of a pulse in people and animals. Inhaling vanadium pentoxide (V_2O_5) causes DNA base oxidation, DNA repair, and the production of micronuclei, nucleoplasmic bridges, and nuclear buds. The workers may be more susceptible to cancer and other disorders linked to DNA instability. In human lymphocytes, it causes single-strand DNA breakage. Vanadium pentoxide (V_2O_5) exposure can weaken lung resistance to infections. Some of the health threats of V_2O_5 for humans include nausea, vomiting, the disappearance of pulse, and the risk of lung cancer [11]. These health issues make the recovery of V_2O_5 from spent catalyst waste an important subject for this research project.

The recovery of vanadium pentoxide (V_2O_5) from spent vanadium catalysts via chemical methods has been reported by many researchers [2–6]. In past investigations, different leaching strategies had been applied to liberate V_2O_5 from the support matrix [8,9]. Sulfuric acid, citric acid, oxalic acid, $(NH_4)S_2O_3$, hydrogen peroxide was used as leaching agents under different experimental conditions [4,12,13].

Several studies have investigated the effectiveness of the recovery of V_2O_5 from spent vanadium catalysts. Li et al. designed the direct acid leaching approach, which is used to extract vanadium from a variety of solid sources. Under pressure from oxygen, the vanadium recovery yield increased to 80% [14]. A similar technique was used by Deng et al. to recover vanadium from unburned stone coal, and they obtained a recovery yield of more than 90% [15]. Shi et al. reported a vanadium recovery of approximately 67.7% in their investigations on the recovery of vanadium from black shale utilizing ultrasonic leaching [16]. On the other hand, it is worth mentioning that some studies provide flotation procedures as a useful technique for the recovery of metals from ores and other sources of metal [17,18].

The quest for new or improved leaching methods for the recovery of V_2O_5 is still a hot topic in chemical research because of its relationship with public health, the environment, and the economy [19–21]. Many studies have been performed on other wastes focusing on environmental issues, such as olive mill wastewater (OMW) [22,23] and textile wastewater [24–26]. In Jordan, to our knowledge, this is the first study performed on spent catalysts and on spent catalyst leaching using a solid spent catalyst as the batch and column scale together with a weak acid or base. The present work is necessary to provide a startup project that utilizes the remediation process of vanadium recovery from a spent catalyst. This work was undertaken with the goal of leaching the maximum amount of V_2O_5 from the spent vanadium catalyst and optimizing the experimental conditions to pursue this task. Moreover, this work aimed to choose an environmentally friendly or nearly environmentally friendly leaching reagent with the fewest adverse effects on the environment.

2. Materials and Methods

2.1. Materials

The following chemicals were used as-received from suppliers without further purification. Oxalic acid ($C_2H_2O_4$) (extra pure) was purchased from Scharlau (Barcelona, Spain); sulfuric acid (H_2SO_4) (99.9% purity) and acetic acid (CH_3COOH) (99.9% purity) were supplied by Merck (Rahway, NJ, USA); potassium hydroxide (KOH) (98% purity) was from Riedel-de Haën (Seelze, Germany); citric acid ($C_6H_8O_7$) (Extra pure) and hydrogen peroxide (H_2O_2) (37 wt.%) were supplied by Labo Chemie (Mumbai, India); vanadium pentoxide (V_2O_5) was supplied by BDH Chemical Ltd. (Poole, UK).

2.2. Preparation and Characterization of Spent Vanadium Catalyst Samples

The spent vanadium catalyst was supplied by the Jordan Phosphate Mines Company (JPMC, Maan, Jordan). The catalyst was provided by the company after the turnover finished, and it was withdrawn as waste. The spent vanadium catalyst samples were composed of 6 mm average diameter with 20 mm length hexagonal prism structures. The samples were ground by using a planetary ball mill and sieved to isolate particles of ≤ 50 μm diameter in the laboratories of Jordan Atomic Energy Commission. Ground samples (mass = 5–25 g) were used in batch experiments, keeping the used volume of leaching acid at 50.0 mL, while 2.40 kg of underground solid samples was used in column experiments with the following dimensions: 150 cm long and 5 cm internal diameter.

Spent catalyst samples were characterized via Zeta potential analysis using Zetasizer, Nano-ZS (Malvern Instruments Ltd., Malvern, UK). Particle size/charge analysis and zeta potential Zetatrac analyzer (Microtrac, York, PA, USA) were used for zeta potential. Particle size analysis and X-ray diffraction XRD (MAXIMA 7000 and X-ray spectrophotometer (Shi-madzu, Tokyo, Japan) were used for the XRD pattern. XRF data were analyzed via X-ray fluorescence XRF, lab center XRF-1800 sequential, and X-ray fluorescence spectrophotometer (Shimadzu, Tokyo, Japan). ICP analysis (ICP-AES GBC E1475, New York, NY, USA) was used for V-ion concentration analysis.

The XRD pattern of the fresh catalyst before industrial use and spent catalysts after the catalytic cycle and leaching analysis were investigated using 7000 Shimadzu 2 kW model X-ray spectrophotometer (Shi-madzu, Tokyo, Japan) with nickel-filtered Cu radiation (CuKα) with $\lambda = 1.54056$ Å.

2.3. Extraction Methods

Extraction strategies involved using an acidic/or basic medium for the extraction of V_2O_5 from the spent vanadium acid. Citric acid, oxalic acid, sulfuric acid, acetic acid, and potassium hydroxide were used for the extraction of V_2O_5. Each leaching experiment was performed (triplet) at the same time under the same conditions by the same person using 250 mL and/or 500 mL clean and dry Erlenmeyer flasks. These flasks were filled with different measured masses of spent catalysts varying from 5.00 to 25.00 g and a specific volume of the leaching acid or base measured by using the graduated cylinder that was added to them; then, these flasks were stirred using multi-stage hot plate with a magnetic stirrer (Witeg, Wertheim, Germany). Then, these flasks were filtrated using vacuum filtration to collect both the solid and the resulting leaching solution. The effect of experimental conditions, such as the concentration of the leaching (2–10%), exposure time (1H–4H), stirring rate (300–900 rpm), temperature (25–75 °C), and solid-to-liquid (S/L) (g of spent catalyst/mL of leaching acid or base) ratio (1/10–4/10), on the percent recovered were investigated. A schematic representation of the steps of sample treatment and the following analysis are shown in Scheme 1.

Concerning the column leaching, the column used in the leaching of the spent catalyst was made of glass (150.0 cm height and 5.0 cm internal diameter) fitted with glass wool plugs at the base to retain materials in the column. About 2.40 kg of the dried, spent catalyst was filled in the column. A total of 4.50 L of 6% oxalic acid solution was poured into the column and allowed to equilibrate for 24 h. Liquid leached samples were drained from the column every 30 min and analyzed for their content of vanadium and some other metals by using ICP. Fresh 6% oxalic acid was added to the column to keep it filled and all the spent catalysts were covered with solution. No fresh water was added.

After eluting the last solution sample from the column, a solid sample from the column was taken and analyzed via XRD and XRF.

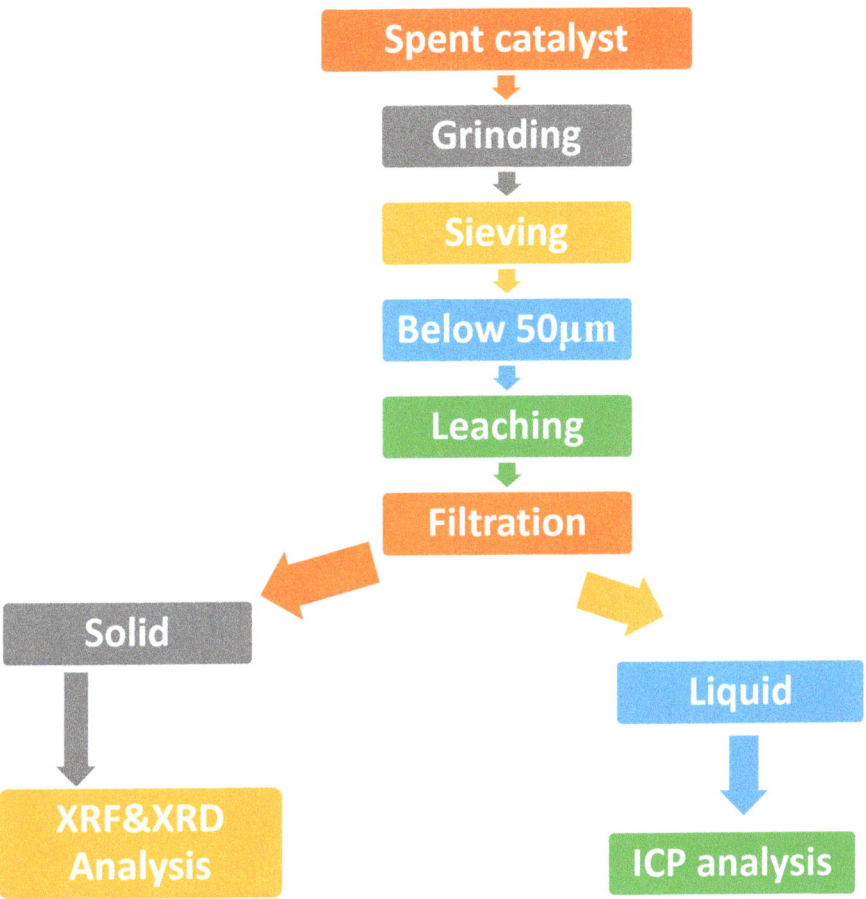

Scheme 1. Schematic representation of the steps of sample treatment and following analysis.

The leaching efficiency was evaluated via the determination of the concentration of vanadium in the solution and relating the recovered vanadium to the nominal value contained in the solid using Equation (1).

$$\%\text{Recovery of Vanadium} = \frac{\text{Vanadium Concentration in the solution} \times \text{V}}{\text{mass of the V} - \text{ion in the original sample}} \times 100\% \quad (1)$$

where V is the volume of the leaching solution (in mL). The concentration of vanadium in the solution was determined via ICP (in ppm).

In batch experiments, ground spent vanadium catalyst was used, but the unground spent vanadium catalyst was used in column experiments.

To minimize errors in the data and to optimize the experimental results, three equivalent runs ($n = 3$) (triplet run) and a data error analysis and statistical analysis with ANOVA test were performed.

3. Results and Discussion

3.1. Characterization of the Spent Vanadium Catalyst

The spent catalyst zeta potential (ζ) was -6.05 ± 0.05 (mV); these low zeta potential values suggest that low coagulation was achieved in the solution, while the XRD data are mentioned in Figures 1 and 2. Figure 1 shows the XRD diffractogram for the active vanadium catalyst (before the industrial process) while Figure 2 shows the diffractogram for the spent vanadium catalyst (after the industrial process).

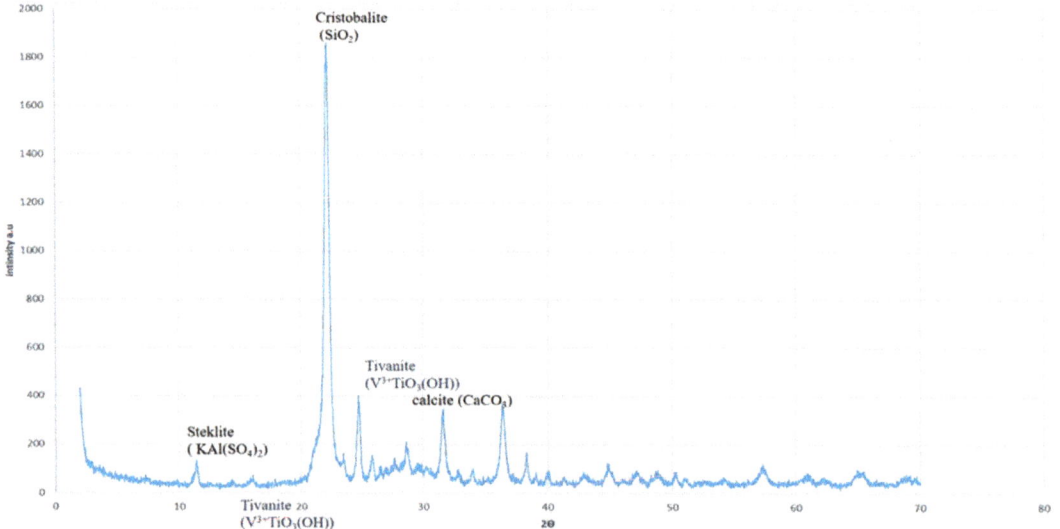

Figure 1. The X-ray powder diffractogram of active vanadium catalyst supplied by the Jordanian Phosphate Mines Company (Maan, Jordan). Experimental conditions: particle size ≤ 50 μm, λ = 1.54056 Å.

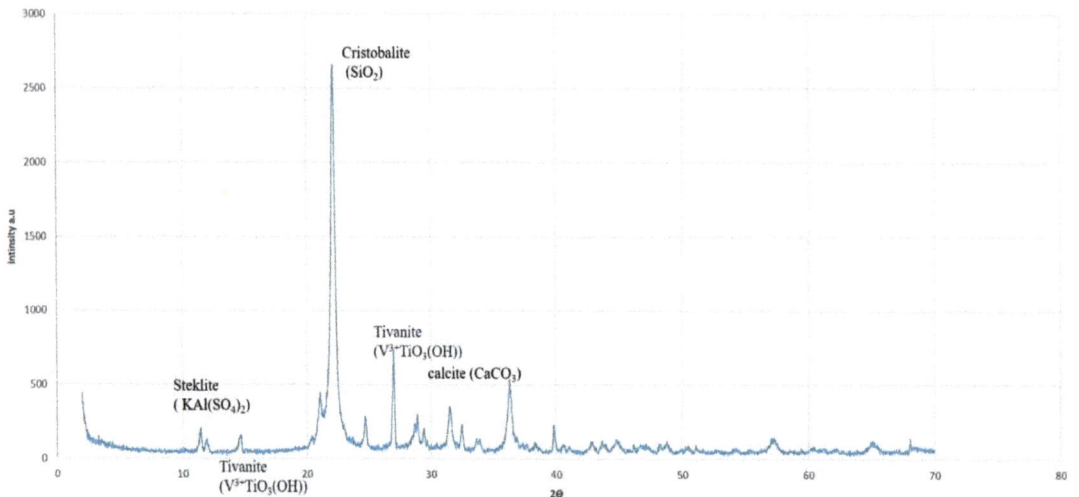

Figure 2. The XRD powder diffractogram of spent vanadium catalyst supplied by the Phosphate Mines Company. Experimental conditions: particle size ≤ 50 μm λ = 1.54056 Å.

The two diffractograms displayed in Figures 1 and 2 for the active and the spent catalyst show that a marginal crystallographic change occurred in the catalyst upon use in the sulfuric acid production industry. The major peaks found at 2Θ 10–12° for steklite ($KAl(SO_4)$), 15–17° and 22–25° for tivanite ($V^{3+}TIO_3(OH)$ reveal the presence of vanadium, 20–23° for cristobalite (SiO_2), and 28–30° for calcite ($CaCO_3$). The absence of any noticeable change in Bragg peaks indicates that there are no major differences between the spent vanadium catalyst and the active catalyst in terms of the crystallographic structure. This result is comparable with the Zhang study that shows the Bragg diffraction of the V_2O_5/TiO_2 catalyst [27]. Apparently, as expected, the catalyst loses its activity because of poisoning without any major change in its chemical structure. This conclusion is reinforced with the results of XRF, displayed in Table 1, which shows that there is a marginal difference between the active and spent vanadium catalysts.

Table 1. Percent (w/w) composition of active and spent vanadium catalyst as determined via X-ray fluorescence spectrometry (XRF). The data were obtained using lab center XRF-1800 (Shimadzu, Tokyo, Japan) (N.D = not detected).

Constituents	Composition (w/w) %		Constituents	Composition (w/w) %	
	Active Vanadium Catalyst	Spent Vanadium Catalyst		Active Vanadium Catalyst	Spent Vanadium Catalyst
SiO_2	64.3229	67.8672	P_2O_5	0.2751	0.2184
SO_3	13.3427	11.4236	MgO	0.2198	0.1091
K_2O	12.1145	11.0965	MnO	0.0132	0.0080
V_2O_5	5.5119	5.3085	NiO	0.0081	0.0080
Al_2O_3	1.5307	1.5182	CuO	0.0067	0.0049
Fe_2O_3	1.1006	0.9865	ZnO	0.0062	0.0055
Na_2O	0.8150	1.0137	SrO	0.0031	0.0040
CaO	0.7265	0.4279	Rb_2O	0.0030	N.D

The XRF data were used as a reference for the succeeding extraction experiments, while the calculation of V_2O_5 recovery depended on Equation (1) and an optimization of the experimental conditions for maximum recovery.

3.2. Effect of Experimental Conditions on Extraction Recovery

3.2.1. The Effect of Acid Type and Concentration on Extraction Recovery

A batch acid extraction experiment was conducted using two concentrations of four different acids (50.00 mL of the acid, 5.00 g of spent catalysts, 300 rpm, 50.0 °C). These acids were citric, oxalic, sulfuric, and acetic acids, while the chosen concentrations of each acid were 2 and 4% (w/w). The acids and the selected concentrations were chosen based on economic and environmental conditions and the expected extraction efficiency.

The results of the effect of the acid type and concentration are displayed in Figures 3 and 4. The figures show the peak time for all investigated acids and the two tested concentrations. The results also indicate that oxalic acid was the best leachant among the tested solutions since it led to a higher % recovery compared to the other acids. It is also a cheap, available, and less vigorous acid. This may be a reason for the strong interaction between V-ion and oxalate ion. The statistical treatment using the ANOVA test (an ANOVA test is a statistical test used to determine if there is a statistically significant difference between two or more categorical groups by testing for differences between means using a variance) of the data showed that there is a significant difference between the results, and the null hypothesis is rejected at $p = 0.5$ if p-values are used. It is worth mentioning that the null hypothesis is a significance test that is used to test the trueness of the variation in experimental data. The null hypothesis assumes with a certain probability that the variation between experimental data is due only to random errors. A positive significance test indicates with a certain probability (p) that the null theory is correct, and the observed variation in data is due to random errors, while a negative test indicates that the null theory is not valid and the observed variation in the data is due to real differences in the measured quantities (The p-value is a number calculated from a statistical test that describes how likely you are to have found a particular set of observations if the null hypothesis were true). This analysis means that oxalic acid gives the best recovery (%) compared to other acids, and oxalic acid is better than the other acids that were used. The decay in leaching efficiency after 3 h might be attributed to another competing mechanism, like the readsorption of the vanadium species on the spent catalyst frame according to the diffusion and concentration differences between the solid and liquid [23]. Figure 5 shows that leaching with 4% oxalic acid is superior to leaching with 2% oxalic acid, and this can easily be explained on a thermodynamic basis where, at a higher concentration of reactants, the equilibrium position will be shifted to the right (i.e., towards a higher degree of completion of the reaction).

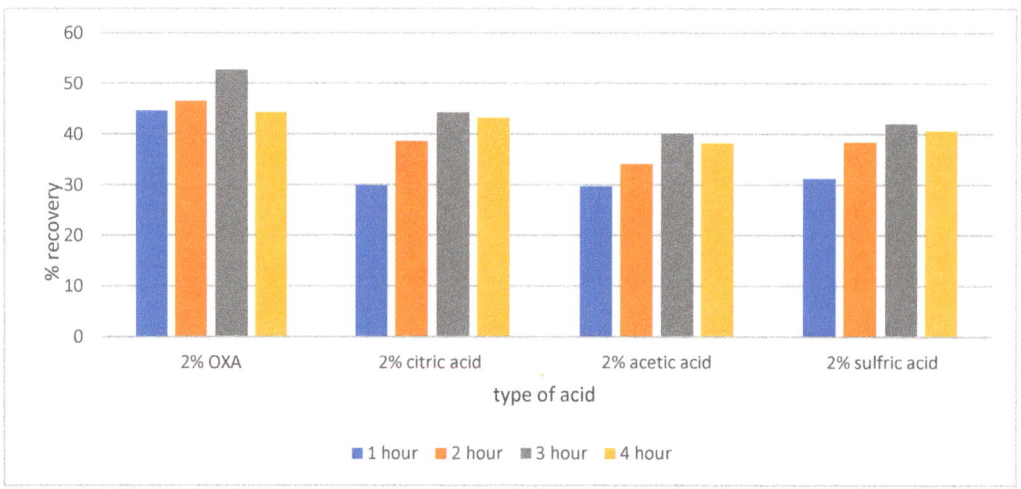

Figure 3. % Recovery of V-ion recovered versus the type of acid (2% w/w) after leaching.

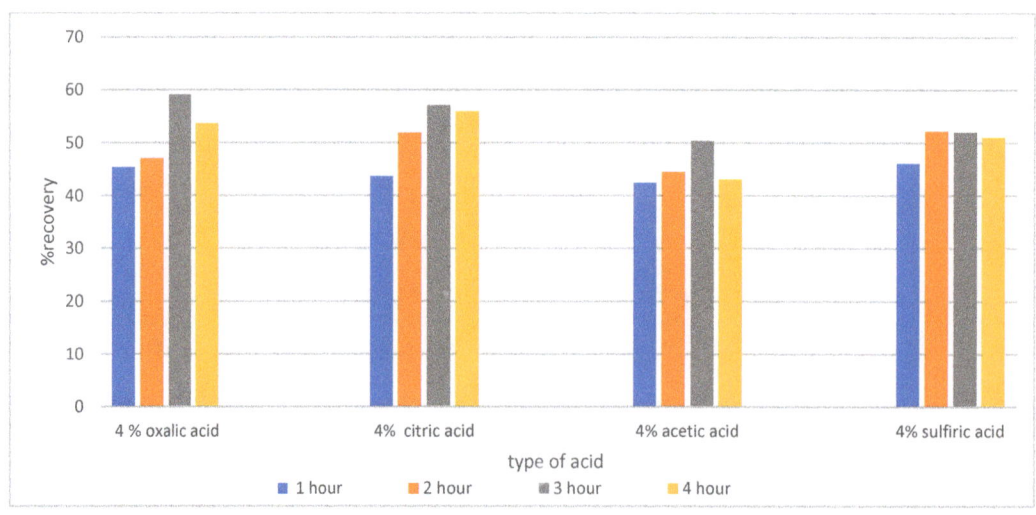

Figure 4. % Recovery of V-ion versus the type of acid (4% w/w) after leaching.

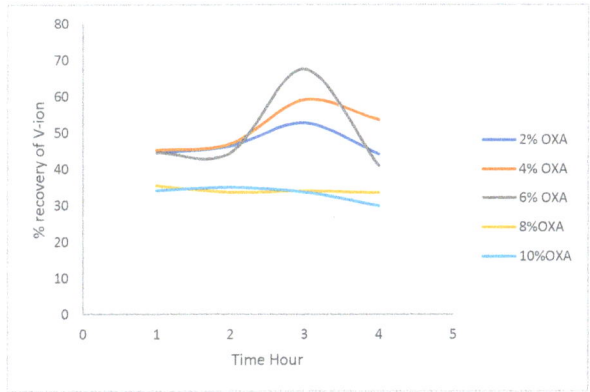

Figure 5. % Recovery of V-ions recovered at different times for different oxalic acid w/w% concentrations. (Color captions are shown on the right of the figure.).

3.2.2. The Effect of Oxalic Acid Concentration on Leaching Efficiency

An additional investigation into the effect of changing the leachant acid concentration on the leaching efficiency was undertaken by evaluating the leaching efficiency across different concentrations of oxalic acid. A succession of oxalic acid concentrations—specifically, 2%, 4%, 6%, 8%, and 10% (w/w)—was meticulously formulated and subsequently employed for the purpose of vanadium leaching from the expended vanadium catalyst using 5.00 g of spent catalysts with a stirring rate of 300 rpm and temperature of 50.0 °C. Figure 5 effectively portrays the extent of leached vanadium in relation to the oxalic acid concentration.

The results displayed in Figure 5 show mostly a peak at 3 h of leaching time, as indicated previously. The highest peak value is for the concentration of the acid of 6% (w/w) oxalic acid concentration. The decline in the concentration of leached vanadium might be attributed to the formation of insoluble oxalic acid–vanadium complexes. This explanation is supported by the observation of a black precipitate (which will be characterized in a future work) in the solution shown in Figure 6 upon using higher concentrations (8% and 10%) of oxalic acid [28]. It is worth mentioning that the ANOVA statistical analysis of these results indicated a significant difference between the results, and that the null hypothesis is rejected at $p = 0.05$ so that 6% of oxalic acid concentration is used as the optimum concentration compared to the other concentrations.

Figure 6. The oxalate complex that precipitated during our work.

3.2.3. The Effect of Stirring Rate

Figure 7 shows the effect of the stirring rate on the vanadium extraction efficiency at a constant mass of spent vanadium catalyst (5.00 g), extraction time (3 h), and volume of oxalic acid solution (50 mL). The results indicate that the optimal stirring rate is 300 rpm for almost all the investigated concentrations.

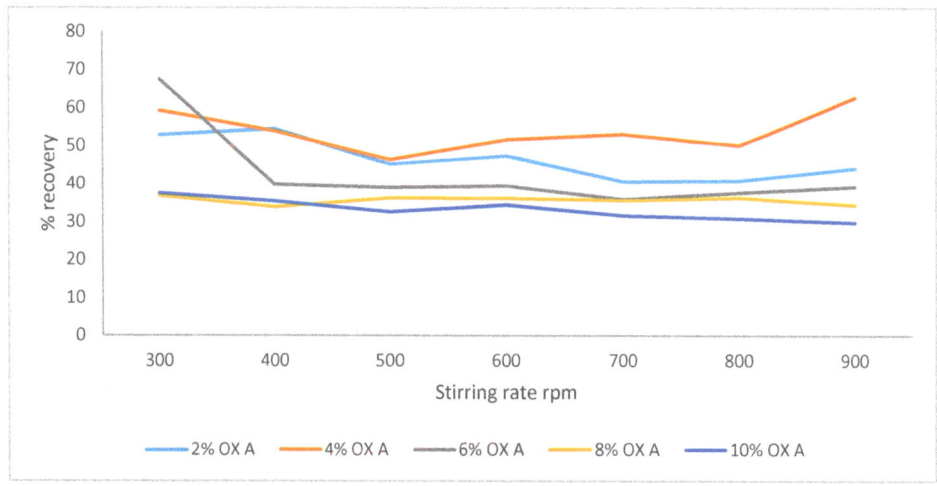

Figure 7. The percent recovery of vanadium versus different stirring rates with different concentrations of oxalic acid.

These results are attributed to the fact that a lower stirring rate allows for better contact between the solvent and the materials inside the pores in the spent porous catalyst. A higher rate of stirring causes the readsorption of the V-ion into the solid from the liquid [29]. In the literature, especially the studies of Mazurek in Poland, there is no mention of variations in stirring rate conditions; they just used 300 rpm as the optimum without listing the values of different rates [29].

3.2.4. The Effect of Solid-to-Leachant Volume (S/L Ratio)

Figure 8 shows the effect of the S/L ratio on the percent recovery of vanadium using five different oxalic acid concentrations (300 rpm, 50.0 °C).

The results indicate that the highest recovery was obtained at the smallest S/L ratio. These results are in general agreement with the leaching efficiency of similar systems [30,31]. In general, the leaching rate depends on several factors, such as the temperature, the size of the solid particles, the solvent used for leaching, and the diffusion (movement) of the leached ions towards the bulk of the solution. Increasing the S/L ratio will, in general, decrease the diffusion due to the increased viscosity. This will result in a hindrance in the movement, leading to a decreased leaching efficiency. This also could be explained using Equation (1), shown above: There is an increase in the mass of V_2O_5 in the spent catalysts as the mass of the spent catalyst increases; however, on the other hand, the mass of V_2O_5 in the spent catalyst is in the denominator of Equation (1). It is clear that the ppm of V recovered from the tested sample is multiplied by the volume of the recovered sample after leaching (Vf) occurs in the nominator, so the percent recovery decreases as a result of the increases in the S/L ratio since the denominator increases as the mass of the spent catalyst increases [30,31].

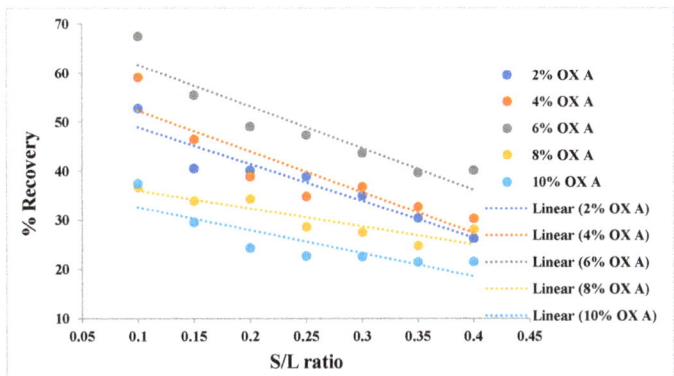

Figure 8. The percent recovery of vanadium ion versus different S/L ratios of spent catalysts with different oxalic acid concentrations.

3.2.5. The Effect of Temperature on Vanadium Recovery from the Spent Catalyst

The effect of temperature on the recovery of vanadium from a spent vanadium catalyst was investigated via the determination of the percent recovery as a function of the extraction temperature between 25 °C and 75 °C under identical experimental conditions. Figure 9 shows that the temperature in the investigated range has little influence on the percent recovery except at 50 °C. ANOVA statistical treatment of the data shows that the null hypothesis is rejected at $p = 0.05$, and there is a significant difference in recovery at 50 °C, while there is an insignificant effect on recovery at other temperatures, and the null hypothesis is valid at all investigated temperatures except 50 °C at $p = 0.05$. This temperature is the same as the temperature chosen in the studies in Poland and Turkey [19,29] where it was mentioned that the effect of temperature on the leaching yield of the vanadium compounds is much smaller when leaching in an acidic environment than in an alkaline environment, while 50 °C was the optimum temperature for both cases without further explanation. From our perspective, this increase in recovery at 50.0 °C may be attributed to the coagulation of the solid spent catalyst in addition to the effect of mass transfer with diffusion factors. This might be explained by two contradicting factors: One is the increased dynamics of the system, which tends to increase the percent recovery with increasing temperature, and the other is the tendency of coagulation and agglomeration of the solution contents, which hinders the leaching of vanadium from the spent catalyst [29,32].

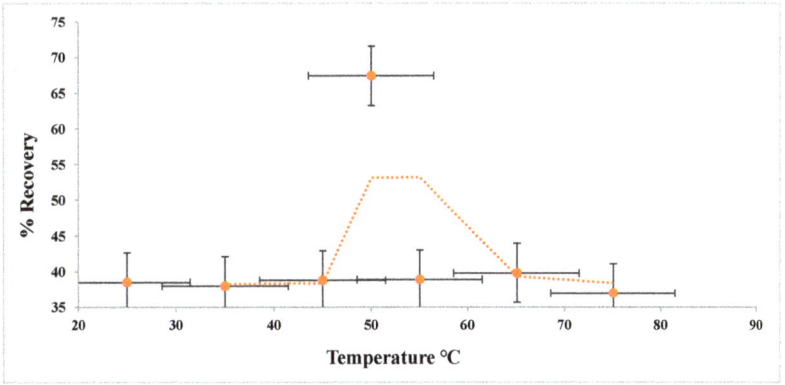

Figure 9. The percent recovery of vanadium ion versus temperature.

3.3. Column Leaching Experiments

Column-leaching experiments were performed on the spent vanadium catalyst (not grinded) packed in the column, as described in the experimental section. The composition and structure of the used spent catalyst were identical to those used in the batch experiments, as evidenced by the XRD and XRF techniques. Figure 10 shows the results of the column extraction of vanadium from the spent catalyst with 6% oxalic acid solution as a function of time. Liquid samples of 50.00 mL were collected at 30 min. intervals after 24 h equilibration time from the time that the eluted solution was introduced to the column. Extraction efficiency shows a peak after 144 h (about 6 days) (Figure 10).

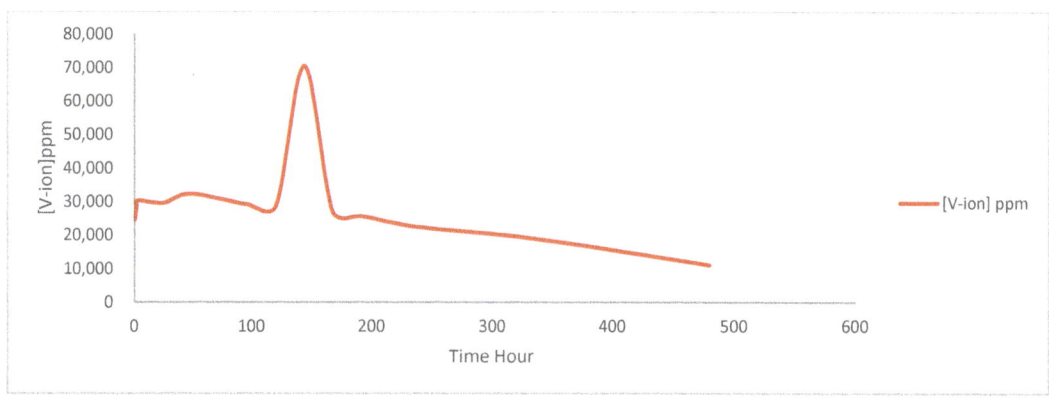

Figure 10. Vanadium ion concentration recovered from column versus different times using 6% oxalic acid solution.

Extraction efficiency was examined via XRD and XRF analysis of the spent vanadium catalyst after 6 days of column extraction. The results are displayed in Figure 11 and Table 2. The XRD diffractogram (Figure 11) shows the absence of the tivanite ($V^{3+}TiO_3(OH)$) peak, which indicates almost total extraction of vanadium from the spent catalyst. XRF (Table 2) shows that the residual V_2O_5 in the spent vanadium catalyst is only 0.30% compared to an initial value of 5.31%. This result attests to the efficiency of the column extraction.

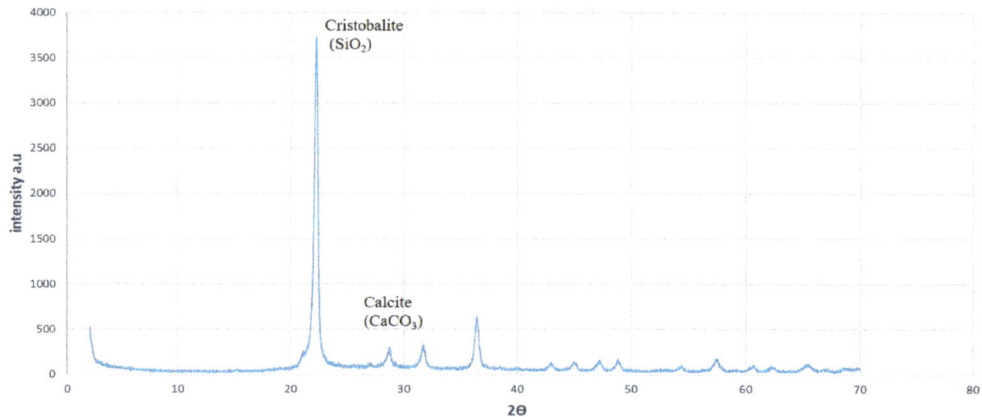

Figure 11. The XRD diffraction pattern for CSC.

Table 2. XRF data for CSC (N.D = not detected).

Constituents	*w/w*% in CSC	Constituents	*w/w*% in CSC
SiO_2	97.5526%	Na_2O	0.0669%
SO_3	0.7799%	CaO	0.0302%
K_2O	0.7623%	P_2O_5	0.0479%
V_2O_5	0.2964%	NiO	0.0106%
Al_2O_3	0.3269%	CuO	0.0028%
Fe_2O_3	0.1104%	MgO, MnO	N.D

3.4. Alkaline Leaching Experiments

The alkaline leaching of vanadium ions from the vanadium-spent catalyst was also investigated. KOH was the only base used to leach vanadium ions from the spent catalyst, noting that alkaline leaching is not preferred because most metals tend to precipitate in basic solutions. In addition, leaching from basic solutions is expensive and economically unfavorable, as shown in Figure 12, which uses 50 mL of 2% KOH and 4%KOH *w/w*%, 5.0000 g of spent catalysts, 300 rpm, 50.0 °C. The alkaline leaching leads to a higher [V] ion recovery compared with some investigated acids; however, in this research, oxalic acid is still superior to KOH as a leacher [17,18].

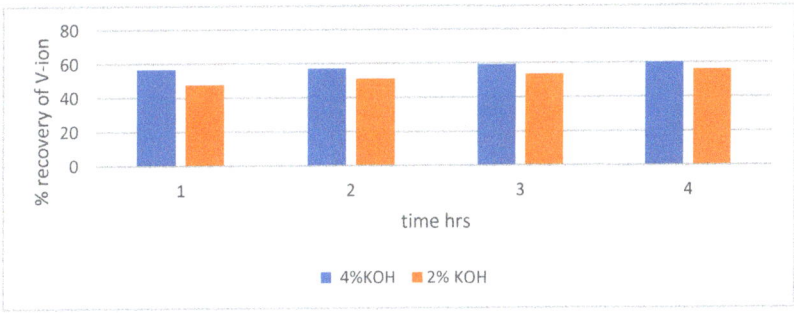

Figure 12. The percent recovery of vanadium ion versus time (leaching using KOH).

4. Conclusions

In the present work, a methodology for the extraction of vanadium from a spent vanadium catalyst used in a sulfuric acid production plant in Jordan was proposed and optimized. Our results provide simple and safe remediation with an almost environmentally friendly procedure for the recovery of V_2O_5 from spent catalysts used in the Jordanian industry. In this approach, oxalic acid was used with 6% (*w/w*) and a 300 rpm stirring rate at 50.0 °C for 3.0 h with 2525 ppm of V-ion recovered in the solution and a percent recovery of 80.58% based on the XRF results. Also, this leaching solution was used for both batch experiments and column experiments. The latter took place over 6 days (144 h), leading to a concentration of recovered vanadium ion equal to 70,620 ppm with a percent recovery of 67.43%. This remediation can be considered a successful approach to the removal of hazardous materials from the environment. Reusing these materials in a beneficial project can open the door to the brand-new application of spent catalyst remediation. The optimization involved the selection of the leaching agent. The leaching agent that showed the highest recovery was a 6% oxalic acid solution. The optimized experimental conditions using 6% oxalic acid solution in batch experiments were 50 °C, a 300 rpm stirring rate, a 0.1 S/L ratio, and 3 hours of extraction time. The highest recovery achieved under the optimized conditions was 67.43%.

Column experiments, on the other hand, lasted for six days, reaching the maximum recovery (94.42%) using 6% oxalic acid under ambient conditions and a gravity-driven flow rate, which is a very encouraging condition for routine leaching and large-scale leaching. Obviously, the extraction efficiency achieved via column extraction requires a longer extraction time and different experimental conditions.

These results lay the foundation for further studies on the extraction of vanadium from hundreds of tons of accumulated spent catalysts and could help to save the environment using an economically acceptable methodology.

Author Contributions: Conceptualization, A.A.B., M.K.H. and M.I.H.; methodology, A.A.B., M.I.H.; validation, A.A.B.; formal analysis, A.A.B., H.H.A.A. and A.K.; investigation, H.H.A.A. and A.K.; resources, A.A.B. and M.I.H.; data curation, H.H.A.A., A.K. and A.A.B.; writing—original draft preparation, H.H.A.A. and A.K.; writing—review and editing, A.A.B., M.I.H., M.K.H.; visualization, A.A.B., M.I.H. and M.K.H.; supervision, M.I.H., A.A.B. and M.K.H.; project administration, A.A.B. and M.I.H.; funding acquisition, A.A.B. and M.I.H. All authors have read and agreed to the published version of the manuscript.

Funding: This research was funded by the Deanship of Academic Research (DAR) at The University of Jordan (UJ) (Grant No. 1679/2019).

Data Availability Statement: All data analyzed during this study are included in this published article.

Acknowledgments: The authors would like to acknowledge the Jordan Atomic Energy Commission (JAC) represented by H.E. Khaled Toukan and Ahmad Al Sabbagh for their scientific and technical support in this project. The authors would like to extend their thanks for Jordan Phosphate Mines Company (JPMC) for providing the raw materials (the spent catalysts), represented by H.E. Mohammed Thneibat, Abdel Wahab Alrowwad, Mohammed Migdadi.

Conflicts of Interest: The authors declare no conflict of interest.

References

1. Imtiaz, M.; Rizwan, M.S.; Xiong, S.; Li, H.; Ashraf, M.; Shahzad, S.M.; Shahzad, M.; Rizwan, M.; Tu, S. Vanadium, recent advancements, and research prospects: A review. *Environ. Int.* **2015**, *80*, 79–88. [CrossRef] [PubMed]
2. Nasimifar, A.; Mehrabani, J. A review on the extraction of vanadium pentoxide from primary, secondary, and co-product sources. *Int. J. Min. Geo. Eng.* **2022**, *56*, 361–382. [CrossRef]
3. Wang, B.; Yang, Q. Recovery of V_2O_5 from spent SCR catalyst by H_2SO_4-ascorbic acid leaching and chemical precipitation. *J. Environ. Chem. Eng.* **2022**, *10*, 108719. [CrossRef]
4. Romanovskaia, E.; Romanovski, V.; Kwapinski, W.; Kurilo, I. Selective recovery of vanadium pentoxide from spent catalysts of sulfuric acid production: Sustainable approach. *Hydrometallurgy* **2021**, *200*, 105568. [CrossRef]
5. Blum, R.-P.; Niehus, H.; Hucho, C.; Fortrie, R.; Ganduglia-Pirovano, M.V.; Sauer, J.; Shaikhutdinov, S.; Freund, H.-J. Surface Metal-Insulator Transition on a Vanadium Pentoxide (001) Single Crystal. *Phys. Rev. Lett.* **2007**, *99*, 226103. [CrossRef]
6. Calderón, H.; Endara, D. Recovery of Vanadium from Acid and Basic Leach Solutions of Spent Vanadium Pentoxide Catalysts. *J. Geol. Resour. Eng.* **2015**, *4*, 213–218. [CrossRef]
7. Ognyanova, A.; Ozturk, A.T.; Michelis, I.D.; Ferella, F.; Taglieri, G.; Akcil, A.; Vegliò, F. Metal extraction from spent sulfuric acid catalyst through alkaline and acidic leaching. *Hydrometallurgy* **2009**, *100*, 20–28. [CrossRef]
8. Yang, B.; He, J.; Zhang, G.; Guo, J. (Eds.) Vanadium and its compounds. In *Vanadium*; Elsevier: Amsterdam, The Netherlands, 2021; pp. 9–32. [CrossRef]
9. Wexler, P.; Judson, R.; de Marcellus, S.; de Knecht, J.; Leinala, E. Health effects of toxicants: Online knowledge support. *Life Sci.* **2016**, *145*, 284–293. [CrossRef]
10. Mohanty, J.; Rath, P.C.; Bhattacharya, I.N.; Paramguru, R.K. The recovery of vanadium from spent catalyst: A case study. *Miner. Process. Extr. Metall.* **2011**, *120*, 56–60. [CrossRef]
11. Bauer, G.; Güther, V.; Hess, H.; Otto, A.; Roidl, O.; Roller, H.; Sattelberger, S. Vanadium and Vanadium Compounds. In *Ullmann's Encyclopedia of Industrial Chemistry*; John Wiley & Sons, Ltd.: Hoboken, NJ, USA, 2000. [CrossRef]
12. Lim, Y.; Choi, J.; Yoo, K.; Lee, S.-H. Feasibility study on the use of magnetic susceptibility for recovery of vanadium component in magnetite. *Geosystem Eng.* **2022**, *25*, 280–284. [CrossRef]
13. Hu, P.; Zhang, Y.; Liu, H.; Liu, T.; Li, S.; Zhang, R.; Guo, Z. High efficient vanadium extraction from vanadium slag using an enhanced acid leaching-coprecipitation process. *Sep. Purif. Technol.* **2023**, *304*, 122319. [CrossRef]
14. Li, M.; Wei, C.; Qiu, S.; Zhou, X.; Li, C.; Deng, Z. Kinetics of vanadium dissolution from black shale in pressure acid leaching. *Hydrometallurgy* **2010**, *104*, 193–200. [CrossRef]

15. Deng, Z.; Wei, C.; Fan, G.; Li, M.; Li, C.; Li, X. Extracting vanadium from stone-coal by oxygen pressure acid leaching and solvent extraction. *Trans. Nonferrous Met. Soc. China* **2010**, *20*, s118–s122. [CrossRef]
16. Jammulamadaka, H.; Pisupati, S.V. A Critical Review of Extraction Methods for Vanadium from Petcoke Ash. *Fuels* **2023**, *4*, 58–74. [CrossRef]
17. Shen, Z.; Wen, S.; Wang, H.; Miao, Y.; Wang, X.; Meng, S.; Feng, Q. Effect of dissolved components of malachite and calcite on surface properties and flotation behavior. *Int. J. Miner. Metall. Mater.* **2023**, *30*, 1297–1309. [CrossRef]
18. Zhao, W.; Yang, B.; Yi, Y.; Feng, Q.; Liu, D. Synergistic activation of smithsonite with copper-ammonium species for enhancing surface reactivity and xanthate adsorption. *Int. J. Min. Sci. Technol.* **2023**, *33*, 519–527. [CrossRef]
19. Erust, C.; Akcil, A.; Bedelova, Z.; Anarbekov, K.; Baikonurova, A.; Tuncuk, A. Recovery of vanadium from spent catalysts of sulfuric acid plant by using inorganic and organic acids: Laboratory and semi-pilot tests. *Waste Manag.* **2016**, *49*, 455–461. [CrossRef]
20. Zhang, Y.; Zhao, Y.; Xiong, Z.; Gao, T.; Xiao, R.; Liu, P.; Liu, J.; Zhang, J. Photo- and thermo-catalytic mechanisms for elemental mercury removal by Ce doped commercial selective catalytic reduction catalyst (V_2O_5/TiO_2). *Chemosphere* **2022**, *287*, 132336. [CrossRef]
21. Tiwari, M.; Bajpai, S.; Dewangan, U.; Tamrakar, R. Suitability of leaching test methods for fly ash and slag: A review. *J. Radiat. Res. Appl. Sci.* **2015**, *8*, 523–537. [CrossRef]
22. Odeh, F.; Abu-Dalo, M.; Albiss, B.; Ghannam, N.; Khalaf, A.; Amayreh, H.H.; Al Bawab, A. Coupling magnetite and goethite nanoparticles with sorbent materials for olive mill wastewater remediation. *Emergent Mater.* **2022**, *5*, 77–88. [CrossRef]
23. Al Bawab, A.; Abu-Dalo, M.; Khalaf, A.; Abu-Dalo, D. Olive Mill Wastewater (OMW) Treatment Using Photocatalyst Media. *Catalysts* **2022**, *12*, 539–555. [CrossRef]
24. Abbas, H.A.; Nasr, R.A.; Khalaf, A.; Bawab, A.A.; Jamil, T.S. Photocatalytic degradation of methylene blue dye by fluorite type $Fe_2Zr_{2-x}W_xO_7$ system under visible light irradiation. *Ecotoxicol. Environ. Saf.* **2020**, *196*, 110518. [CrossRef] [PubMed]
25. Nasr, R.A.; Abbas, H.A.; Khalaf, A.; Bozeya, A.; Jamil, T.S. Nano-sized $Ga_{2-x}CuxZr_{2-x}W_xO_7$ for Malachite green decolorization under visible light. *Desalination Water Treat.* **2020**, *183*, 389–403. [CrossRef]
26. Abu-Zurayk, R.; Khalaf, A.; Abbas, H.A.; Nasr, R.A.; Jamil, T.S.; Al Bawab, A. Photodegradation of Carbol Fuchsin Dye Using an $Fe_{2-x}Cu_xZr_{2-x}W_xO_7$ Photocatalyst under Visible-Light Irradiation. *Catalysts* **2021**, *11*, 1473–1489. [CrossRef]
27. Zhao, Y.; Chen, L.; Yi, H.; Zhang, Y.; Song, S.; Bao, S. Vanadium Transitions during Roasting-Leaching Process of Vanadium Extraction from Stone Coal. *Minerals* **2018**, *8*, 63. [CrossRef]
28. Lee, M.; Le, M.N. Selective dissolution of vanadium (V) from spent petroleum catalysts by oxalic acid solution and its kinetic study. *J. Min. Metall. Sect. B Metall.* **2020**, *56*, 127–133. [CrossRef]
29. Mazurek, K. Recovery of vanadium, potassium, and iron from a spent vanadium catalyst by oxalic acid solution leaching, precipitation, and ion exchange processes. *Hydrometallurgy* **2013**, *134*, 26–31. [CrossRef]
30. Irannajad, M.; Meshkini, M.; Azadmehr, A.R. Leaching of zinc from low grade oxide ore using organic acid. *Physicochem. Probl. Miner. Process.* **2013**, *49*, 547–555.
31. Chen, Y.; Wang, Y.; Bai, Y.; Feng, M.; Zhou, F.; Lu, Y.; Mu, T. Mild and efficient recovery of lithium-ion battery cathode material by deep eutectic solvents with natural and cheap components. *Green Chem. Eng.* **2022**, *4*, 303–311. [CrossRef]
32. Luo, H.; Cheng, Y.; He, D.; Yang, E.-H. Review of leaching behavior of municipal solid waste incineration (MSWI) ash. *Sci. Total Environ.* **2019**, *668*, 90–103. [CrossRef]

Article

The Use of Lightweight Aggregates in Geopolymeric Mortars: The Effect of Liquid Absorption on the Physical/Mechanical Properties of the Mortar

Emilia Vasanelli [1,*], Silvia Calò [2], Alessio Cascardi [3] and Maria Antonietta Aiello [2]

[1] CNR-ISPC (National Research Council—Institute of Heritage Science), University Campus, Prov. le Lecce Monteroni, 73100 Lecce, Italy

[2] Department of Innovation Engineering, University of Salento, University Campus, Prov. le Lecce Monteroni, 73100 Lecce, Italy; silvia.calo@unisalento.it (S.C.); antonietta.aiello@unisalento.it (M.A.A.)

[3] Department of Civil Engineering, University of Calabria, Via P. Bucci, 87036 Rende, Italy; alessio.cascardi@unical.it

* Correspondence: emilia.vasanelli@cnr.it

Citation: Vasanelli, E.; Calò, S.; Cascardi, A.; Aiello, M.A. The Use of Lightweight Aggregates in Geopolymeric Mortars: The Effect of Liquid Absorption on the Physical/Mechanical Properties of the Mortar. *Materials* **2024**, *17*, 1798. https://doi.org/10.3390/ma17081798

Academic Editors: F. Pacheco Torgal and Maria Harja

Received: 20 March 2024
Revised: 10 April 2024
Accepted: 11 April 2024
Published: 14 April 2024

Abstract: Geopolymers have been proposed as a green alternative to Portland cement with lowered carbon footprints. In this work, a geopolymeric mortar obtained using waste materials is studied. Fly ash, a waste generated by coal combustion, is used as one of the precursors, and waste glass as lightweight aggregates (LWAs) to improve the thermal performance of the mortar. The experimental study investigates the effect of varying the alkali activating solution (AAS) amount on the workability, compressive strength, and thermal conductivity of the mortar. Indeed, AAS represents the most expensive component in geopolymer production and is the highest contributor to the environmental footprint of these materials. This research starts by observing that LWA absorbs part of the activating solution during mixing, suggesting that only a portion of the solution effectively causes the geopolymerization reactions, the remaining part wetting the aggregates. Three mixes were investigated to clarify these aspects: a reference mix with a solution content calibrated to have a plastic consistency and two others with the activating solution reduced by the amount absorbed by aggregates. In these cases, the reduced workability was solved by adding the aggregates in a saturated surface dry state in one mix and free water in the other. The experimental results evidenced that free water addiction in place of a certain amount of the solution may be an efficient way to improve thermal performance without compromising the resistance of the mortar. The maximum compressive strength reached by the mortars was about 10 MPa at 48 days, a value in line with those of repair mortars. Another finding of the experimental research is that UPV was used to follow the curing stages of materials. Indeed, the instrument was sensitive to microstructural changes in the mortars with time. The field of reference of the research is the rehabilitation of existing buildings, as the geopolymeric mortars were designed for thermal and structural retrofitting.

Keywords: geopolymer mortars; lightweight aggregates; sustainable mortars; thermal conductivity

1. Introduction

Geopolymer concretes and mortars have received high attention from researchers in the last twenty years [1–3]. They have been proposed as a green and sustainable alternative to ordinary Portland cement with lowered carbon footprints [4–6]. Their use to replace Portland cement products generally results in vast energy and virgin materials savings, resulting in sustainable concrete production. Geopolymers reuse the solid waste generated in the industrial and manufacturing sectors, following the circular economy principles [7]. They are aluminosilicate-based amorphous inorganic materials obtained through a polymerization process, starting from natural or waste materials with a high content of aluminum or silicon (precursors), such as slag from blast furnaces from steel mills, clays, flying and volcanic ashes, etc. [8,9]. The reaction of the precursor powder with an alkaline activating solution (AAS), consisting of sodium and/or potassium hydroxides and silicates, at a temperature below 100 °C, produces this alkali-activated material, a robust polymer used as the binder in concretes and mortars. It has been proved that in the construction industry, the production of ordinary Portland cement is a high greenhouse gas emitter, with almost 8% of total CO_2 production in the world [6]. Thus, the use of geopolymers represents a challenge to producing new building materials or optimizing the existing ones, reducing consumption during production, the emission of greenhouse gases, and environmental impact [10].

Besides higher sustainability, geopolymers are considered optimal candidates to substitute cement because they overcome some drawbacks. Depending on employed precursors, activating solutions, and aggregates, geopolymers may have better properties than traditional cement-based materials, like improved durability, fire resistance, and reduced thermal conductivity [11]. In particular, the literature results proved that geopolymers may be used as thermal insulating mortars and lightweight structural concrete with even better performances than cement-based materials [12–14]. Several methods have been reported to improve the thermal properties of geopolymers [15]: the incorporation of lightweight aggregates, the omission of fine aggregates (the no-fines concrete), and the use of foaming and air-entraining admixtures that create large bubbles and voids in concrete (the foamed or aerated concrete) [16,17]. Lightweight aggregates (LWAs) have many benefits, the most important being a high strength-to-weight ratio, but they also enhance tensile strength, and heat and sound insulation, and lower thermal expansion [15]. On the other hand, their high porosity poses some drawbacks in the mixing procedure and internal curing of the material, affecting its mechanical and physical properties. This aspect is well known from the literature on lightweight concrete mix design [18,19]. Lightweight aggregates influence more than normal weight aggregates the internal curing condition of the material, due to water absorption and interaction between lightweight aggregates and pastes. Indeed, the state of moisture of aggregates before mixing strongly influences slump loss and shrinkage of the material [20,21]. This aspect has been poorly investigated in the literature for geopolymers employing lightweight aggregates. Huiskes et al. [22] analyzed the effect of pre-soaking LWAs on the properties of geopolymer concrete. They found that pre-soaked LWAs in the mixture generated more reliable characteristics than non-pre-soaking mixtures in terms of stability, compaction, air content, porosity, and particle distribution. Kupaei et al. [23] added oil palm shells (OPS) as lightweight aggregates in two moisture conditions: the air dry (AD) and saturated surface dry (SSD) condition. They found that the compressive strength development was dependent on the conditions of OPS. The mixes with air-dry OPS produced a lower strength of about 7% compared to the mixes with OPS in the SSD condition.

The present study wants to deepen the effects of liquid absorption by LWA on the final properties of a geopolymeric mortar. It is part of extensive research aiming at developing a geopolymeric mortar for the seismic/thermal retrofitting of existing buildings, from the perspective of the circular economy. The mortar employs waste materials as aggregates and precursors. The precursor is fly ash, a waste generated by coal combustion and widely used as an additive in the building sector. The aggregates were expanded waste glass particles of different grades. They were used as LWAs to improve the thermal insulating properties of the geopolymeric mortar. The LWAs usually absorb water when employed in dry/ambient conditions in concrete mixes. In the case of a geopolymer, the liquid part of the mix is the alkaline activating solution, which is indispensable to start the geopolymerization reactions. When LWAs are added to the mix at ambient conditions, part of the AAS is absorbed by aggregates [22,23]. Thus, it can be supposed that only part of the solution effectively causes the geopolymerization reactions to take place as the remaining part wets the aggregates and ensures the necessary workability of the mix. This circumstance is antieconomic if it is considered that the alkaline solution is expensive and has a high impact on the total cost of the material. Furthermore, it has been proved that AAS has a higher environmental impact regarding categories other than global warming, namely freshwater, terrestrial, and marine aquatic ecotoxicity potential, ozone depletion potential, and human toxicity potential [4,6,24]. Based on these considerations, this paper wants to investigate the effects of reducing the content of the AAS by subtracting the amount that aggregates absorbed during mixing. To this scope, three formulations were made: a reference mix in which the content of the AAS was sufficient to obtain a plastic consistency of the mortar and two other mixes with a reduced content of the activating solution. In these cases, the subsequent reduced workability was solved in two ways: by adding free water (FW mix) or aggregates in the SSD state (SSD mix). The workability, compressive strength, and thermal conductivity were measured to compare the properties of the SSD and FW mixes with the reference one.

2. Materials and Methods

2.1. Materials

Three geopolymeric mortar formulations were developed. All the mixes were obtained by mixing the precursors (fly ash and metakaolin) with the alkali-activating solution and then adding LWAs.

The composition obtained by SEM-EDS analysis and the loss of ignition (LOI) of fly ash and metakaolin are reported in Table 1.

Table 1. Fly ash and metakaolin compositions obtained by SEM-EDS.

	\multicolumn{9}{c	}{Oxides Concentration %}	LOI %
	SiO_2	CaO	Al_2O_3	Fe_2O_3	K_2O	TiO_2	SO_3	Na_2O	MgO	
Fly ash	59.54	1.76	27.26	2.91	2.71	0.59	1.02	2.56	1.65	3.20
Metakaolin	56.50	2.51	27.92	0.77	1.83	0.44	0.29	9.03	0.71	1.29

The alkali-activating solution was made of sodium silicate (SS) solution (Extra Pure by Merck KGaA, Darmstadt, Germany) and sodium hydroxide solution (SH). The last was obtained by dissolving sodium hydroxide in distilled water to have a twelve-molar solution (12 M).

Lightweight aggregates employed in the study were commercially available (by Poraver, Schlüsselfeld, Germany) and made of expanded waste glass particles. Five different grain sizes were used (0.04–0.125 mm, 0.1–0.3 mm, 0.25–0.5 mm, 0.5–1 mm, 1–2 mm) and graded according to the Fuller's distribution. The LWA dry loose bulk densities vary between 230 kg/m^3 (for the coarsest fraction) and 530 kg/m^3 (for the finest fraction).

The geopolymer mortars were formulated maintaining the following fixed proportions among the ingredients, based on previous research by the authors and the literature findings [25,26].

- The weight of metakaolin was 10% of the total weight of precursors.
- The weight ratio SS/SH was 2.5.
- The aggregate-to-binder volume ratio was 0.95 corresponding to 40% in weight.

The three geopolymer mixes differed in the alkali-activating solution and water contents. The reference mix is called SOL because the liquid part of the mix was only the activating solution without water addiction. In the SOL mix, the AAS-to-binder ratio was 0.67. This value was obtained after preliminary tests to guarantee the plastic consistency of the mortar.

The other two mixes were developed to reduce the amount of the alkaline solution by the quantity absorbed by aggregates during mixing. It was estimated starting from the technical datasheet of aggregates. The supplier furnished the water absorption of each aggregate fraction to reach a saturated surface dry condition (SSD). Considering the amounts of each fraction given by Fuller's distribution, a water absorption of about 20% of the total weight of aggregates was calculated. The corresponding amount of activating solution was estimated to be 30% of the LWA weight, accounting for the difference in density between water and activating solution (1000 kg/m^3 and 1325 kg/m^3, respectively). Based on this value, a solution-to-binder ratio equal to 0.55 was obtained. Thus, the other two mixes (SSD and FW) were designed considering this ratio. In the SSD mix, the aggregates were added in SSD conditions; in the FW mix the aggregates were at ambient conditions, and free water was added. The water content was equal to the quantity necessary to bring aggregates in the saturated surface dry condition according to the values reported in the datasheet. The resulting mix proportions are reported in Table 2.

Table 2. Mix proportioning of the geopolymeric mortars. FA: fly ash; MK: metakaolin; SS: sodium silicate solution, SH: sodium hydroxide solution; LWA: lightweight aggregate; W: water in the mix, added as free water in FW mix and absorbed by aggregates in SSD mix.

Mix Nomenclature	FA + MK (kg/m^3)	SS + SH (kg/m^3)	LWA (kg/m^3)	W (kg/m^3)
SOL	504.84	337.35	202.73 *	-
SSD	529.13	289.84	361.22 **	148.73
FW	507.56	278.02	203.82 *	40.76

* Ambient conditions; ** SSD condition.

2.2. Mixing and Curing

The AAS was prepared 24 h before mixing. The solution was added gradually to dry precursors in a Hobart mixer at 140 rpm. After the solution addiction, the mixer continued to work for another five minutes at the same speed. The lightweight aggregates were then added. They were at ambient conditions for the SOL and FW mix, while for the SSD mix, aggregates were preliminarily conditioned to have a saturated surface dry condition. They were immersed in tap water for 24 h before mixing and drained on a sieve to eliminate water. Then, they were dabbed on the surface with filter paper to reach the SSD condition. The cone procedure described in [27] was carried out to verify the SSD state. Free water was added to the FW mix after aggregate addiction. After adding aggregates and free water to the FW mix, the mixer was run for another five minutes. The mortars were manually mixed with a spatula before being placed into molds to ensure the homogeneity of the mixture. Specimens were de-molded after two to five days, depending on their consistency, and then cured at laboratory conditions (23 °C, 50% RH).

2.3. Test Methods

The consistency of the mortars was measured immediately after mixing by a flow-table test [28]. The mold was filled in two layers 25 mm thick. Each layer was consolidated with 25 times tamping of the flow table before measurements. The mold was removed to let the mortar flow. Two perpendicular diameters were measured to determine the mean diameter of the sample (flowability) expressed in millimeters.

Apparent density was monitored during curing up to 48 days from casting. It was determined on four specimens 40 × 40 × 80 mm for each mix by dividing their mass by the volume. The last was calculated by measuring each side of the specimens at four different points using a caliper. All the measurements were carried out under laboratory conditions and not after oven conditioning to avoid the influence of temperature on the kinetic of geopolymerization reactions and the water vapor evaporation from specimens.

Ultrasonic pulse velocity (UPV) measurements were carried out in direct transmission mode using 150 kHz transducers (Pundit PL 200, Screening Eagle Technologies AG, Schwerzenbach, Zurich). UPVs were measured at the timescale and on the specimens used for density measurements. Four measurements were taken, and the average value was calculated for each specimen (along the 40 × 80 faces).

After 48 days from casting, all the specimens were broken under compression test to determine the compressive strength of the materials [29]. The compressive strength was calculated as the mean value of the four measurements at a loading rate of 200 N/s.

The thermal conductivity of the mortars was measured on disk-like specimens, 50 mm in diameter (FOX 50 Laser Comp—TA Instruments, New Castle, DE, USA) according to [30,31]. The two-thickness method was used. It had the best precision guaranteed by the instrument producer (3%) and allowed for estimating the specimen's thermal conductivity and contact thermal resistance. The two thicknesses were 15 mm and 6 mm. Specimen surfaces were lapped to minimize the surface roughness and ensure good contact with the instrument plates. The upper plate temperature was set to 15 °C and the lower plate to 25 °C. Two couples of specimens were tested for each batch, and the average value was calculated. The specimens were in ambient conditions when tested.

3. Results and Discussion

3.1. Consistency of the Mortars

It is known from the literature that the content of water and activating solution strongly affects the workability of the geopolymers: generally, the higher the liquid-to-binder ratio, the higher the workability of the mix [32–35]. The results of the present study followed the results of the literature, as the consistencies of the three mixes were very different depending on the content of the activating solution and water.

The consistency of the SOL mix was plastic, with a flow value of 180 mm. The value is comparable to those of commercially available mortars for composite-reinforced mortar application, and it is suitable for application on building vertical surfaces. The SSD mix had a higher flowability (>300 mm) than the SOL mix. The result can be explained by the saturated surface dry condition of aggregates, which caused the activating solution to be not absorbed by aggregates during mixing and be available to provide workability [22]. To effectively compare the liquid-to-binder ratio of SOL and SSD mixes, the aggregates should be considered in the same conditions. Thus, the liquid content of the SSD mix should be calculated as the sum of the activating solution and water absorbed by aggregates after dabbing. In this case, the liquid-to-binder ratio of the SSD mix would be 0.83 in contrast with 0.67 of the SOL mix. The values explain the higher flowability of the mix. The amount of water absorbed by aggregates with the adopted SSD procedure was much higher than that reported in the datasheet of aggregates, probably due to the difficulties in effectively drying the surface of the finest aggregate fraction [36]. More research is needed to clarify this point and optimize the SSD procedure for the finest fractions of the mix.

As regards the FW mix, a 245 mm flow was registered. This value was higher than that of the SOL mix, even if the total liquid content (SS + SH + W) and the ratio of liquid to binder ((SS + SH + W)/(FA + MK)) were slightly lower (Table 2). The difference in viscosity between water and the activating solution may explain the result. It is reported in the literature that at the same liquid content, the ratio of SS solution on SH strongly influences workability [32,35,37]. It is explained by the higher viscosity of the SS solution compared to that of SH and water, which reduced the workability [36,38]. The SOL and FW mixes had the same SS-to-SH ratio, but the content of SS in the SOL mix was sensibly higher than in the FW mix, as it was 72% of the total liquid content instead of 62% for the FW mix.

3.2. Density and UPV Monitoring

The apparent density of the geopolymers decreases with curing time, especially during the first ten days (Figure 1). Indeed, water contained in the activating solution, free water, or water produced during the geopolymerization reactions evaporates with time [33,39,40] in all the mixes, causing a density reduction. At the age of 32 days, the density of the mixes became constant. Before this time, it decreased with different speeds depending on the mix.

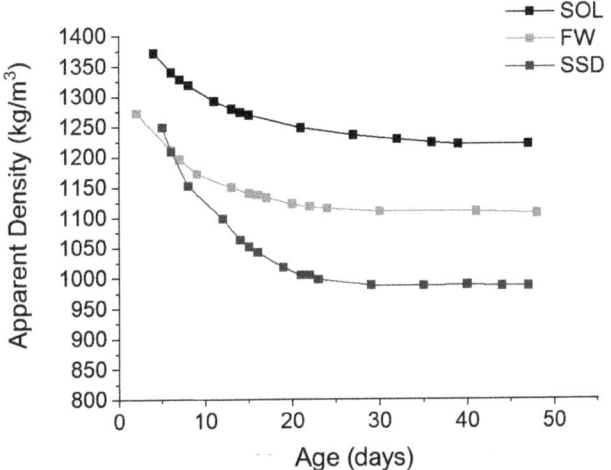

Figure 1. Apparent density vs age of the mortars (days), measured in SOL, FW, and SSD mixes.

SOL and FW behaviors with time were similar. Up to ten days, the rate of density decrease was higher, and then it diminished in both the mixes to become constant after 30–32 days from casting. The SOL density was higher than the FW density because of its higher content of activating solution, following the results of consistency tests.

The SSD behavior was different in the first days of curing. Because of its very fluid state, it was possible to demold specimens only after five days from casting. Up to 20 days, the mix lost about 20% of the initial density, contrasting with 12% and 10% of FW and SOL mixes, respectively. The SSD mix had a lower density than the other two mixes, following the difference in consistency among mixes.

The residual water inside the specimens was estimated using the following equation [39]:

$$Residual\ water\ (\%) = \frac{W_{H_2O} - W'_{H_2O}}{W_n} \tag{1}$$

where W_{H_2O} was the initial weight of the water in the specimen, W'_{H_2O} was the total water loss by the specimen on the day of measurement starting from the day of demolding, and W_n was the weight of the specimen on the day of measurement. The initial weight of water was determined as the sum of the content of water contained in SS and SH, the free water in the case of the FW mix, and the water absorbed by aggregates in the case of the SSD mix.

The results (Figure 2) were interesting as they showed that different mixes reached a similar value of the residual water in the material independently of the initial amount of water and solution in the mix.

The ultrasonic pulse velocity measurements were carried out frequently during the curing period, especially in the first days after demolding (Figure 3). The measurements highlight some topics that, to the author's knowledge, were not evidenced in the literature. The literature on the UPV application on geopolymers shows that UPV increases with time due to the progression of geopolymerization reactions [41,42]. The measurements were generally performed on fixed dates, like 7, 14, and 28 days from casting. In the present study, measurements were performed on several dates after demolding, evidencing different stages in UPV variation with time. Indeed, in all the mixes UPV first increased and then decreased. UPV increased again in the FW and SSD mixes while they remained constant in the SOL mix. In the last stage of measurements, UPV also remained constant for the FW and SSD mix.

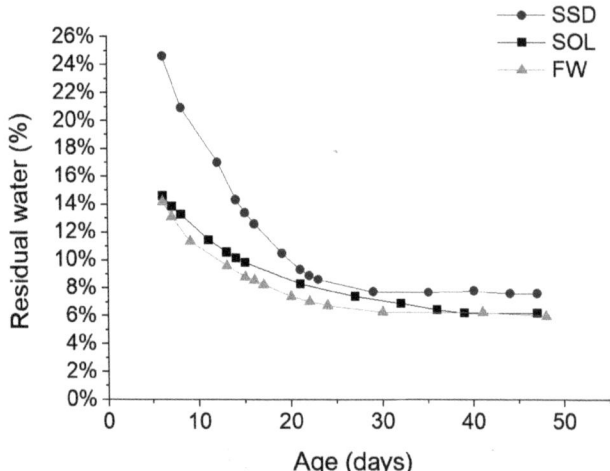

Figure 2. Residual water (%) vs mortar age (days) calculated for SOL, FW, and SSD mixes.

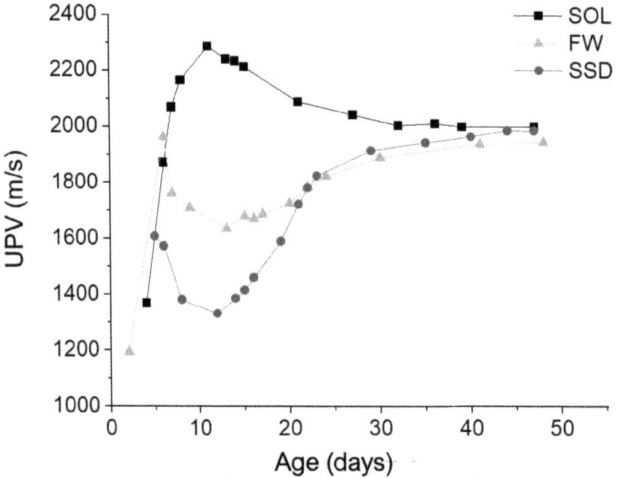

Figure 3. UPV values vs mortar age (days) calculated for SOL, FW, and SSD mixes.

The UPV behavior with time may be explained by considering two contemporaneous effects in the geopolymers curing. 1. The decrease in density with time due to water evaporation; 2. the hardening of the material with time due to the progression of the geopolymerization reactions. Water evaporation caused a substitution of water with air within the material pores, determining a UPV reduction. In contrast, the hardening of the material with the geopolymerization reaction caused UPV to rise. These two effects and the prevalence of one over the other determined the trend of UPV measurements.

Four stages can be distinguished (Table 3, Figure 3).

Table 3. Four stages of UPV behavior: periods of duration. In brackets are the dates in which there was a change in the rates of UPV increase/decrease.

Mix	Stage 1	Stage 2	Stage 3	Stage 4
FW	Up to 6th day	6th–(7th)–13th day	13th–(30th)–41st day	41st day
SSD	Up to 5th day	5th–(8th)–12th day	12th–(23rd)–44th day	44th day
SOL	Up to 11th day	11th–(21st)–32nd day		32nd day

- Stage 1: in the first days from casting, the increase in stiffness prevailed, even if density decreased faster in this period (Figure 1). In fact, during this stage, a UPV increase can be observed, characterized by a longer duration for SOL (11 days) than for FW and SSD (6 and 5 days, respectively).
- Stage 2: the effect of density reduction prevailed, causing a UPV lowering. During this stage, the geopolymerization reactions slowed down, accompanied by prevailing water evaporation. This stage lasted longer for the SOL mix (about 20 days) than for FW and SSD mixes (about one week). The rate of UPV reduction of the SOL mix slowed down after the 21st day, following the density behavior (see Figure 1). The FW and SSD mixes behave differently. The duration of this stage was shorter and followed by stage three, not experienced by the SOL mix.
- Stage 3: This stage was visible only for SSD and FW mixes. The UPV increased in both of the mixes due to the material hardening. The increment had two phases with different rates, first higher and then lower (Table 3). For the SSD mix, the UPV increase after the 12th day was considerable: 50%, in contrast with 20% for the FW mix.
- Stage 4: UPVs remained constant in all the mixes. This stage started earlier for the SOL mix (32nd day) and later in FW (41st day) and SSD (44th day) mixes.

Considering the UPV trends in the mixes, it can be deduced that most of the hardening of the SOL mix was until the 11th day from casting. In the other two mixes, hardening of the materials happened in the first week, then it stopped for some days and started again after about 12–13 days from casting. The higher ratio of activating solution to binder ratio of the SOL mix may have contributed to the faster development of the geopolymerization reactions [2,35,36]. It is reported in the literature that an excess of water may cause a delay in the progression of the geopolymerization reactions [36,40,43]. It probably happened in the FW and especially in the SSD mix following the higher flowability of the two mixes compared to the SOL mix.

At the end of UPV monitoring, the three mixes had similar UPV values.

3.3. Compressive Strength and Thermal Conductivity

It is well known from the literature that the compressive strength of geopolymers depends on several factors: type of precursors, curing conditions, molarity of NaOH solution, SS/SH ratio, alkaline solution-to-binder ratio, the content of free water, and aggregate-to-binder content are the most cited [22,37,40,44,45]. Most of these factors, namely the aggregates' content, the NaOH solution's molarity, and the SS/SH ratio, were constant in this study. In contrast, the alkaline solution content was varied among the mixes.

The compressive strength of the mortars was measured at the age of 48 days when density and UPV were constant for all the mortars. Testing mortars at this date and not at 28 days, as usual, ensured a proper comparison of the mechanical properties of different mixes because most of the geopolymerization reactions occurred in all of them. Indeed, after 28 days, while the SOL mix had almost stable values of UPV, both the FW and SSD mixes still had UPV rising with time. The compressive strengths of the FW and SOL mixes were comparable (Figure 4a), even if the content of the alkaline solution was different. The alkaline solution-to-binder ratio was 0.67 and 0.55, respectively, for the SOL and the FW mixes. Thus, the compressive strength results showed that a ratio of 0.55 was sufficient to cause geopolymerization reactions and the consequent material stiffening. Replacing part of the solution with water in the FW mix guaranteed a higher flowability than the SOL mix without compromising the compressive strength. Anyway, the threshold of water content in substitution of the activating solution that does not compromise strength should be assessed by further investigations. Indeed, it is reported in the literature that increasing water instead of the solution may bring adverse effects in resistance [35,37].

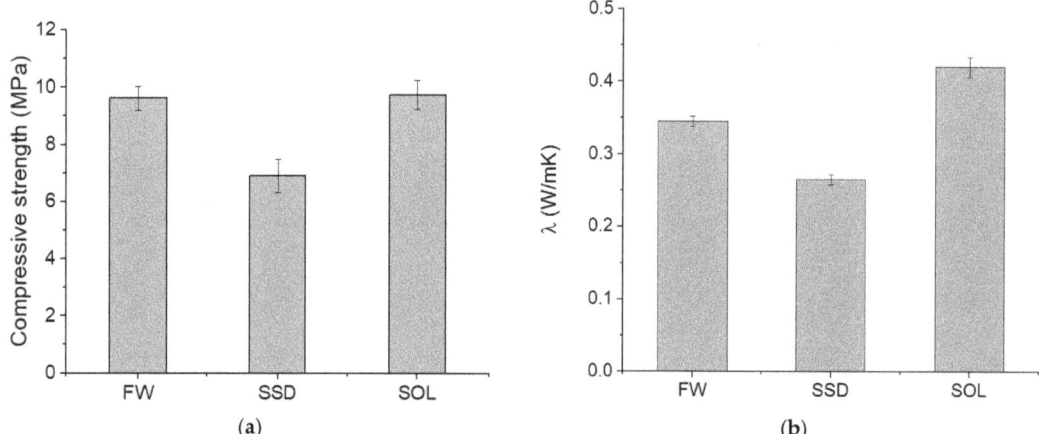

Figure 4. Compressive strength (**a**) and thermal conductivity λ (**b**) of the investigated mortars.

The SSD mix had a lower compressive strength than the other two mixes. This result cannot be attributed to the content of the activating solution, as it was the same as that of the FW mix. The lower compressive strength result followed the higher flowability and lower density results, and it is due to the different behavior of aggregates in the SSD state. As stated before, the total liquid-to-binder ratio of the SSD mix was higher than those of the other two mixes, causing a higher flowability, lower density, and lower compressive strength of the mix. Thus, the optimization of AAS content or the procedure to obtain LWAs in SSD conditions should be performed to reduce the flowability of the SSD mix and eventually reach the resistance of the other two mixes. More research is needed to deepen this aspect.

Thermal conductivity as compressive strength was affected by the content of the activating solution as the results varied with mixes (Figure 4b). SSD had the lowest thermal conductivity (0.27 W/mK), and the SOL mix the highest (0.42 W/mK). The thermal conductivity of the FW mix was between the values of the other two mixes (0.34 W/mK).

Both thermal conductivity and compressive strength strongly depend on the density of the mixes, following the literature findings [15,22,32,46]. In Figure 5, linear correlations between the density (Figure 1), compressive strength (Figure 4a), and thermal conductivity (Figure 4b) results are reported (Figure 5). The high coefficient of determination of the obtained correlations evidenced a strong dependence among results. The best linear fit was between the thermal conductivity and the density results. All these properties depend on the porosity of the mixes, which in turn depends on the content of the liquid part of the mix and consistency. Thus, the higher flowability of the mix caused a higher porosity and, consequently, lower density, compressive strength, and thermal conductivity. The obtained correlations are not exhaustive due to the low number of data but are indicative of the dependence among the investigated properties of the mortars.

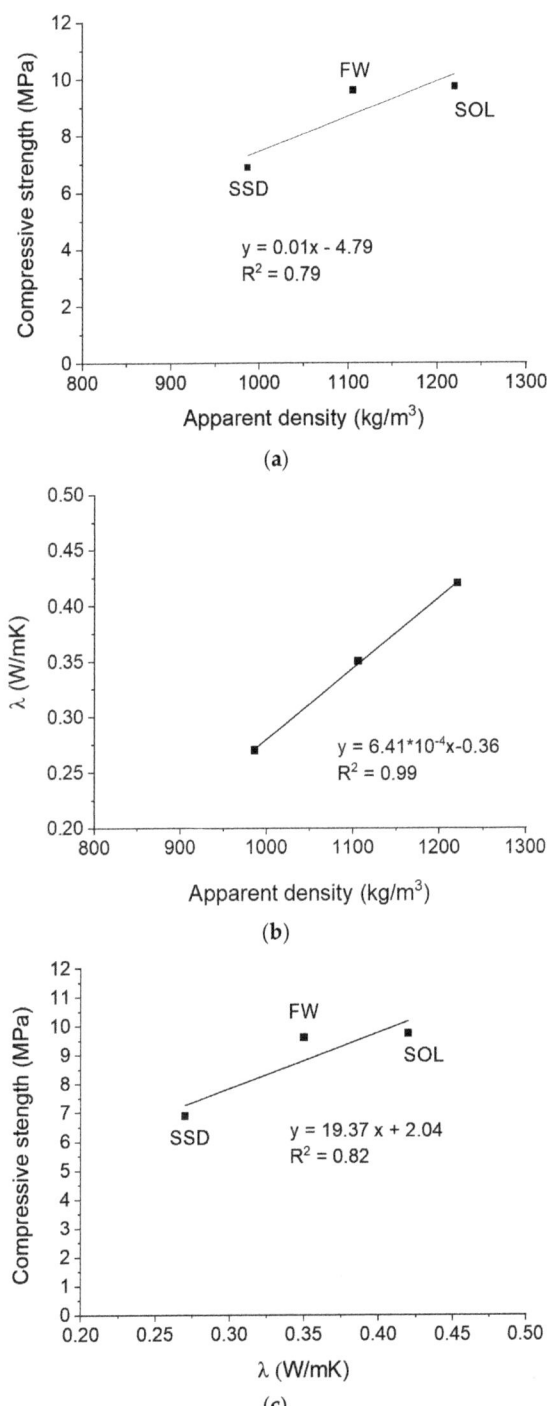

Figure 5. Linear correlations between compressive strength and density (**a**), thermal conductivity and density (**b**), and compressive strength and thermal conductivity (**c**).

4. Conclusions

This work is part of extensive experimental research that aims to optimize a geopolymer mortar with good mechanical and thermal performances. The mortar is formulated to be a sustainable material, as it employs waste aggregates and precursors, and it does not contain cement, which is considered the largest gas emitter in the construction industry. Fly ash, a waste of the coal industry, was employed as a precursor and waste glass as lightweight aggregates to improve the thermal performance of the mortar.

This experimental study investigates the effect of varying the content of the activating solution on the workability, compressive strength, and thermal conductivity of the material. This work starts from the observation that the expanded glass aggregates absorbed a consistent part of the activating solution during mixing. Thus, it can be supposed that only part of the solution effectively causes the geopolymerization reactions while the remaining part is needed to wet the aggregates, guaranteeing workability. This circumstance is antieconomic if it is considered that the alkaline solution is expensive and increases the environmental impact of the material.

Three mixes were investigated to clarify these aspects. The SOL mix was the reference mix in which the AAS content was calibrated to have a plastic consistency of the mortar. The other two mixes (SSD and FW mixes) had their AAS lowered by the quantity absorbed by aggregates. In these cases, two strategies were considered to overcome the reduced workability: in the SSD mix, aggregates were added in a saturated surface dry state; in the FW mix, the aggregates were at ambient conditions, and free water was added.

The three mixes had different properties due to the differences in the contents of water and AAS. The SOL mix had the highest density, compressive strength, and thermal conductivity. The FW mix had a compressive strength comparable to the SOL mix, a lower density, and a lower thermal conductivity. The SSD mix had the lowest compressive strength, density, and thermal conductivity.

The comparable mechanical results of FW and SOL mixes mean that replacing part of the activating solution with free water may be a way to obtain better thermal performances without compromising the resistance of the mortar. On the other hand, more research is needed to find the optimal water content to compromise the mechanical performance and thermal efficiency of the material.

Regarding the SSD mix results, the mix cannot be considered optimized as its consistency was too fluid. It was probably due to difficulties obtaining the finest fractions of LWA in SSD states as they absorbed too much water. Optimizing the procedure or reducing the AAS content may allow for better performances of the SSD mix in flowability and mechanical resistance. More research is needed to investigate this aspect.

Another finding of the experimental research is using UPV to follow the curing stages of materials. Indeed, the instrument was sensitive to microstructural changes with time. The result is noteworthy because UPV reduces the number of samples that would otherwise be required for destructive testing to monitor the hardening progress of the material over time.

The obtained experimental results concern a limited number of samples, so this research is not intended to be exhaustive of the topics covered; it is considered exploratory for the subsequent experimental investigations. Indeed, the experimental results should be further improved by increasing the number of tested specimens and deepening other aspects that have not been faced, like the influence of the investigated parameters on the shrinkage, porosity, and durability of the different mixes. These aspects will be the object of future research.

Author Contributions: Conceptualization, E.V.; validation, E.V., S.C., A.C. and M.A.A.; formal analysis, E.V. and S.C.; investigation, E.V., S.C. and A.C.; resources, M.A.A.; data curation, E.V. and S.C.; writing—original draft preparation, E.V.; writing—review and editing, E.V., S.C., A.C. and M.A.A.; supervision, M.A.A.; funding acquisition, M.A.A. All authors have read and agreed to the published version of the manuscript.

Funding: The research presented in this article was funded by the Italian Ministry of University and Research in the framework of the national research project PRIN (Progetti Di Ricerca Di Rilevante Interesse Nazionale—Bando 2022)—Project PRIN sTructurAl and energy rEnovation for sustainable buildingS—TARGETS.

Institutional Review Board Statement: Not applicable.

Informed Consent Statement: Not applicable.

Data Availability Statement: Data are contained within the article.

Acknowledgments: The authors would like to thank Sipre Srl for its valuable support in sample making and workability testing.

Conflicts of Interest: The authors declare no conflicts of interest.

References

1. Singh, N.B.; Middendorf, B. Geopolymers as an alternative to Portland cement: An overview. *Constr. Build. Mater.* **2020**, *237*, 117455. [CrossRef]
2. Zhang, P.; Zheng, Y.; Wang, K.; Zhang, J. A review on properties of fresh and hardened geopolymer mortar. *Comp. Part B* **2018**, *152*, 79–95. [CrossRef]
3. Provis, J.L. Alkali-activated materials. *Cem. Concr. Res.* **2018**, *114*, 40–48. [CrossRef]
4. Passuello, A.; Rodríguez, E.D.; Hirt, E.; Longhi, M.; Bernal, S.A.; Provis, J.L.; Kirchheim, A.P. Evaluation of the potential improvement in the environmental footprint of geopolymers using waste-derived activators. *J. Clean. Prod.* **2017**, *166*, 680–689. [CrossRef]
5. Habert, G.; De Lacaillerie, J.D.E.; Roussel, N. An environmental evaluation of geopolymer based concrete production: Reviewing current research trends. *J. Clean. Prod.* **2011**, *19*, 1229–1238. [CrossRef]
6. Imtiaz, L.; Kashif-ur-Rehman, S.; Alaloul, W.S.; Nazir, K.; Javed, M.F.; Aslam, F.; Musarat, M.A. Life cycle impact assessment of recycled aggregate concrete, geopolymer concrete, and recycled aggregate-based geopolymer concrete. *Sustainability* **2021**, *13*, 13515. [CrossRef]
7. Nodehi, M.; Taghvaee, V.M. Alkali-Activated Materials and Geopolymer: A Review of Common Precursors and Activators Addressing Circular Economy. *Circ. Econ. Sust.* **2022**, *2*, 165–196. [CrossRef]
8. Duxson, P.; Fernández-Jiménez, A.; Provis, J.L.; Lukey, G.C.; Palomo, A.; Van Deventer, J.S.J. Geopolymer technology: The current state of the art. *J. Mater. Sci.* **2007**, *42*, 2917–2933. [CrossRef]
9. Davidovits, J. Geopolymers: Ceramic-like inorganic polymers. *J. Ceram. Sci. Technol.* **2017**, *8*, 335–350. [CrossRef]
10. Provis, J.L.; Duxson, P.; van Deventer, J.S.J. The role of particle technology in developing sustainable construction materials. *Adv. Powder Technol.* **2010**, *21*, 2–7. [CrossRef]
11. Davidovits, J. Geopolymers and geopolymeric materials. *J. Therm. Anal.* **1989**, *35*, 429–441. [CrossRef]
12. Medri, V.; Papa, E.; Mazzocchi, M.; Laghi, L.; Morganti, M.; Francisconi, J.; Landi, E. Production and characterization of lightweight vermiculite/geopolymer-based panels. *Mater. Des.* **2015**, *85*, 266–274. [CrossRef]
13. Wongsa, A.; Sata, V.; Nuaklong, P.; Chindaprasirt, P. Use of crushed clay brick and pumice aggregates in lightweight geopolymer concrete. *Construct. Build. Mater.* **2018**, *188*, 1025–1034. [CrossRef]
14. Nematollahi, B.; Ranade, R.; Sanjayan, J.; Ramakrishnan, S. Thermal and mechanical properties of sustainable lightweight strain hardening geopolymer composites. *Arch. Civ. Mech. Eng.* **2017**, *17*, 55–64. [CrossRef]
15. Masoule, M.S.T.; Bahrami, N.; Karimzadeh, M.; Mohasanati, B.; Shoaei, P.; Ameri, F.; Ozbakkaloglu, T. Lightweight geopolymer concrete: A critical review on the feasibility, mixture design, durability properties, and microstructure. *Ceram. Int.* **2022**, *48*, 10347–10371. [CrossRef]
16. Jaya, N.A.; Yun-Ming, L.; Cheng-Yong, H.; Abdullah, M.M.A.B.; Hussin, K. Correlation between pore structure, compressive strength and thermal conductivity of porous metakaolin geopolymer. *Constr. Build. Mater.* **2020**, *247*, 118641. [CrossRef]
17. Degefu, D.M.; Liao, Z.; Berardi, U.; Labbé, G. The effect of activator ratio on the thermal and hygric properties of aerated geopolymers. *J. Build. Eng.* **2022**, *45*, 103414. [CrossRef]
18. *ACI 211.2-98*; Standard Practice for Selecting Proportions for Structural Lightweight Concrete. American Concrete Institute: Farmington Hills, MI, USA, 1998.
19. Yu, Q.L.; Spiesz, P.; Brouwers, H.J.H. Development of cement-based lightweight composites–Part 1: Mix design methodology and hardened properties. *Cem. Con. Comp.* **2013**, *44*, 17–29. [CrossRef]
20. Collins, F.; Sanjayan, J.G. Strength and shrinkage properties of alkali-activated slag concrete containing porous coarse aggregate. *Cem. Concr. Res.* **1999**, *29*, 607–610. [CrossRef]
21. Weber, S.; Reinhardt, H.W. A new generation of high-performance concrete: Concrete with autogenous curing. *Adv. Cem. Based Mater.* **1997**, *6*, 59–68. [CrossRef]
22. Huiskes, D.M.A.; Keulen, A.; Yu, Q.L.; Brouwers, H.J.H. Design and performance evaluation of ultra-lightweight geopolymer concrete. *Mater. Des.* **2016**, *89*, 516–526. [CrossRef]

23. Kupaei, R.H.; Alengaram, U.J.; Jumaat, M.Z.B.; Nikraz, H. Mix design for fly ash based oil palm shell geopolymer lightweight concrete. *Construct Build Mat* **2013**, *43*, 490–496. [CrossRef]
24. Ouellet-Plamondon, C.; Habert, G. Life cycle analysis (LCA) of alkali-activated cements and concretes. In *Handbook of Alkali-Activated Cements, Mortars and Concretes*; Woodhead Publishing-Elsevier: Cambridge, UK, 2014; pp. 663–686.
25. Longo, F.; Lassandro, P.; Moshiri, A.; Phatak, T.; Aiello, M.A.; Krakowiak, K.J. Lightweight geopolymer-based mortars for the structural and energy retrofit of buildings. *Energy Build.* **2020**, *225*, 110352. [CrossRef]
26. Alanazi, H.; Hu, J.; Kim, Y.R. Effect of slag, silica fume, and metakaolin on properties and performance of alkali-activated fly ash cured at ambient temperature. *Constr. Build. Mater.* **2019**, *197*, 747–756. [CrossRef]
27. *UNI EN 1097-6:2022*; Test for Mechanical and Physical Properties of Aggregates. Part 6: Determination of Particle Density and Water Absorption. iTeh Standards: Etobicoke, ON, Canada, 2022.
28. *UNI EN 1015-3:2007*; Methods of Test for Mortar for Masonry—Part 3: Determination of Consistence of Fresh Mortar (by Flow Table). iTeh Standards: Etobicoke, ON, Canada, 2007.
29. *UNI EN 1015-11:2019*; Methods of Test for Mortar for Masonry—Part 11: Determination of Flexural and Compressive Strength of Hardened Mortar. iTeh Standards: Etobicoke, ON, Canada, 2020.
30. *ASTM C518-21*; Standard Test Method for Steady-State Thermal Transmission Properties by Means of the Heat Flow Meter Apparatus. ASTM International: West Conshehoken, PA, USA, 2021.
31. *ISO 8301:1991*; Thermal Insulation—Determination of Steady-State Thermal Resistance and Related Properties—Heat Flow Meter Apparatus. International Organization for Standardization: Geneva, Switzerland, 2019.
32. Wongsa, A.; Zaetang, Y.; Sata, V.; Chindaprasirt, P. Properties of lightweight fly ash geopolymer concrete containing bottom ash as aggregates. *Constr. Build. Mater.* **2016**, *111*, 637–643. [CrossRef]
33. Bondar, D.; Nanukuttan, S.; Provis, J.L.; Soutsos, M. Efficient mix design of alkali activated slag concretes based on packing fraction of ingredients and paste thickness. *J. Clean. Prod.* **2019**, *218*, 438–449. [CrossRef]
34. Pouhet, R.; Cyr, M. Formulation and performance of flash metakaolin geopolymer concretes. *Constr. Build. Mater.* **2016**, *120*, 150–160. [CrossRef]
35. Hadi, M.N.; Zhang, H.; Parkinson, S. Optimum mix design of geopolymer pastes and concretes cured in ambient condition based on compressive strength, setting time and workability. *J. Build. Eng.* **2019**, *23*, 301–313. [CrossRef]
36. Nath, P.; Sarker, P.K. Effect of GGBFS on setting, workability and early strength properties of fly ash geopolymer concrete cured in ambient condition. *Constr. Build. Mater.* **2014**, *66*, 163–171. [CrossRef]
37. Waqas, R.M.; Butt, F.; Zhu, X.; Jiang, T.; Tufail, R.F. A comprehensive study on the factors affecting the workability and mechanical properties of ambient cured fly ash and slag based geopolymer concrete. *Appl. Sci.* **2021**, *11*, 8722. [CrossRef]
38. Chindaprasirt, P.; Chareerat, T.; Sirivivatnanon, V. Workability and strength of coarse high calcium fly ash geopolymer. *Cem. Concr. Compos.* **2007**, *29*, 224–229. [CrossRef]
39. Wang, H.; Wu, H.; Xing, Z.; Wang, R.; Dai, S. The Effect of Various Si/Al, Na/Al Molar Ratios and Free Water on Micromorphology and Macro-Strength of Metakaolin-Based Geopolymer. *Materials* **2021**, *14*, 3845. [CrossRef] [PubMed]
40. Xie, J.; Kayali, O. Effect of initial water content and curing moisture conditions on the development of fly ash-based geopolymers in heat and ambient temperature. *Constr. Build. Mater.* **2014**, *67*, 20–28. [CrossRef]
41. Sitarz, M.; Hager, I.; Choińska, M. Evolution of Mechanical Properties with Time of Fly-Ash-Based Geopolymer Mortars under the Effect of Granulated Ground Blast Furnace Slag Addition. *Energies* **2020**, *13*, 1135. [CrossRef]
42. Ghosh, R.; Sagar, S.P.; Kumar, A.; Gupta, S.K.; Kumar, S. Estimation of geopolymer concrete strength from ultrasonic pulse velocity (UPV) using high power pulser. *J. Build. Eng.* **2018**, *16*, 39–44. [CrossRef]
43. Zuhua, Z.; Xiao, Y.; Huajun, Z.; Yue, C. Role of water in the synthesis of calcined kaolin-based geopolymer. *Appl. Clay Sci.* **2009**, *43*, 218–223. [CrossRef]
44. Ohno, M.; Li, V.C. An integrated design method of Engineered Geopolymer Composite. *Cem. Concr. Compos.* **2018**, *88*, 73–85. [CrossRef]
45. Reddy, V.S.; Krishna, K.V.; Rao, M.S.; Shrihari, S. Effect of molarity of sodium hydroxide and molar ratio of alkaline activator solution on the strength development of geopolymer concrete. In *E3S Web of Conferences*; EDP Sciences: Hulis, France, 2021; Volume 309, p. 01058. [CrossRef]
46. Agustini, N.K.A.; Triwiyono, A.; Sulistyo, D. Effects of water to solid ratio on thermal conductivity of fly ash-based geopolymer paste. *IOP Conf. Ser. Earth Environ. Sci.* **2020**, *426*, 012010. [CrossRef]

materials

Article

Sugar-Free, Vegan, Furcellaran Gummy Jellies with Plant-Based Triple-Layer Films

Anna Stępień [1,*], Joanna Tkaczewska [2], Nikola Nowak [3], Wiktoria Grzebieniarz [3], Urszula Goik [1], Daniel Żmudziński [1] and Ewelina Jamróz [4]

[1] Department of Engineering and Machinery for Food Industry, Faculty of Food Technology, University of Agriculture, Balicka Street 122, PL-30-149 Cracow, Poland; urszula.goik@urk.edu.pl (U.G.); daniel.zmudzinski@urk.edu.pl (D.Ż.)

[2] Department of Animal Products Processing, University of Agriculture, Balicka Street 122, PL-30-149 Cracow, Poland; joanna.tkaczewska@urk.edu.pl

[3] Department of Chemistry, University of Agriculture, Balicka Street 122, PL-30-149 Cracow, Poland; nikola.nowak@urk.edu.pl (N.N.); wiktoria.grzebieniarz@urk.edu.pl (W.G.)

[4] Department of Product Packaging, Cracow University of Economics, Rakowicka Street 27, PL-31-510 Cracow, Poland; ewelina.jamroz@urk.edu.pl

[*] Correspondence: anna.stepien@urk.edu.pl

Abstract: Increasing consumer awareness of the impact of nutrition on health and the growing popularity of vegan diets are causing a need to look for new plant-based formulations of standard confectionery products with high energy density and low nutritional value, containing gelatin. Therefore, the aim of this study was to develop vegan and sugar-free gummy jellies based on an algae-derived polysaccharide—furcellaran (FUR). Until now, FUR has not been used as a gel-forming agent despite the fact that its structure-forming properties show high potential in the production of vegan confectionery. The basic formulation of gummy jellies included the addition of soy protein isolate and/or inulin. The final product was characterized regarding its rheological, antioxidant, mechanical and physicochemical properties. Eco-friendly packaging for the jellies composed of a three-layer polymer film has also been developed. It was observed that the highest values of textural parameters were obtained in jellies containing the addition of soy protein isolate, whose positive effect was also found on antioxidant activity. Before drying, all furcellaran-based gel systems showed G' and G'' values characteristic of strong elastic hydrogels. Storing jellies for a week under refrigeration resulted in an increase in hardness, a decrease in moisture content and reduced water activity values. Overall, our study indicates the high potential of furcellaran both as a gelling agent in confectionery products and as a base polymer for their packaging.

Keywords: confectionery gels; low-sugar candy; gummy jelly; furcellaran; biodegradable films

Citation: Stępień, A.; Tkaczewska, J.; Nowak, N.; Grzebieniarz, W.; Goik, U.; Żmudziński, D.; Jamróz, E. Sugar-Free, Vegan, Furcellaran Gummy Jellies with Plant-Based Triple-Layer Films. *Materials* **2023**, *16*, 6443. https://doi.org/10.3390/ma16196443

Academic Editor: Daniela Fico

Received: 23 August 2023
Revised: 20 September 2023
Accepted: 25 September 2023
Published: 27 September 2023

1. Introduction

Furcellaran is an anionic polysaccharide obtained by the extraction of red algae *Furcellaria lumbricalis*, composed of units consisting of a fragment of (1 → 3) β-D-galactopyranose with a sulfate group at C-4 and (1 → 4)-3,6-anhydro-α-D-galactopyranose, which contains 16–20% sulfate content. Structurally, furcellaran is largely similar to kappa-carrageenan [1,2]. According to European Union legislation, FUR has been approved for use as a food additive functioning as a gel-forming substance and thickener. As interest in algae-derived substances has been growing steadily in recent times, there are reports of explorations for new industrial applications of furcellaran. Studies mainly focus on the use of FUR as a component of biodegradable films to replace synthetic packaging in the food industry [3]. However, due to its gelling properties, the ability to form stable complexes with other polymers and the low energy value resulting from the presence of bonds resistant to human amylases, furcellaran is a promising component that can find an application in the production of confectionary.

Oenothera biennis L. as a biennial plant belongs to *Onagraceae* family, which also has many useful biological properties. As a commercial medicinal plant, its oil is used in the pharmaceutical industries, nutraceutical, cosmetic, and feed sectors. It traditionally contains a high content of protein, fiber, vitamin (A and D) and minerals (mainly calcium, potassium and magnesium) [4]. Within the protein composition, it is rich in sulfur amino acids. Oenothera oil is very popular among consumers and is often used as herbal supplements for dietary additions or in the cosmetic and skincare industry. Obtaining oenothera oil generates a lot of protein waste, which can be a valuable raw material with a high nutritional value with active components [5]. So, we have made an assumption that the oenothera by-product obtained after oil extraction, due to its high content of protein as well as complexation/enrichment with polyphenols, can be recommended for use as a promising active ingredient in engineering multi-layer or emulsified films.

The growing interest in a plant-based diet has meant that new vegan alternatives to products containing animal ingredients are constantly appearing on the food market. Therefore, also within the candy segment, new formulations are being searched for to produce jelly beans and gummies without the addition of gelatin [6]. Nevertheless, it is a significant challenge to achieve the resilient structure provided by gelatin using vegan hydrocolloids, so there is a need to find new solutions in this area. Plant-based substitutes for gelatin used as gelling agents in confectionary products include agar [7], carrageenan [6], alginate [8], and pectin [9], among others. However, there are no reports as of yet on the application of furcellaran in this type of food.

Gummy jellies, classified often as so-called confectionery gels containing, among other things, gelling agents, sweeteners and fruit acids, are generally considered to have low nutritional value due to their lack of health-promoting substances and high proportion of simple sugars [10,11]. In addition, the fact that gummies are popular especially among children and adolescents intensifies the need to look for more high-value formulations. Therefore, considering the health of potential confectionary consumers, increasing attention is being given to the development of products that are reduced in calories or incorporate health-promoting additives. As an excessive supply of sugar in the diet contributes to the development of many metabolic diseases [12], one of the directions for modifying gummy jellies is the elimination of sucrose or sugar syrups. For example, confectionery products with sorbitol [13], sucralose and erythritol [14], mannitol [15] and xylitol [16] were reported. Increasing the value of gummy candies can be carried out by adding various types of bioactive substances such as probiotics [17,18], vitamins [9,19] or plant extract [20,21].

Since gummy jellies are a product categorized as a ready-to-go snack, they are sold in small unit packs made of synthetic films, which are categorized as single-use plastic (SUP). It is assumed that synthetic packaging materials represent the highest share of waste classified as SUP [22]. Therefore, the main task implemented as part of rational waste management is the search for alternative biodegradable packaging plastics. One of the most promising solutions, within environmentally friendly packaging technologies, is the deployment of biopolymers. Incorporating waste from food processing into biopolymer films can effectively reduce waste and promote sustainability in the food industry. However, the selection of materials and components for biopolymer films should consider desired properties and the type of food product being packaged. While some biopolymers have shown promise for food packaging, their limitations in physicochemical properties have restricted their application [23]. To overcome these limitations, researchers have explored the combination of various biopolymers, such as proteins, lipids and polysaccharides, in the development of reinforced biopolymer-based films with improved mechanical properties and biological activity [24].

Proteins and polysaccharides are commonly used as feedstock for biopolymer films [25]. It was found that furcellaran, used as a gelling agent in our study, can be also used as a matrix for many biopolymer films due to its ability to form film form [26–28]. The incorporation of biopolymer films into extracts with antioxidant or antimicrobial activity can lead to active packaging, extending product quality.

The objectives of this research include the following: (1) to develop a formulation of furcellaran-based sugar-free jellies and to characterize their basic physicochemical properties, (2) to develop technology to produce biodegradable film as packaging for jellies; and (3) to determine the effect of packaging on the basic physical and mechanical characteristics of the product during storage. We aimed to develop a basic gummy formulation that takes into account an FUR, acidity regulator, sweetener, natural color and flavor, and two potential structuring compounds: inulin and soy protein isolate (SPI). Gummy jellies in three different variants were characterized with respect to their textural, rheological and antioxidant properties. Their nutritional value was also estimated, and storage analyses were performed to check the changing of basic qualities over time. Furcellaran was also used to produce a three-layer film enriched with oenothera oil, which was the packaging for jellies. The physical characteristics of gummy jellies stored for a week in polymer films and without packaging were compared. Therefore, the research resulted in an innovative product with packaging, using the same biopolymer—furcellaran as a gelling agent and component of biodegradable films.

2. Materials and Methods

2.1. Materials

Oenothera by-products were obtained from a local processor Olinii (Nowa Wieś, Poland). This was waste from the production of oenothera cold-pressed oil. To isolate the protein from oenothera, hexane (Sigma Aldrich, St. Louis, MO, USA), hydrochloric acid and sodium (Chemland, Stargard, Poland) were used. The following were employed to make the gummy jellies and films: spirulina powder (Green Essence, Pyrzyce, Poland), furcellaran type 7000 (Est-Agar AS, Karla Village, Estonia), soy protein isolate (Bene Vobis, Gdańsk, Poland), inulin HPX (Beneo Orafti®, Mannheim, Germany), citric acid (Biomus, Lublin Poland), cold-pressed oenothera (*Oenothera biennis*) oil from (Oleofarm, Wrocław, Poland), glycerol and Tween 80 (Chemland, Stargard, Poland).

2.2. Preparations of Furcellaran Gummy Jellies

The four variants of gummy jellies were prepared according to the formulation shown in Table 1. The weighed amount of furcellaran powders was mixed with distilled water at room temperature and stirred using a magnetic stirrer with simultaneous heating set at 70 °C and 500 rpm. Once the polysaccharide was completely dissolved, the rest of the ingredients were added with the exception of spirulina and flavor, which was dissolved in 10 mL of distilled water and inserted into the solution as soon as it had cooled to 60 °C. The solutions were poured into hexagonal prism molds and left to solidify in the refrigerator for 24 h. Individual gels removed from the molds were spread evenly on perforated trays and dried by convection at 45 °C for 4 h, rotating after half the drying time. For the purposes of further analysis, samples after gelling but before drying will be referred to as gels. Meanwhile, jellies/gummies/candies are the dried samples which are final products.

Table 1. Furcellaran gummy jellies formulation.

Component (g/100 g)	Sample Code			
	GJ-I	GJ-SPI	GJ-I+SPI	GJ-C
Furcellaran	4.00	4.00	4.00	4.00
Erythritol	10.00	10.00	10.00	10.00
Soy protein isolate	0.00	2.00	1.00	0.00
Inulin	2.00	0.00	1.00	0.00
Citric acid	0.20	0.20	0.20	0.20
Spirulina	0.25	0.25	0.25	0.25
Aroma	0.01	0.01	0.01	0.01
Water	83.55	83.55	83.55	85.55

2.3. Rheological Characteristics of the Gels

The rheological properties of the furcellaran sample at the temperatures 6 °C, 20 °C and 40 °C were tested. Rheological tests were performed using an RS-6000 rotational rheometer (ThermoFischer, Karlsruhe, Germany), cone-plate system with an angle of 2°, 35 mm diameter and gap of 1.4 mm. Measurements in the range of linear viscoelasticity (LVE) are the standard criterion when selecting the value of the vibration amplitude γ. The measurements are carried out at a constant frequency value, determining the relationship between the storage modulus G' and the loss modulus G'' as a function of the strain amplitude. The area of experimental curves is independent of the applied strain amplitude and defines the range of linear viscoelasticity. The appropriate value determination usually consists of determining the maximum amplitude in the LVE range, which translates into the largest possible deformation, and thus increases the accuracy of the measurements [29]. Determination of the relationship between the storage modulus—G' (Pa) and the loss modulus—G'' (Pa) at a constant frequency of 1 Hz determines the range of linear viscoelasticity. A frequency sweep test, for frequencies from 0.1 to 50 Hz in the range of linear viscoelasticity was carried out. To determine the strength of the gels, the $\tan(\delta)$ phase angle tangent values as the ratio G''/G' were calculated.

2.4. Physicochemical Properties and Nutritional Labeling of Gummy Jellies

The moisture content of the samples at each stage of the technological process (after gelling, drying and storage) was determined by the gravimetric method in a laboratory dryer at 60 °C until a constant weight was obtained. The water activity was measured in triplicate using AquaLAB 4TE equipment (Decagon Devices Inc., Pullman, WA, USA). The pH values of the gummy jelly solutions were determined using a pH-meter CP-505 (Elmetron, Zabrze, Poland). The values of basic macronutrients of the four variants of gummy jellies were calculated using Cronometer (Revelstoke, BC, Canada) software (https://cronometer.com/, accessed on 20 September 2023).

2.5. Antioxidant Activity

The jelly extracts (20 mg/mL) were prepared by adding jelly in the amount of 0.2 g to 10 mL of distilled water. The tubes containing the jelly extracts were put in a water bath at a temperature of 50 °C with shaking action for 10 min to ensure complete dissolution of the jelly. The extracts prepared in such a manner were then used for Ferric Reducing Antioxidant Power Assay (FRAP) and metal-chelating activity.

The FRAP assay was completed according to the method of Benzie and Strain [30]. The FRAP reagent comprised an acetate buffer (pH 3.6), a ferric chloride solution (20 mM), and a 2,4,6-tripyridyl-s-triazine solution (10 mM TPTZ in 40 mM HCl) at the ratio of 10:1:1 $(v/v/v)$, respectively. Firstly, the FRAP reagent was incubated at a temperature of 37 °C in the dark for 30 min, and then, it was mixed with a jelly extract in the ratio of 0.4:3.6 (v/v). The solution was incubated one more time at 37 °C for 10 min in dark conditions, and the following was absorbance measurement at 593 nm via the Helios Gamma UV-1601 spectrophotometer (Thermo Fisher Scientific, Waltham, MA, USA). The metal-chelating ability of jelly extract was assessed according to the method described by Tkaczewska, Zając, Jamróz, and Derbew [31]. The analyses were performed in duplicate for 3 samples of the jelly (n = 2 × 3).

2.6. Texture Profile Analysis

The texture properties of gels and gummy jellies were evaluated by a two-bite compression test using a Shimadzu EZ Test EZ-LX (Shimadzu, Kyoto, Japan) universal testing machine equipped with a load cell of 500 N. Before analyses, jellies were rested at room temperature 30 min before the test. Each sample was compressed twice to 50% of its original height with a 36 mm diameter cylindrical press jig, which allowed the jelly to deform without penetration. The test speed was 1 mm/s, and the trigger point was 0.05 N. The obtained reports for this test included hardness, springiness and cohesiveness. Textural parameters values were determined using Trapezium X software (https://www.shimadzu.eu/trapezium-x-software, accessed on 20 September 2023) (Shimadzu, Kyoto, Japan). Analyses were carried out in 10 replicates.

2.7. Film Preparations, Packing and Storage Analysis

2.7.1. Isolation of Protein from Post-Production Oenothera Pomace

Isolation of the protein was performed according to Trigui et al. [32]. Oenothera pomace was ground using a laboratory mill. It was then defatted by placing them in hexane ((1:5 w/v), pomace: hexane) for 4 h. The procedure was repeated twice. The defatted pomace was placed under a fume hood for 24 h at room temperature to evaporate the hexane. The as-prepared pomace was dispersed in distilled water and raised to pH 11 using a previously prepared 1.0 M NaOH solution. The pH value was controlled continuously using a CP-505 pH-meter (Elmetron, Zabrze, Poland). The suspension was stirred on a magnetic stirrer (MR Hei-Tec, Heidolph Instruments GmbH & Co. KG, Schwabach, Bayern) for 2.5 h at 30 °C. After this time, the suspension was centrifuged 3000× g at 25 °C for 20 min, and the supernatant was collected. The process was repeated twice, and all supernatants thus obtained were combined. Subsequently, a pH value of 4.5 was set to precipitate the protein using a previously prepared 1 M HCl solution. The solutions were centrifuged again, the supernatant was discarded and the precipitate was washed with water, lyophilized and stored at −20 °C until analysis.

2.7.2. Preparation of the Emulsion and Spirulina Solutions

About 2 g of protein isolated from oenothera pomace was dissolved in buffer pH 11. Then, 5 mL of oenothera oil and 5 mL of Tween 80 were added. The mixture was homogenized using a Polytron PT2500E homogenizer (Dan-Lab, Białystok, Poland). Aqueous spirulina solution was prepared by dissolving 2 g of spirulina in 80 g of water. It was then stirred using a magnetic stirrer for 30 min (400 rpm, temp. 30 °C).

2.7.3. Preparation of the Multilayer Film

A solution of 1% furcellaran was prepared and glycerol (1% w/w) was added as a plasticizer. The so-obtained solution was stirred under continuous heating (temperature 100 °C) to dissolve the polymer. Then, the temperature of the solution was lowered to 60 °C, and 250 mL was taken and poured onto a previously prepared form to turn into a gel. This formed the 1st layer of the biopolymer film. Then, 200 mL of the solution was taken and 50 mL of the previously prepared emulsion was added to it, which constituted the 2nd layer of the biopolymer film. After thorough mixing, the solution was poured over the gelled 1st layer and again left until it turned into a gel. The 3rd layer was made up of 240 mL of furcellaran solution and 10 mL of spirulina aqueous solution. After spilling the third layer, the composite was left to dry under a laboratory fume hood at room temperature. The procedure for obtaining three-layer films and their active ingredients is shown in Figure 1.

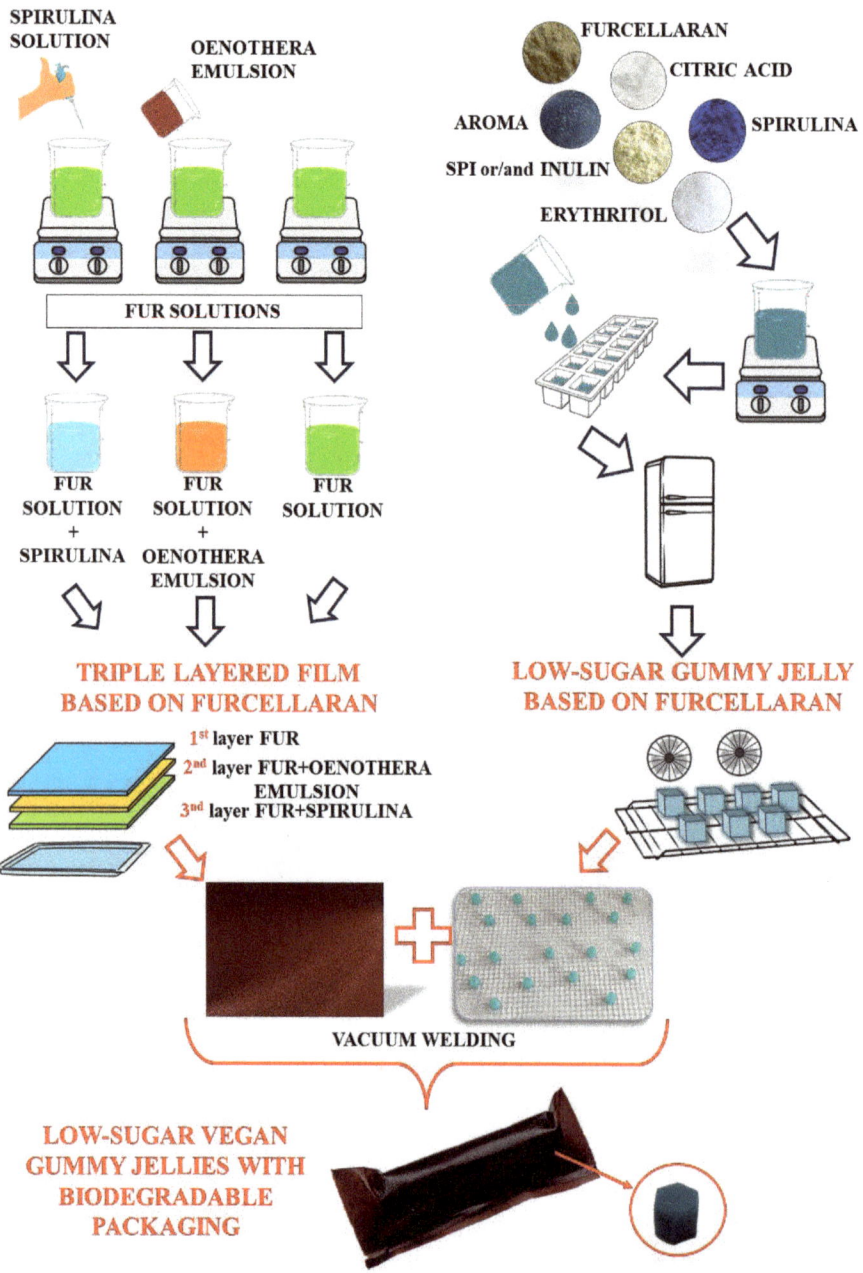

Figure 1. Production scheme for furcellaran-based jelly and packaging films.

2.7.4. Packing and Storage

After drying, the gummy jellies were cooled to room temperature, piled one on top of the other in 10-piece packets, and wrapped in biopolymer film, which was then sealed in tube form using a food vacuum sealer Freshpack Pro-QH(QH-01) (SaengQ, Wenzhou, China). Gummies packaged in film, and jellies without packaging as reference samples, were stored for 7 days in a climate chamber KBK-140W (Wamed, Warsaw, Poland) at 25 °C under 60% relative humidity. After one week, the samples were removed from the film, weighed and examined for textural properties and water activity according to the methodology described in earlier paragraphs.

2.8. Statistical Analyses

Averages and standard deviations of samples were reported based on the triplicate measurement. Statistica (StatSoft, Inc., Tulsa, OK, USA) software version 13.3 was used to perform multifactor Analyses of Variance (ANOVAs) in order to determine whether the effect of the variables (formulation and stage of the technological process) on the product was significant. For the other variables, significant differences ($p < 0.05$) were evaluated using one-way ANOVA with the Duncan test.

3. Results and Discussion

3.1. Rheological Characteristic of Furcellaran Gel Systems

The rheological characteristics of furcellaran gels allow studying the physical stability, esthetic effect, quality and usability of obtained preparations. The rheological studies allow researchers to understand and evaluate the response of a material to various inputs, such as stress or strain at different frequencies and at different temperatures. Results of rheological analyses are also useful for evaluating product consistency as well as the effect of the additives on the product itself. The rheological properties of hydrogels allow them to be classified as viscoelastic materials because they exhibit both elastic and viscous behavior and exhibit physical properties between liquid and solid states. The frequency sweep test allows researchers to observe how the relationship between the viscous and elastic moduli as a function of frequency was changed. The frequency sweep test is performed within the linear viscoelastic region (LVER), which is defined by the initial constant frequency amplitude sweep test to avoid sample damage. In this range, G', G'' and tan (δ) do not change with the applied strain [33].

Figures 2–4 shows the storage modulus G' and the loss modulus G'' as a function of strain for the tested furcellaran-based gels. The tested samples did not show quantitative similarity. At low strain amplitudes, G' was almost constant, which suggest a linear viscoelastic range in which there is no permanent damage to the sample structure. Then, both modules become strain dependent, with G' decreasing and G'' passing through the peak which is located just before or at the intersection point whose strain is defined as destructive [33]. A similar behavior of the samples was observed in different systems where bacterial alginate-like exopolymers (ALE) gels were investigated [34,35].

The results of the frequency sweep test obtained at temperatures of 6 °C, 20 °C and 40 °C are presented graphically in Figures 5–7. For all samples, G' is always about 10 times greater than G'' at the full frequency range at all tested temperatures 6 °C, 20 °C and 40 °C. G' and G'' run almost parallel without an intersection point, which indicates the elastic nature of the cross-linked gels prevailing over the viscous nature in the tested conditions. This behavior is a typical feature of a strong gel system, which suggest strong gel formation for all furcellaran hydrogels; similar behavior was reported by Yang et al. [36].

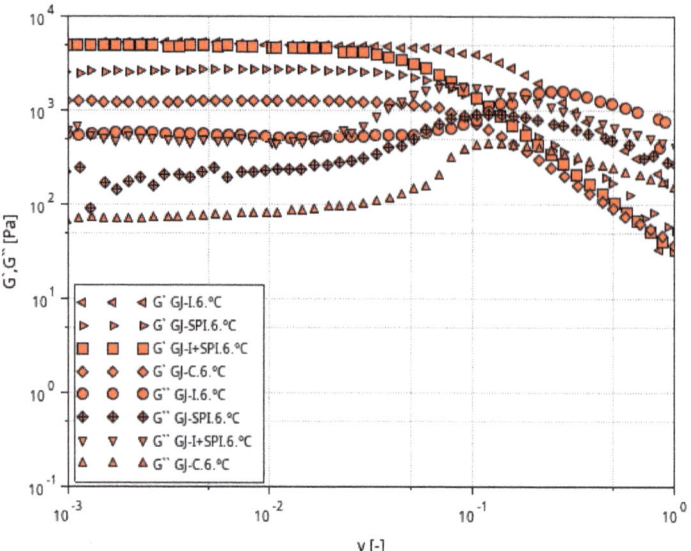

Figure 2. Relationship between the G′ and G″ modulus values as a function of strain amplitude at constant oscillation frequency for furcellaran-based gels at a temperature of 6 °C.

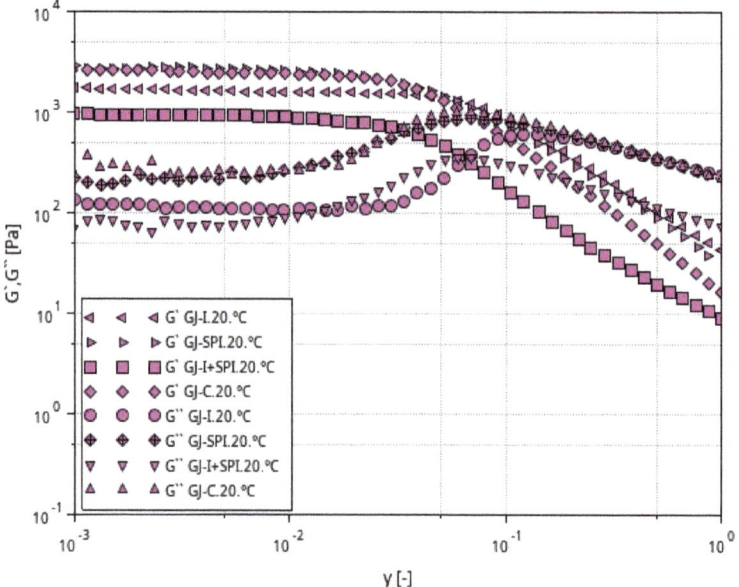

Figure 3. Relationship between the G′ and G″ modulus values as a function of strain amplitude at constant oscillation frequency for furcellaran-based gels at a temperature of 20 °C.

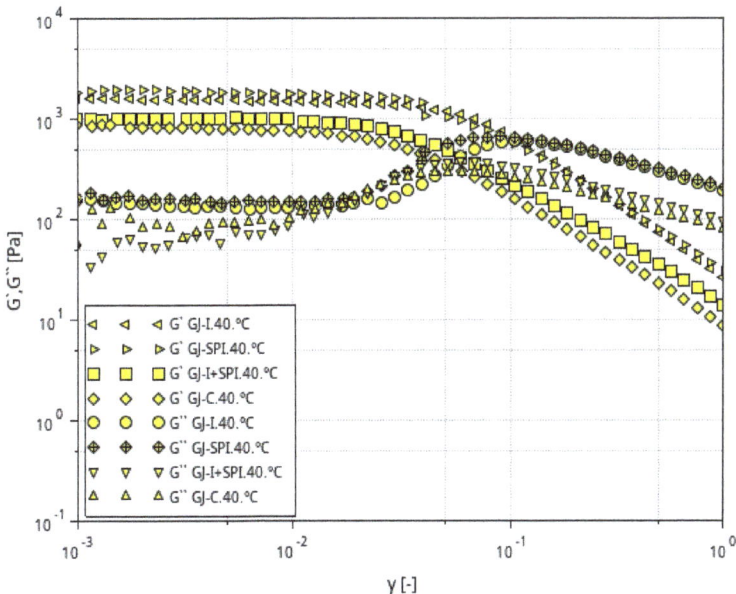

Figure 4. Relationship between the G′ and G″ modulus values as a function of strain amplitude at constant oscillation frequency for furcellaran-based gels at a temperature of 40 °C.

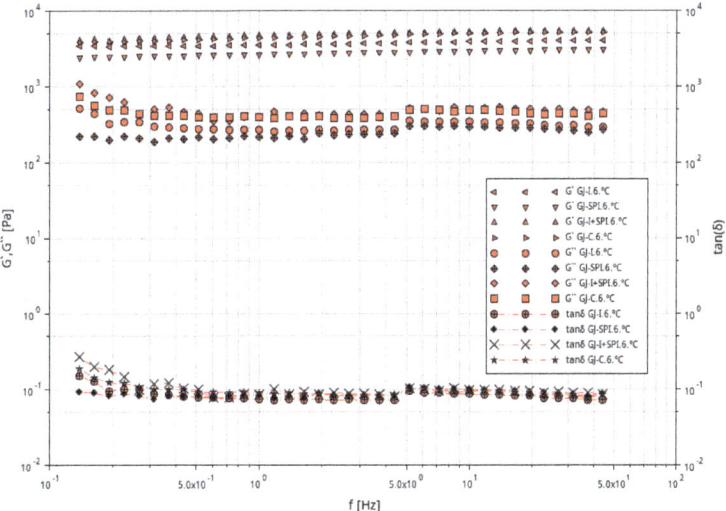

Figure 5. Dependence of G′ and G″ modulus on frequency for furcellaran-based gels at 6 °C and the dependence of the tan (δ) phase shift angle tangent on frequency for examined systems.

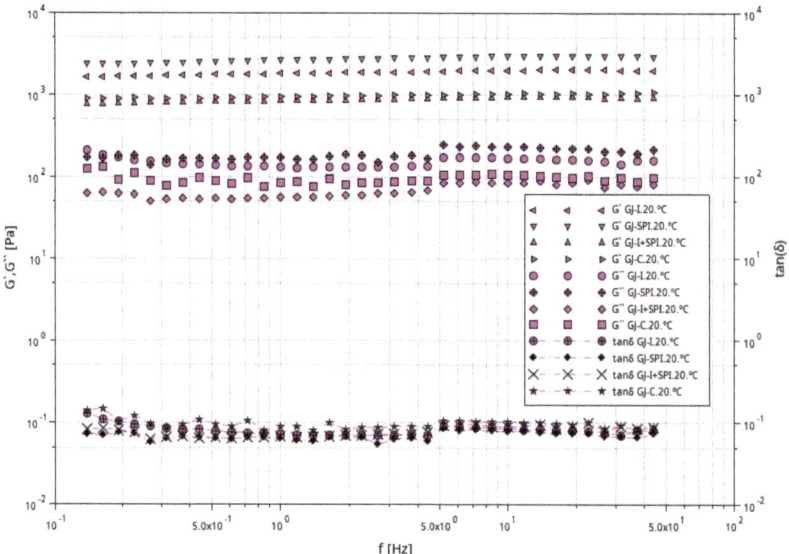

Figure 6. Dependence of G′ and G″ modulus on frequency for furcellaran gels at the temperature of 20 °C and the dependence of the tan (δ) phase shift angle tangent on frequency for examined systems.

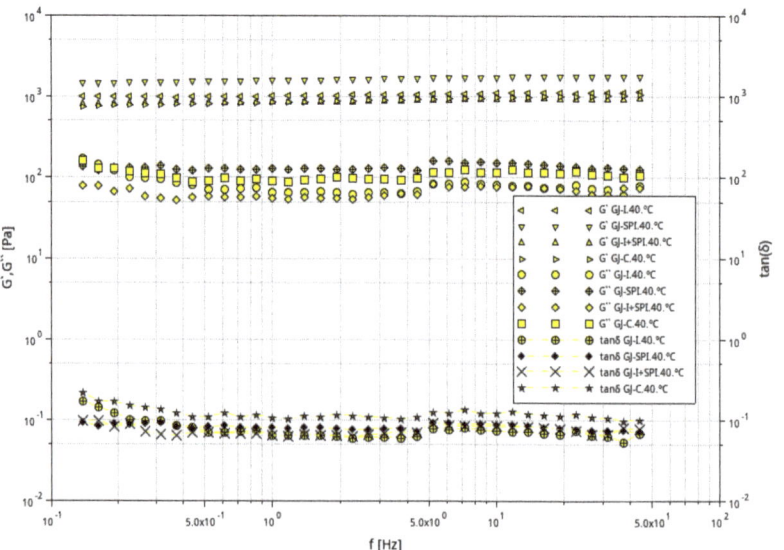

Figure 7. Dependence of G′ and G″ modulus on frequency for furcellaran-based gels at the temperature of 40 °C and the dependence of the tan (δ) phase shift angle tangent on frequency for examined systems.

The response of the elasticity modulus is independent of the frequency in the range from 0.1 to 50 Hz. The G″ moduli were almost frequency-independent in the range of 0.3–50 Hz, with the exception of samples GJ-I, GJ-SPI and GJ-C at 60 °C, where a clear decrease in G″ in the range of 0.1–0.3 Hz was observed. A characteristic slight increase in G″ at the frequency of 5 Hz was also observed for all tested systems. Similar behavior of the G′ and G″ modules was reported for 2% kappa-carrageenan gels [37]. The lack of literature about the rheology of furcellaran-based gels encourages further research on these systems. The ratio of the lost energy to the stored energy in each cycle is described as tan(δ), which informs about the physical behavior of the analyzed material. The phase angle tangent ($\tan(\delta) = G''/G'$) showed the relative importance of the viscosity modulus G'' and the elastic modulus G' [38]. The obtained values of tan(δ) were always lower than one, which confirms the constant behavior of the gels at all range of the frequency values and confirms a high degree of cross-linking of the gels. At lower frequency values of ~0.1–1 Hz, the examined systems have a tan(δ) value between 0.3 and 0.1 at 6 °C and 20 °C, while at the higher frequencies, tan(δ) values are below 0.1. All the evaluated furcellaran-based gels showed a strong gel network in the entire tested frequency and temperature range [39]. However, at the temperature of 40 °C, the control (GJ-C) gel has values higher than 0.1 in the entire frequency range, which causes the weakening of the gel network. Therefore, it can be concluded that the tested systems were strong gels, and the additions of inulin and soy protein isolate did not change the nature of the strength of the gels; it can even be assumed that they keep it constant. Analyzing the evaluated systems, it can also be seen that the GJ-SPI sample with the addition of SPI was the most stable, because the values of the G' and G'' modules are constant over the entire range of tested frequencies and temperatures.

3.2. Physicochemical Properties and Nutritional Labeling of Gummy Candies

Table 2 shows basic physicochemical data: sample weights, water activity and pH of low-sugar jellies during the following manufacturing stages. The experiments included four sample formulations and characteristics: gels before drying, jellies after drying, jellies stored in film (obtained as described in Section 2.7.1) and jellies stored without packaging. The weight of the samples after cooling and gelling varied between 9.54 g for the GJ-SPI sample and 10.03 g obtained for the variant with the addition of 2% inulin. These differences are caused by different formulations. However, except for the control sample, all versions were characterized by the same dry matter content; thus, these differences turned out to be statistically insignificant. The transformation of gels into gummy jellies includes drying in which excess water is removed, thus changing the structural properties of the product and stabilizing its structure. Therefore, the selection of proper process conditions including time and temperature is essential for the quality of the final product. During the production of commercial gummy candies, water is removed so as to ensure not only their high quality but also to provide optimum efficiency and continuity of production while taking into account the energy requirements important for economic reasons. For all variants of furcellaran-based gel systems, approximately twofold weight loss was observed during convection drying. It is assumed that the standard moisture content of jellies varied from 8 to 22% [11]. In our study, drying was carried out for four hours at a relatively low temperature so as to possibly preserve all thermolabile components such as volatile aromatic compounds as a result of which the final jellies had a significantly higher (64–66%) than recommended moisture content. The choice of such a method for removing moisture from furcellaran gels was predicated on the fact that priority was given to evaluating the macroscopic appearance of the sample during drying, especially homogeneous surface without signs of crystallization for erythritol.

Table 2. Physicochemical parameters of the developed low-sugar gummy jellies.

		Weight (g)	**Water Activity**	**pH**
after gelling	GJ-I	10.03 ± 0.19 [f]	0.982 ± 0.001 [b]	3.94 ± 0.09 [b]
	GJ-SPI	9.54 ± 0.14 [f]	0.980 ± 0.001 [b]	4.24 ± 0.12 [cd]
	GJ-I+SPI	9.55 ± 0.09 [f]	0.980 ± 0.002 [b]	4.02 ± 0.06 [b]
	GJ-C	9.75 ± 0.20 [f]	0.981 ± 0.001 [b]	4.13 ± 0.12 [cd]
after drying (4 h/45 °C)	GJ-I	4.80 ± 0.20 [e]	0.819 ± 0.004 [a]	3.29 ± 0.11 [a]
	GJ-SPI	4.45 ± 0.18 [de]	0.837 ± 0.004 [a]	3.71 ± 0.15 [e]
	GJ-I+SPI	4.53 ± 0.24 [de]	0.843 ± 0.006 [a]	3.34 ± 0.12 [a]
	GJ-C	4.26 ± 0.26 [d]	0.842 ± 0.003 [a]	3.41 ± 0.02 [a]
after storage without packing	GJ-I	3.15 ± 0.17 [a]	0.816 ± 0.008 [a]	
	GJ-SPI	3.63 ± 0.16 [b]	0.828 ± 0.008 [a]	
	GJ-I+SPI	3.61 ± 0.21 [bc]	0.841 ± 0.012 [a]	
	GJ-C	3.49 ± 0.22 [bc]	0.837 ± 0.036 [a]	
after storage with packing	GJ-I	4.17 ± 0.21 [d]	0.815 ± 0.004 [a]	
	GJ-SPI	4.22 ± 0.10 [d]	0.822 ± 0.003 [a]	
	GJ-I+SPI	4.38 ± 0.31 [d]	0.837 ± 0.004 [a]	
	GJ-C	3.83 ± 0.22 [c]	0.836 ± 0.007 [a]	

Values with different letters in the same row present significant differences.

In order to test the effect of biopolymer packaging on the properties of jellies, they were stored under refrigeration (temperature: 5 °C, relative humidity: 60%) for one week. The reference sample consisted of candies stored without packaging. As a result of controlling the weight of jellies after storage, we observed a decrease in the average weight for both wrapped and unwrapped samples. As a result of controlling the weight of jellybeans after storage, we observed a decrease in the average weight for both wrapped and unwrapped samples with more water lost by the unpackaged samples. Thereby, it can be concluded that the water content of the material was too high to store it in an unchanged state under the refrigeration conditions. The average weight of jellies unit after 7 days of packaged candies varied from 3.83 to 4.38 g, while for unpackaged candies, the values were in the range of 3.15–3.63 g. The greatest weight loss after storage (about 18%) was observed for samples containing inulin. This means that furcellaran alone binds and cross-links water better in gel systems than furcellaran with added inulin. The negative effect of inulin on the structure of furcellaran gels may also be due to its tendency toward crystallization, which is determined by the polymer chain length and its initial concentration in the system [40]. Water loss in jellies stored in triple-layer films ranged from 1.2 to 5.2%, indicating that despite the sealed packaging, free water migrated from the product structure into the film or optionally to the environment. Since, depending on the formulation, the decrease in moisture content during sample storage varied, it can be concluded that the physical stability of investigated furcellaran-based gel systems primarily depends on their composition and the solids–water interaction.

An attempt was made to produce a multi-layer packaging based on furcellaran, which was enriched with oenothera emulsion and spirulina. The packaging was designed to meet the needs of consumers who are environmentally conscious and want to reduce synthetic waste from food packaging. The material designed in this way was intended to contribute to extending the shelf life of gummy jellies. The appearance of a portion of the jellies in a biopolymer package is shown in Figure 8.

Figure 8. Example photo of the gels and the final packaging on the product.

Despite the fact that moisture level has a significant influence on texture and shelf life, water content by itself is not sufficient to completely characterize gummy jelly quality and stability during storage. A parameter often used to describe microbial stability, texture and water migration over time is water activity defined as the ratio of the vapor pressure of the material to the vapor pressure of pure water [11]. From a practical point of view, the a_w value indicates the amount of water not associated with other components in the structure of the material. This type of water is considered available for various chemical, physical, enzymatic or microbiological processes that in food cause deteriorative reaction during storage. In our study, water activity after drying decreased about 0.15 and ranged from 0.819 to 0.843, which is relatively high compared to literature reports. Gok et al. [15] obtained an a_w value that ranged between 0.76 and 0.84 for low-calorie jellies with mannitol addition, while Periche et al. [41] determined an a_w level from 0.721 to 0.908 for gummies based on isomaltulose. The high a_w values obtained for furcellaran jellies are due to insufficient moisture removal during drying. It is believed that the safe-for-consumers water activity value of this type of range should be between 0.55 and 0.75 [11]. Hence, in order to improve the formulation of the proposed product, it would be necessary to consider changing the drying parameters or increasing the content of solutes, especially those that tend to bind water. It was observed that after 7 days of storage, the water activity of all variants of the gels decreased slightly with a slightly greater decrease for the packaged samples. This tendency can be explained by the fact that every physical system tends toward thermodynamic equilibrium. Since the water activity of the gels was high, non-bonded moisture migrated. The phenomenon was more intense for samples in film with an a_w equal to 0.433 than for gummies without package stored in an environment with relative humidity 60% (a_w = 0.6).

The main goal of this work was to design vegan gummies, while the secondary goal was to develop a packaging that would be an alternative to synthetic packaging. Multi-layer films based on furcellaran are characterized by improved functional properties in relation to single-layer films, which is certainly due to their complex structure [26,42,43]. The addition of spirulina and oenothera emulsion to a specific layer was intended to give the films barrier properties.

Acidity (pH) is an important factor affecting the safety and quality of food products. As expected, drying the gels to make the final product resulted in an increase in pH. The production of standard confectionary products usually includes the addition of fruit acids such as citric and tartaric. The purpose is to lower the pH value but also to give the product its characteristic sour taste. As expected, the original acidity (3.94–4.24) of furcellaran gels decreased (3.29–3.71) as a result of drying. Acidity is a resultant of material composition; therefore, depending on the applied ingredients, it varies significantly even within a product of the same type. Lekahena and Boboleha [13] obtained seaweed jellies with pH between 4.64 and 4.74, while the acidity of gelatin gummies developed by Delgado and Banon [44] was equal to 3.07.

Standard commercially available gummy jellies are mostly characterized by low nutritional value and high energy density, making them widely considered an "unhealthy consumer choice". Depending on the type and manufacturer, the energy value of gummies can range from 300 to 400 kcal in 100 g of product. Thus, assuming an average daily caloric intake of 2000 kcal, the consumption of the entire package can correspond to as much as one-fifth of the daily energy requirements. The energetic value of furcellaran-based jellies ranged from 29.3 to 37.3 kcal/100 g (Table 3). Assuming that, according to the results described above, the developing formulation requires the removal of more water, the calculated caloric value of the developed product is low. Potentially reducing the moisture content of the product, even by half, will still result in an energy value of less than 100 kcal/100 g. The highest protein content at the level of 4.07 g was, of course, shown by the sample with the addition of SPI, while in order to increase the value of the product in future research, it is planned to increase the addition of soy protein isolate. Dietary fiber ranging from 5.50 to 8.7 g/100 g could also be higher due to its beneficial effects on health. As expected, the gummies had a low fat content. The carbohydrates, resulting mainly from the erythritol and furcellaran presence, ranged from 28.66 to 30.27 g/100 g, while according to the assumptions, the product did not contain any simple sugars at all. As stated by regulation (EU) 1924/2006 [45] on health and nutrition, sugar-free claims may only be used if the product contains no more than 0.5 g of sugar per 100 g. The gummy jellies developed in this study therefore can be included into this group of food products.

Table 3. Nutritional values of the developed low-sugar gummy jellies.

	GJ-I (100 g)	GJ-SPI (100 g)	GJ-I+SPI (100 g)	GJ-C (100 g)
Energy (kcal/kJ)	33.0/138.95	33.2/138.95	37.3/156.06	29.3/122.63
Protein (g)	0.21	4.07	2.11	0.23
Total carbohydrate (g)	30.27	27.21	28.66	29.08
Dietary fiber (g)	8.77	5.14	6.95	5.50
Sugars (g)	0.00	0.00	0.00	0.00
Fat (g)	0.00	0.21	0.21	0.00

3.3. Antioxidant Activity

Table 4 shows the results of investigating the antioxidant activity of furcellaran-based samples at two different stages of the production process. The highest antioxidant activity, expressed both as Ferric Reducing Antioxidant Power and metal-chelating activity, is found in freshly prepared jellies containing soy protein isolate (0.68 uM Trolox equivalent/mg and 20.19%, respectively). As the SPI content of the jellies decreases, their antioxidant activity also decreases. Soy protein isolates (SPIs) are a highly refined form of protein derived from soybeans. They are commonly used in various food products, including beverages, protein bars, meat alternatives, and baked goods. One of the notable properties of soy protein isolates is their antioxidant activity, which can have several health benefits [46]. According to literature data [47], soy protein isolates commonly contain a lot of phenolic compounds, including isoflavonoids that could stabilize free radicals. Both the native soy protein structure and also possibly the presence of residual antioxidant phenolic compounds could cause an antioxidant effect. Furthermore, nonhydrolyzed SPI has been reported to be able to inhibit lipid oxidation and retard rancidity odor development in various model products [48,49].

No changes in the antioxidant activity of the jelly were noted during the 7-day storage. There was also no negative effect of packaging the jelly in biopolymer films on their antioxidant properties. Polyphenols themselves can undergo oxidation when exposed to oxygen in the air. This process can lead to a decrease in their antioxidant capacity. The loss of antioxidant activity due to oxidation can impact the overall effectiveness of the polyphenol-rich product in neutralizing free radicals [50]. Therefore, the lack of change in the antioxidant activity of stored samples is a positive development. However, it would be necessary to conduct further studies in which the storage time of the samples is extended.

Table 4. Antioxidant activity of the developed low-sugar gummy jellies.

	Sample	FRAP uM Trolox Equivalent/mg	Metal-Chelating Activity (%)
after drying (2 h/35 °C)	GJ-I	0.56 ± 0.04 [ad]	0 ± 0.00 [a]
	GJ-SPI	0.86 ± 0.06 [b]	20.19 ± 2.47 [c]
	GJ-I+SPI	0.58 ± 0.04 [abd]	7.51 ± 1.08 [b]
	GJ-C	0.50 ± 0.10 [ac]	0 ± 0.00 [a]
after storage without packing	GJ-I	0.60 ± 0.04 [abd]	0 ± 0.00 [a]
	GJ-SPI	0.69 ± 0.05 [b]	29.11 ± 1.47 [c]
	GJ-I+SPI	0.60 ± 0.02 [abd]	8.69 ± 2.03 [b]
	GJ-C	0.44 ± 0.01 [c]	0 ± 0.00 [a]
after storage with packing	GJ-I	0.52 ± 0.11 [ac]	0 ± 0.00 [a]
	GJ-SPI	0.66 ± 0.04 [d]	29.11 ± 5.65 [c]
	GJ-I+SPI	0.52 ± 0.03 [ac]	6.34 ± 2.54 [b]
	GJ-C	0.42 ± 0.05 [c]	0 ± 0.00 [a]

Values with different letters in the same row present significant differences.

The impact of biopolymer packaging on the active properties of products stored within them can vary depending on several factors such as type of biopolymer used, the nature of the product, and storage conditions [24]. Proper selection of biopolymer materials, design considerations, and testing under relevant storage conditions are essential to ensure that the packaging maintains or enhances the active properties of the products it contains.

3.4. Textural Properties of Gummy Jelly

The textural properties of food products classified as confectionery have a major impact on consumer perception and acceptance. Our study applied an instrumental TPA technique in order to determine the hardness (N), springiness, cohesiveness and gumminess (N) of furcellaran low-sugar jellies with inulin and/or soy protein isolate additions. The obtained results and statistical effect of the formulations and technological process stage are shown in Table 5.

Four different stages of jelly production were considered in the study: after gelling, after drying, and after one week of refrigerated storage in bioactive film or without packaging. The most significant change in all examined parameters, independently of the sample composition, was observed after drying the gels, which is clearly related to the significant loss of water. In addition, jellies stored in active furcellaran film for seven days showed higher values of hardness, cohesiveness, gumminess and springiness than analogous samples stored in the refrigerator without packaging. This trend is also related to differences in moisture content. As shown in the previous paragraph, the water activity of the storage environment was 0.55, while the a_w value of furcellaran films was equal to 0.43. Therefore, jellies hermetically sealed in film, according to the base of thermodynamic equilibrium, gave up more moisture than samples without packaging. There are almost no reports in the literature on the textural properties of confectionary gels based on furcellaran. In addition, it is worth noting that the values of individual parameters determined by TPA depend on the geometry of the tested sample, so it is necessary to take into account not only its composition but also its shape and size [10]. Regardless of the stage of production, hardness values, defined as the maximum force required for the sample deformation in the first compression, were the highest for jellies with 2% soy protein isolate additions and ranged from 8.264 N determined for the sample after gelling to 33.542 N obtained for packing jellies. The increased hardness of the system as a result of the addition of proteins is due to their structure-forming properties. There are no reports on the properties of composite hydrogel systems containing furcellaran and protein, while the increase in hardness of gels based on carrageenan and soy protein has been proven [51]. However, interestingly, both furcellaran and SPI chains are negatively charged, so the thermodynamic incompatibility between the protein and polysaccharide fractions had a positive effect on the textural properties of the samples [52]. Slightly lower hardness than the GJ-SPI samples, at all stages of production, were found for the jellies with the addition of both protein isolate and inulin. Meanwhile, the values determined for the GJ-I+SPI systems are more similar to those obtained for gels with the addition of protein alone. This shows that furcellaran–SPI interactions are definitely stronger than those between inulin and protein. Compatibility between protein–polysaccharide systems, which determines good mechanical properties, is a result of many factors including their type, composition, molecular weight, structure and chain branching [10]. Obtained values of cohesiveness, expressed as internal bonds in the sample and springiness connected with elastic recovery that occurs in the material after compressive force removal, showed the same tendency regarding the hardness. This therefore confirms the key effect of the addition of the protein fraction on the mechanical characteristics desired in confectionary products. Gumminess is a measure representing the energy required to break down a semi-solid product ready for swallowing. The highest values of this texture variable were determined at each stage of the production process for gels containing soy protein isolate, while gumminess is far more determined by the stage of the production process than by the formulation. It was noted that for all systems tested except the GJ-C control sample, drying the gels resulted in a more than threefold increase in gumminess values. It has been demonstrated that the increase in gumminess values as a result of drying can be explained by both gelation and dehydration phenomena. The simultaneous actions of both gelling agents on the colloidal system containing water and other food components resulted in the typical firm and chewy structure of gummy candies [44].

Statistical analysis (multi-factor ANOVA) shows that the technological process step was the variable with the most influence (the highest values of F-ratio) on every textural parameter with the most significant impact on gumminess. The formulation of the jellies exhibited a remarkable effect on the obtained values of cohesiveness and gumminess. A statistically significant effect of both individual variables, process step and formulation, was found for each of the textural parameters, while the F-ratio value for springiness was the highest. The effect of the samples' formulation on hardness and springiness was not statistically significant.

Table 5. Comparison of texture parameters of furcellaran low-sugar gummy jellies.

	After Gelling			
	GJ-I	GJ-SPI	GJ-I+SPI	GJ-C
Hardness (N)	3.582 ± 0.30 [a]	8.827 ± 0.56 [c]	8.264 ± 0.57 [bc]	2.743 ± 0.27 [a]
Cohesiveness (N)	0.148 ± 0.021 [ac]	0.346 ± 0.026 [d]	0.227 ± 0.016 [bc]	0.139 ± 0.029 [a]
Springiness	0.079 ± 0.014 [a]	0.103 ± 0.004 [a]	0.094 ± 0.006 [a]	0.058 ± 0.007 [a]
Gumminess (N)	3.241 ± 0.633 [a]	5.723 ± 0.278 [b]	4.424 ± 0.844 [ab]	3.537 ± 0.191 [a]
	After drying [4 h/45 °C]			
	GJ-I	GJ-SPI	GJ-I+SPI	GJ-C
Hardness (N)	8.495 ± 0.81 [bc]	16.483 ± 0.88 [e]	15.138 ± 0.81 [d]	7.313 ± 0.30 [b]
Cohesiveness (N)	0.251 ± 0.025 [b]	0.489 ± 0.079 [e]	0.444 ± 0.054 [e]	0.198 ± 0.010 [ab]
Springiness	0.482 ± 0.057 [c]	0.607 ± 0.040 [ef]	0.508 ± 0.075 [cd]	0.236 ± 0.031 [b]
Gumminess (N)	11.734 ± 0.971 [d]	16.321 ± 1.010 [f]	15.681 ± 0.852 [f]	7.681 ± 0.852 [c]

Table 5. *Cont.*

	After 7 days without packing			
	GJ-I	GJ-SPI	GJ-I+SPI	GJ-C
Hardness (N)	16.476 ± 0.84 [e]	20.680 ± 0.93 [g]	19.073 ± 1.41 [f]	15.144 ± 0.82 [d]
Cohesiveness (N)	0.383 ± 0.012 [bc]	0.625 ± 0.083 [f]	0.437 ± 0.029 [e]	0.286 ± 0.027 [cd]
Springiness	0.533 ± 0.017 [cde]	0.626 ± 0.069 [f]	0.570 ± 0.024 [def]	0.273 ± 0.019 [b]
Gumminess (N)	12.863 ± 0.861 [de]	21.087 ± 0.995 [h]	19.265 ± 0.935 [g]	8.392 ± 0.950 [c]
	After 7 days with packing			
	GJ-I	GJ-SPI	GJ-I+SPI	GJ-C
Hardness (N)	26.114 ± 0.98 [g]	33.546 ± 0.60 [i]	30.566 ± 0.77 [h]	18.286 ± 1.16 [f]
Cohesiveness (N)	0.584 ± 0.060 [f]	0.914 ± 0.051 [g]	0.850 ± 0.053 [g]	0.348 ± 0.021 [d]
Springiness	0.730 ± 0.090 [g]	0.965 ± 0.089 [g]	0.825 ± 0.088 [h]	0.557 ± 0.053 [cdef]
Gumminess (N)	13.980 ± 0.779 [e]	28.170 ± 1.275 [i]	20.642 ± 1.023 [h]	8.506 ± 0.779 [c]
	ANOVA (F-ratio)			
	Hardness	Cohesiveness	Springiness	Gumminess
Formulation	1.378	9.992 *	2.324	19.161 *
Process step	22.117 *	24.221 *	20.916 *	51.126 *
Interaction	2.207 *	4.030 *	10.286 *	1.894 *

* Statistical significance ≥ 95% (*p*-values ≤ 0.005). Values with different letters in the same row present significant differences.

4. Conclusions

The results of the current research indicate that furcellaran has high potential as a gelling agent used in the production of sugar-free gummy jellies. The performed rheological characterization showed that hydrogels based on this polysaccharide are characterized by high stability and elasticity, which confirms furcellaran's compatibility with other ingredients used in the formulation: erythritol, soy protein isolate and inulin. After drying, the gels transformed into gummies by removing about 50% of the water showed water activity ranging from 0.819 to 0.843 and pH between 3.29 and 3.71. The most significant metal-chelating activity of 29.11% was determined for the variant of jellies with soy protein isolate. The hardness of the dried jellies, which is the most important textural parameter, ranged from 7.31 to 16.38 N, and its value increased during storage regardless of the packaging applied. The use of three-layer furcellaran film as packaging for jellies did not have any negative effect on the antioxidant activity, texture or physicochemical properties of the product. During 7 days of storage, jellies lost only a small amount of water, which is not due to the use of packaging but rather due to the insufficient drying of the product. Therefore, further research will include formulation improvements that take into account an increase in the final dry matter content. Since the use of erythritol as a sweetener did not in any way reduce the mechanical stability or properties of the jellies, future research will be conducted in the direction of increasing the proportion of protein in the product in order to obtain low-sugar confectionery with high nutritional value, which is suitable for consumers who limit sugar in their diets. In order to make the product available at a larger scale, it is also necessary to investigate the effect of biopolymer films on the microbiological safety of jellies under different storage conditions. Nevertheless, based on the results obtained so far, it can be concluded that the use of furcellaran as both a gelling agent and a component of the film makes it possible to obtain a complete vegan confectionary product with environmentally friendly packaging.

Author Contributions: Conceptualization: A.S. and E.J.; methodology: A.S., U.G., E.J., J.T., W.G. and N.N.; software: A.S. and D.Ż.; investigation: A.S., U.G., E.J., J.T. and A.S.; writing—original draft preparation: A.S., U.G., E.J. and J.T.; writing—review and editing: A.S.; visualization: A.S. and E.J. All authors have read and agreed to the published version of the manuscript.

Funding: This research received no external funding.

Institutional Review Board Statement: Not applicable.

Informed Consent Statement: Not applicable.

Data Availability Statement: Raw data available on request to authors.

Acknowledgments: Not applicable.

Conflicts of Interest: The authors declare no conflict of interest.

References

1. Laos, K.; Ring, S.G. Note: Characterisation of Furcellaran Samples from Estonian Furcellaria Lumbricalis (Rhodophyta). *J. Appl. Phycol.* **2005**, *17*, 461–464. [CrossRef]
2. Imeson, A.P. Carrageenan and Furcellaran. In *Handbook of Hydrocolloids*; Elsevier: Amsterdam, The Netherlands, 2009; pp. 164–185. ISBN 978-1-84569-414-2.
3. Marangoni, L., Jr.; Vieira, R.P.; Jamróz, E.; Anjos, C.A.R. Furcellaran: An Innovative Biopolymer in the Production of Films and Coatings. *Carbohydr. Polym.* **2021**, *252*, 117221. [CrossRef] [PubMed]
4. Hudson, B.J.F. Evening Primrose (*Oenothera* Spp.) Oil and Seed. *J. Am. Oil Chem. Soc.* **1984**, *61*, 540–543. [CrossRef]
5. Hadidi, M.; Ibarz, A.; Pouramin, S. Optimization of Extraction and Deamidation of Edible Protein from Evening Primrose (*Oenothera Biennis* L.) Oil Processing by-Products and Its Effect on Structural and Techno-Functional Properties. *Food Chem.* **2021**, *334*, 127613. [CrossRef] [PubMed]
6. Song, X.; Chiou, B.; Xia, Y.; Chen, M.; Liu, F.; Zhong, F. The Improvement of Texture Properties and Storage Stability for Kappa Carrageenan in Developing Vegan Gummy Candies. *J. Sci. Food Agric.* **2022**, *102*, 3693–3702. [CrossRef] [PubMed]
7. Utomo, B.S.B.; Darmawan, M.; Hakim, A.R.; Ardi, D.T. Physicochemical properties and sensory evaluation of jelly candy made from different ratio of k-carrageenan and konjac. *Squalen Bull. Mar. Fish. Postharvest Biotechnol.* **2014**, *9*, 25. [CrossRef]
8. De Avelar, M.H.M.; Efraim, P. Alginate/Pectin Cold-Set Gelation as a Potential Sustainable Method for Jelly Candy Production. *LWT* **2020**, *123*, 109119. [CrossRef]
9. Ghiraldi, M.; Franco, B.G.; Moraes, I.C.F.; Pinho, S.C. Emulsion-Filled Pectin Gels for Vehiculation of Vitamins D $_3$ and B $_{12}$: From Structuring to the Development of Enriched Vegan Gummy Candies. *ACS Food Sci. Technol.* **2021**, *1*, 1945–1952. [CrossRef]
10. Burey, P.; Bhandari, B.R.; Rutgers, R.P.G.; Halley, P.J.; Torley, P.J. Confectionery Gels: A Review on Formulation, Rheological and Structural Aspects. *Int. J. Food Prop.* **2009**, *12*, 176–210. [CrossRef]
11. Ergun, R.; Lietha, R.; Hartel, R.W. Moisture and Shelf Life in Sugar Confections. *Crit. Rev. Food Sci. Nutr.* **2010**, *50*, 162–192. [CrossRef]
12. Rippe, J.; Angelopoulos, T. Relationship between Added Sugars Consumption and Chronic Disease Risk Factors: Current Understanding. *Nutrients* **2016**, *8*, 697. [CrossRef] [PubMed]
13. Lekahena, V.N.J.; Boboleha, M.R. The Effects of Sucrose Substitution with Sorbitol on Physicochemical Properties and Sensory Evaluation of Seaweed Jelly Candy. In *Proceedings of the 5th International Conference on Food, Agriculture and Natural Resources (FANRes 2019)*; Atlantis Press: Ternate, Indonesia, 2020.
14. Riedel, R.; Böhme, B.; Rohm, H. Development of Formulations for Reduced-Sugar and Sugar-Free Agar-Based Fruit Jellies. *Int. J. Food Sci. Technol.* **2015**, *50*, 1338–1344. [CrossRef]
15. Gok, S.; Toker, O.S.; Palabiyik, I.; Konar, N. Usage Possibility of Mannitol and Soluble Wheat Fiber in Low Calorie Gummy Candies. *LWT* **2020**, *128*, 109531. [CrossRef]
16. Bartkiene, E.; Sakiene, V.; Bartkevics, V.; Wiacek, C.; Rusko, J.; Lele, V.; Ruzauskas, M.; Juodeikiene, G.; Klupsaite, D.; Bernatoniene, J.; et al. Nutraceuticals in Gummy Candies Form Prepared from Lacto-Fermented Lupine Protein Concentrates, as High-Quality Protein Source, Incorporated with *Citrus Paradise* L. Essential Oil and Xylitol. *Int. J. Food Sci. Technol.* **2018**, *53*, 2015–2025. [CrossRef]
17. Lele, V.; Ruzauskas, M.; Zavistanaviciute, P.; Laurusiene, R.; Rimene, G.; Kiudulaite, D.; Tomkeviciute, J.; Nemeikstyte, J.; Stankevicius, R.; Bartkiene, E. Development and Characterization of the Gummy–Supplements, Enriched with Probiotics and Prebiotics. *CyTA—J. Food* **2018**, *16*, 580–587. [CrossRef]
18. Silva, J.R.; Silva, J.B.D.; Costa, G.N.; Santos, J.S.D.; Castro-Gomez, R.J.H. Probiotic Gummy Candy with Xylitol: Development and Potential Inhibition of Streptococcus Mutans UA 159. *RSD* **2020**, *9*, e7369108942. [CrossRef]
19. Zhou, X.; Yu, J.; Yu, H. Effect of Gelatin Content and Oral Processing Ability on Vitamin C Release in Gummy Jelly. *J. Food Sci. Technol.* **2022**, *59*, 677–685. [CrossRef]
20. Cedeño-Pinos, C.; Martínez-Tomé, M.; Murcia, M.A.; Jordán, M.J.; Bañón, S. Assessment of Rosemary (*Rosmarinus Officinalis* L.) Extract as Antioxidant in Jelly Candies Made with Fructan Fibres and Stevia. *Antioxidants* **2020**, *9*, 1289. [CrossRef]

21. Gramza-Michalowska, A.; Regula, J. Use of Tea Extracts (*Camelia Sinensis*) in Jelly Candies as Polyphenols Sources in Human Diet. *Asia Pac. J. Clin. Nutr.* **2007**, *16*, 43–46.
22. Chen, Y.; Awasthi, A.K.; Wei, F.; Tan, Q.; Li, J. Single-Use Plastics: Production, Usage, Disposal, and Adverse Impacts. *Sci. Total Environ.* **2021**, *752*, 141772. [CrossRef]
23. Basumatary, I.B.; Mukherjee, A.; Katiyar, V.; Kumar, S. Biopolymer-Based Nanocomposite Films and Coatings: Recent Advances in Shelf-Life Improvement of Fruits and Vegetables. *Crit. Rev. Food Sci. Nutr.* **2022**, *62*, 1912–1935. [CrossRef] [PubMed]
24. Abdullah; Cai, J.; Hafeez, M.A.; Wang, Q.; Farooq, S.; Huang, Q.; Tian, W.; Xiao, J. Biopolymer-Based Functional Films for Packaging Applications: A Review. *Front. Nutr.* **2022**, *9*, 1000116. [CrossRef] [PubMed]
25. Azeredo, H.M.C.; Waldron, K.W. Crosslinking in Polysaccharide and Protein Films and Coatings for Food Contact—A Review. *Trends Food Sci. Technol.* **2016**, *52*, 109–122. [CrossRef]
26. Grzebieniarz, W.; Tkaczewska, J.; Juszczak, L.; Krzyściak, P.; Cholewa-Wójcik, A.; Nowak, N.; Guzik, P.; Szuwarzyński, M.; Mazur, T.; Jamróz, E. Improving the Quality of Multi-Layer Films Based on Furcellaran by Immobilising Active Ingredients and Impact Assessment of the Use of a New Packaging Material. *Food Chem.* **2023**, *428*, 136759. [CrossRef]
27. Jamróz, E.; Kulawik, P.; Guzik, P.; Duda, I. The Verification of Intelligent Properties of Furcellaran Films with Plant Extracts on the Stored Fresh Atlantic Mackerel during Storage at 2 °C. *Food Hydrocoll.* **2019**, *97*, 105211. [CrossRef]
28. Jancikova, S.; Jamróz, E.; Kulawik, P.; Tkaczewska, J.; Dordevic, D. Furcellaran/Gelatin Hydrolysate/Rosemary Extract Composite Films as Active and Intelligent Packaging Materials. *Int. J. Biol. Macromol.* **2019**, *131*, 19–28. [CrossRef] [PubMed]
29. Dziubiński, M.; Kiljański, T.; Sęk, J. *Podstawy Teoretyczne i Metody Pomiarowe Reologii*; Wydawnictwo Politechniki Łódzkiej: Łódź, Poland, 2014; ISBN 83-7283-641-8.
30. Benzie, I.F.F.; Strain, J.J. The Ferric Reducing Ability of Plasma (FRAP) as a Measure of "Antioxidant Power": The FRAP Assay. *Anal. Biochem.* **1996**, *239*, 70–76. [CrossRef]
31. Tkaczewska, J.; Zając, M.; Jamróz, E.; Derbew, H. Utilising Waste from Soybean Processing as Raw Materials for the Production of Preparations with Antioxidant Properties, Serving as Natural Food Preservatives—A Pilot Study. *LWT* **2022**, *160*, 113282. [CrossRef]
32. Trigui, I.; Yaich, H.; Sila, A.; Cheikh-Rouhou, S.; Krichen, F.; Bougatef, A.; Attia, H.; Ayadi, M.A. Physical, Techno-Functional and Antioxidant Properties of Black Cumin Seeds Protein Isolate and Hydrolysates. *Food Meas.* **2021**, *15*, 3491–3500. [CrossRef]
33. Ahmed, J.; Basu, S. *Advances in Food Rheology and Its Applications*; Woodhead Publishing: Thorston, UK, 2016; ISBN 0-08-100432-X.
34. Eshtiaghi, N.; Markis, F.; Baudez, J.-C.; Slatter, P. Proxy Model Materials to Simulate the Elastic Properties of Digested Municipal Sludge. *Water Res.* **2013**, *47*, 5557–5563. [CrossRef]
35. Pfaff, N.M.; Dijksman, J.A.; Kemperman, A.J.B.; Van Loosdrecht, M.C.M.; Kleijn, J.M. Rheological Characterisation of Alginate-like Exopolymer Gels Crosslinked with Calcium. *Water Res.* **2021**, *207*, 117835. [CrossRef] [PubMed]
36. Yang, Z.; Yang, H.; Yang, H. Effects of Sucrose Addition on the Rheology and Microstructure of κ-Carrageenan Gel. *Food Hydrocoll.* **2018**, *75*, 164–173. [CrossRef]
37. Yang, Z.; Yang, H.; Yang, H. Characterisation of Rheology and Microstructures of κ-Carrageenan in Ethanol-Water Mixtures. *Food Res. Int.* **2018**, *107*, 738–746. [CrossRef] [PubMed]
38. Owczarz, P.; Ryl, A.; Modrzejewska, Z.; Dziubiński, M. The Influence of The Addition of Collagen on The Rheological Properties of Chitosan Chloride Solutions. *PCACD* **2017**, *XXII*, 176–189. [CrossRef]
39. Lapasin, R. *Rheology of Industrial Polysaccharides: Theory and Applications*; Springer: Berlin/Heidelberg, Germany, 2012; ISBN 1-4615-2185-8.
40. Bot, A.; Erle, U.; Vreeker, R.; Agterof, W.G.M. Influence of Crystallisation Conditions on the Large Deformation Rheology of Inulin Gels. *Food Hydrocoll.* **2004**, *18*, 547–556. [CrossRef]
41. Periche, A.; Heredia, A.; Escriche, I.; Andrés, A.; Castelló, M.L. Optical, Mechanical and Sensory Properties of Based-Isomaltulose Gummy Confections. *Food Biosci.* **2014**, *7*, 37–44. [CrossRef]
42. Jamróz, E.; Tkaczewska, J.; Zając, M.; Guzik, P.; Juszczak, L.; Kawecka, A.; Turek, K.; Zimowska, M.; Wojdyło, A. Utilisation of Soybean Post-Production Waste in Single- and Double-Layered Films Based on Furcellaran to Obtain Packaging Materials for Food Products Prone to Oxidation. *Food Chem.* **2022**, *387*, 132883. [CrossRef]
43. Jamróz, E.; Cabaj, A.; Tkaczewska, J.; Kawecka, A.; Krzyściak, P.; Szuwarzyński, M.; Mazur, T.; Juszczak, L. Incorporation of Curcumin Extract with Lemongrass Essential Oil into the Middle Layer of Triple-Layered Films Based on Furcellaran/Chitosan/Gelatin Hydrolysates—In Vitro and in Vivo Studies on Active and Intelligent Properties. *Food Chem.* **2023**, *402*, 134476. [CrossRef]
44. Delgado, P.; Bañón, S. Determining the Minimum Drying Time of Gummy Confections Based on Their Mechanical Properties. *CyTA—J. Food* **2015**, *13*, 329–335. [CrossRef]
45. EU. *Regulation (EC) No 1924/2006—Nutrition and Health Claims on Foods*; European Union: Maastricht, The Netherlands, 2006.
46. Rachman, A.; Brennan, M.A.; Morton, J.; Torrico, D.; Brennan, C.S. In-Vitro Digestibility, Protein Digestibility Corrected Amino Acid, and Sensory Properties of Banana-Cassava Gluten-Free Pasta with Soy Protein Isolate and Egg White Protein Addition. *Food Sci. Hum. Wellness* **2023**, *12*, 520–527. [CrossRef]
47. Penta-Ramos, E.A.; Xiong, Y.L. Antioxidant Activity of Soy Protein Hydrolysates in a Liposomal System. *J. Food Sci.* **2002**, *67*, 2952–2956. [CrossRef]

48. Zhang, L.; Li, J.; Zhou, K. Chelating and Radical Scavenging Activities of Soy Protein Hydrolysates Prepared from Microbial Proteases and Their Effect on Meat Lipid Peroxidation. *Bioresour. Technol.* **2010**, *101*, 2084–2089. [CrossRef] [PubMed]
49. Kang, H.-J.; Kim, S.-J.; You, Y.-S.; Lacroix, M.; Han, J. Inhibitory Effect of Soy Protein Coating Formulations on Walnut (*Juglans Regia* L.) Kernels against Lipid Oxidation. *LWT—Food Sci. Technol.* **2013**, *51*, 393–396. [CrossRef]
50. Synge, R.L.M. Interactions of Polyphenols with Proteins in Plants and Plant Products. *Plant Foods Hum. Nutr.* **1975**, *24*, 337–350. [CrossRef]
51. Zhang, Q.; Gu, L.; Su, Y.; Chang, C.; Yang, Y.; Li, J. Development of Soy Protein Isolate/κ-Carrageenan Composite Hydrogels as a Delivery System for Hydrophilic Compounds: Monascus Yellow. *Int. J. Biol. Macromol.* **2021**, *172*, 281–288. [CrossRef] [PubMed]
52. Baeza, R.I.; Carp, D.J.; Pérez, O.E.; Pilosof, A.M.R. κ-Carrageenan—Protein Interactions: Effect of Proteins on Polysaccharide Gelling and Textural Properties. *LWT* **2002**, *35*, 741–747. [CrossRef]

Article

Carbon-Fiber-Recycling Strategies: A Secondary Waste Stream Used for PA6,6 Thermoplastic Composite Applications

Marco Valente [1,2,*], Matteo Sambucci [1,2], Ilaria Rossitti [1,2], Silvia Abruzzese [1], Claudia Sergi [1,2], Fabrizio Sarasini [1,2] and Jacopo Tirillò [1,2]

[1] Department of Chemical Engineering, Materials, Environment, Sapienza University of Rome, 00184 Rome, Italy; matteo.sambucci@uniroma1.it (M.S.); ilaria.rossitti@uniroma1.it (I.R.); abruzzese.1638778@studenti.uniroma1.it (S.A.); claudia.sergi@uniroma1.it (C.S.); fabrizio.sarasini@uniroma1.it (F.S.); jacopo.tirillo@uniroma1.it (J.T.)

[2] INSTM Reference Laboratory for Engineering of Surface Treatments, Department of Chemical Engineering, Materials, Environment, Sapienza University of Rome, 00184 Rome, Italy

* Correspondence: marco.valente@uniroma1.it

Abstract: With a view to achieving sustainable development and a circular economy, this work focused on the possibility to valorize a secondary waste stream of recycled carbon fiber (rCF) to produce a 3D printing usable material with a PA6,6 polymer matrix. The reinforcing fibers implemented in the research are the result of a double-recovery action: starting with pyrolysis, long fibers are obtained, which are used to produce non-woven fabrics, and subsequently, fiber agglomerate wastes obtained from this last process are ground in a ball mill. The effect of different amounts of reinforcement at 5% and 10% by weight on the mechanical properties of 3D-printed thermoplastic composites was investigated. Although the recycled fraction was successfully integrated in the production of filaments for 3D printing and therefore in the production of specimens via the fused deposition modeling technique, the results showed that fibers did not improve the mechanical properties as expected, due to an unsuitable average size distribution and the presence of a predominant dusty fraction ascribed to the non-optimized ball milling process. PA6,6 + 10 wt.% rCF composites exhibited a tensile strength of 59.53 MPa and a tensile modulus of 2.24 GPa, which correspond to an improvement in mechanical behavior of 5% and 21% compared to the neat PA6,6 specimens, respectively. The printed composite specimens loaded with the lowest content of rCF provided the greatest improvement in strength (+9% over the neat sample). Next, a prediction of the "optimum" critical length of carbon fibers was proposed that could be used for future optimization of recycled fiber processing.

Keywords: short-carbon-fiber-reinforced PA6,6 thermoplastic composites; recycled carbon fibers; 3D printing; recoverability; critical length

Citation: Valente, M.; Sambucci, M.; Rossitti, I.; Abruzzese, S.; Sergi, C.; Sarasini, F.; Tirillò, J. Carbon-Fiber-Recycling Strategies: A Secondary Waste Stream Used for PA6,6 Thermoplastic Composite Applications. *Materials* **2023**, *16*, 5436. https://doi.org/10.3390/ma16155436

Academic Editor: Daniela Fico

Received: 8 July 2023
Revised: 28 July 2023
Accepted: 31 July 2023
Published: 3 August 2023

1. Introduction

In recent years, carbon fibers (CFs) have found applications in a wide variety of fields, including automotive, aircraft, electronic equipment, and sporting goods industries [1]. Their use as reinforcement in composite materials is constantly growing even in fields more oriented toward mass production by replacing traditional metallic components, providing highly specific mechanical properties that result in lower CO_2 emissions [2]. Data analyses of this trend suggest that the world production of CFs already doubled between 2009 and 2014 from 27 to 53 ktons and reached a maximum in 2022 with 117 ktons [3].

Such a boost in the use of CFs affects the generation of wastes that derive both from the production process (process wastes representing about 30–40 wt.% of the total) and from the end of life of the products [1]. Currently, the main ways for the disposal of composite waste are incineration and landfilling, but these strategies will no longer be adequate due to environmental pollution and loss of CFs with high added value [2]. Recently, several recycling technologies have been proposed to treat composite materials with CFs and recover the CFs to reuse them in other applications for both economic and environmental reasons, aiming at the production of sustainable composites made with recycled reinforcement and a recyclable matrix [4]. The main technologies explored so far include mechanical recycling (shredding, crushing, milling), chemical recycling (solvent, catalyst, or supercritical fluids), and thermal recycling (pyrolysis, oxidation, steam thermolysis), and the objective of these techniques is to recover CFs under conditions as close to their initial state as possible to facilitate reuse in other applications [5]. Mechanical recycling is currently one of the consolidated methods and involves several steps to reduce the size of waste. First, the composites are ground to a size of about 50–100 mm, and then, further grinding is applied to obtain recycled materials with different dimensions, which can range from powder to fibrous agglomerates. The materials obtained from this type of recycling can be used as reinforcement in short-fiber composite materials, such as those used for extrusion and injection molding. Because of the resultant limited fiber aspect ratio, these materials have a low market value [6]. Due to the friction caused by materials during recycling, damage to equipment can occur, which directly increases the cost of various operations, and this decreases the economic margin of recycled materials, often making this choice less feasible. Dust produced by the recycling system is a major safety and health hazard, but it can be easily reduced with engineering controls and good ventilation [1,6]. In the case of chemical recycling, called solvolysis, the polymeric matrix is decomposed with a solution of acids, bases, and solvents whose composition must be adjusted according to the nature of the matrix. To increase the contact surface with the solution and aid in the dissolution of the matrix, solid composites are ground first. At the end of the process, the CFs are washed to remove the decomposed polymer residues and the solvent residues.

Recycled CFs obtained in this way can be longer than mechanically ground ones and have been shown to maintain their tensile strength, with only a small percentage less than virgin CFs. However, the use of hazardous and concentrated chemicals has a significant environmental impact [7]. In thermal recycling, high temperatures are used to degrade the polymeric matrix and leave the CFs as a residue, and it can be divided into different types. For thermal processes, the operating parameters must be carefully controlled to avoid the loss of valuable products or a change in the chemistry of the recovered materials. If the process temperature is too low, the fiber surface is covered by a layer of amorphous carbon following the limited degradation of the matrix, whereas if the temperature is too high, the CF surfaces oxidizes, with consequent reduction in the diameter of the fibers and mechanical properties in general [2,8,9]. Therefore, the best way is first to recover long fibers through the process of pyrolysis, which will be later used to produce semi-finished products, such as non-woven fabrics, to be impregnated in a subsequent step. This step, in turn, generates waste, namely short fibers. The latter, through further processing, could be used in fused deposition modeling (FDM) technology. The mechanical properties of composite materials depend on many factors, and the length of the fibers must certainly be mentioned among the main ones [10]. Normally, short-CF-reinforced composite materials are prepared using extrusion compounding and injection molding techniques, and in addition to the initial length of the fibers, it should be remembered that during the production of these composites, fiber breakage occurs due to fiber–fiber interaction [11].

The manufacturing of filaments for 3D printing also falls within the scope of extrusion, which is the focus of this study. Three-dimensional printing has already been successfully applied in the manufacture of polymeric components ranging from prototypes to final products, but the main problem is that the resulting parts have inferior mechanical performance compared to parts fabricated using conventional techniques, such as injection molding. To solve this problem, it has been seen that the addition of fibers to the polymer matrix forming a composite produces a significant improvement in the resistance of the molded parts [12]. At present, thermoplastic polymers are the ones mostly used in these processes, including acrylonitrile butadiene styrene (ABS), polylactic acid (PLA), polycarbonate (PC), and polyamide (PA).

In recent years, many authors have studied the addition of short fibers to a thermoplastic polymer to create filaments of the composite material used as raw material [13]. All these studies have analyzed the effect of fiber content, fiber orientation, and fiber length [14] on the processability and performance of the manufactured components. In addition, some studies have reported comparisons between the properties of 3D-printed composites and those manufactured with traditional molding techniques [12,15]. Among them, the work of mechanical and morphological characterization and of PLA reinforced with 15 wt.% of short CFs carried out by Ferreira et al. [14], the study of the effects of process parameters on the tensile properties of parts fabricated with ABS and CFs by Ning et al. [16], and the investigation of engineering applications of PEEK composites and short CFs by Wang et al. [17] can be worth mentioning. In their work, Ferreira et al., however, show problems in the adhesion between PLA and CFs, and embrittlement of the reinforced material due to the addition of short CFs is reported [14]. In the study by Ning et al., the effects of process parameters on the tensile properties of the composites made with FDM are described, which have much in common with the composites made in the context of this study with the additive manufacturing technique [16]. Wang et al. in their research instead deal with the production of composite materials with PEEK and CFs with a variable fiber content between 5 and 15% by weight successfully produced with extrusion [17]. In addition, the work by Giani et al. [18], in which the applicability of recycled carbon fibers to produce CF-reinforced PLA composites is demonstrated, is also worth mentioning. In this case, the comparison between the neat and reinforced samples showed about a doubling of the elastic modulus and maximum stress. All these results are particularly interesting, given the ease of manufacturing filaments for FDM and 3D printing, and confirm the possibility of applying CFs in these areas, also by exploiting different matrices as environmental and economically sustainable alternatives.

The carbon fibers used in this study fall into the category of waste generated directly by the production process that uses waste material, and the main aim lies in the identification of a potential application for this by-product considered a processing waste up to now. Given this fact and to fall within the scope of eco-sustainable and circular composite materials, the objective of this work is to study the characteristics and mechanical properties of composite materials obtained by combining a PA6,6 thermoplastic matrix with doubly recycled carbon fiber reinforcement by exploiting the additive manufacturing process. The result of this work contributes some knowledge about converting by-products to added-value 3D-printable composites with improved performance. To the best of the authors' knowledge, investigations on the influence of fiber size on CF-reinforced filaments for 3D printing are still insufficient. Few studies have dealt with the effect of fiber length downsizing due to recycling processing on the mechanical behavior of filaments and 3D-printed parts. This article addresses the issues related to the size of the fibers used and the presence of a predominant dusty fraction resulting from the milling treatment of pyrolyzed carbon fibers to obtain fillers suitable for the 3D-printable compounding. Filaments made of neat PA6,6, as a reference material, and filaments loaded with 5 wt.% and 10 wt.% of rCFs were manufactured. All filaments were studied in terms of mechanical properties, microstructure, fiber–polymer composition, and fiber size distribution. Next, 3D-printed specimens were produced and characterized via mechanical and morphological analysis. The findings of the research aim to establish a link between recycler and FDM user by providing information to optimize the processing of rCFs for maximizing the mechanical performance of the final product.

2. Materials and Methods

2.1. Materials

The materials used in this study were polyamide BASF Ultramid® 1000-11 NF2001 PA6,6 (BASF, Ludwigshafen, Germany) used as a matrix and carbon microfibers supplied by Carbon Task Srl (Biella, Italy) used as reinforcement for the composite. The characteristics of the raw polymer material are listed in Table 1.

Table 1. Physical and mechanical properties of BASF Ultramid® 1000-11 NF2001 PA6,6 from the technical data sheet.

Properties	Values	Comments
Density	1.14 g/cc	ISO 1183-1
Water absorption	8.5%	ISO 62
Moisture absorption at equilibrium	2.5%	23 °C/50% R.H.; ISO 62
Tensile strength, yield	83.0 MPa	50 mm/min; ISO 527-1
Elongation at break	25%	50 mm/min, normal strain; ISO 527-1
Elongation at yield	5.0%	50 mm/min; ISO 527-1
Tensile modulus	3.00 GPa	1 mm/min; ISO 527-1
Flexural strength	117 GPa	ASTM D790 test
Flexural modulus	2.90 GPa	ASTM D790 test
Melting point	257 °C	10 K/min ASTM D3418 test

The carbon microfibers (7 μm diameter) were supplied by Carbon Task Srl (Biella, Italy), a company that deals with the weaving of dry CFs from production waste and that, through its own production line designed over the past four years, has created a process that allows the production of non-woven fabrics of various weights.

These microfibers are the result of a double-recovery action, starting from long-fiber recycled carbon waste deriving from the pyrolysis of fiber-reinforced composites, subsequently sent to the production of non-woven fabric. The conversion of the fibers into woven-non-woven tissue carbon, including carding, fiber pile laying, and woven bonding/fixation, produces processing scrap in the form of agglomerates of microfibers (average length 550 μm), which are separated using cyclone filters. The agglomerates are collected and then ground using ball milling, obtaining the rCF filler used as reinforcement for the PA6,6 matrix in this work. Figure 1 shows a schematization of the double-recovery action that leads to the formation of carbon microfibers.

Figure 1. Schematization of the double-use action.

The physical properties and chemical composition of these microfibers were obtained experimentally. Their density was determined using a helium pycnometer (ASTM UOP851) and was found to be 1.917 g/cm³. Subsequently, scanning electron microscopy (MIRA 3 FEG-SEM, TESCAN, Brno, Czech Republic) and energy-dispersive spectroscopy (Octane Elect EDS system, Edax, Mahwah, NJ, USA) analyses of the microfibers were carried out to study their morphology and elemental composition. As can be seen in Figure 2, the sample was characterized by numerous fragments of CFs with a varied granulometry and a predominant fraction of pulverulent material, due to the ball milling process to which the raw materials were subjected.

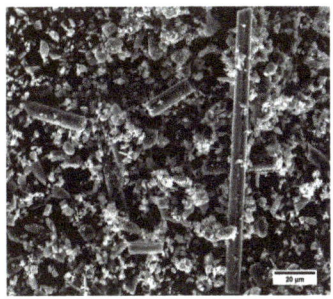 

Figure 2. SEM micrographs of a sample of carbon microfibers (magnifications 3.75 k× (**left**) and 17.5 k× (**right**)).

Preliminary measurements were carried out using the SEM analysis program to evaluate the average size of the fiber fragments, which reported an average length of 30 ± 12 μm.

Regarding the EDS analysis, traces of aluminum and oxygen (Figure 3) were identified.

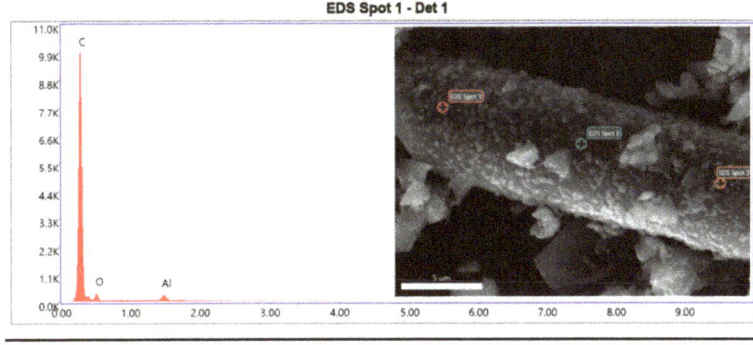

Element	Atomic %
O	2.06
Al	1.32

Figure 3. EDS analysis spectrum and atomic percentage of the elements.

When analyzing the relative amounts of aluminum and oxygen, it was noted that they were present in a ratio of about 2:3 (Table 2), and therefore, it could be assumed that they were traces of alumina (Al_2O_3) probably deriving from the mill balls, which are made of alumina.

Table 2. Temperature profiles used during the extrusion of PA6,6 filaments (neat and composites).

	PA6,6 Neat Filament	PA6,6 + rCF
Zone 1 temperature, °C	255	260
Zone 2 temperature, °C	255	260
Zone 3 temperature, °C	260	265
Zone 4 temperature, °C	260	265
Zone 5 temperature, °C	260	265
Zone 6 temperature, °C	255	260
Zone 7 temperature, °C	250	255
Die temperature, °C	235	240
Screw speed, rpm	150	150

2.1.1. Filament Extrusion for 3D Printing

Three types of filaments with different amounts of fibers were produced:

- Neat PA6,6 filament
- Filament of PA6,6 + 5% rCF, i.e., with the addition of 5 wt.% of recycled CFs
- Filament of PA6,6 + 10% rCF, i.e., with the addition of 10 wt.% of recycled CFs

A die system was connected to a Thermo Scientific Process 11 twin-screw extruder (Thermo Fisher Scientific, Waltham, MA, USA), in which the material leaving the extruder is passed through a series of rotating cylinders that allow the filament to be given a constant pre-established cross section to reach the final phase of winding on the reel.

Different extrusion temperature profiles were selected for neat and composite filaments, given the presence of carbon microfibers that could lead to alterations in the rheology and thermal properties of the mixture. Table 2 reports the temperature programs inside the extruder (from feed to die) used in this work.

In all cases, the diameter of the filaments was periodically checked during the winding phase to be sure that it fell within the dimensional range allowed for feeding to the 3D printer, i.e., a diameter between 1.60 and 1.80 mm, with an optimum of 1.75 mm [19,20]. The filaments' diameters were measured using a digital caliper (sensitivity of 0.05 mm). Figure 4 illustrates the PA6,6 filaments obtained from the extrusion process.

Figure 4. 3D printing filaments obtained from extrusion processing.

2.1.2. 3D Printing

After the in-depth study of the properties and characteristics of the 3D printing filaments, the 3D printing of dumbbell-shaped samples with the three formulations chosen was carried out in order to perform tensile tests. The machine used was a PRUSA i3 3D printer (Prusa, Prague, Czech Republic) (Figure 5a). The standard dumbbell-shaped specimens were 7 mm × 3 mm, with a useful length of 45 mm, with curved fittings and no perimeter (Figure 5b). As far as the operating temperatures are concerned, a printing temperature of 265 °C and a bed temperature of 90 °C were used for the neat PA6,6 and 5 wt.% rCF specimens, while for the 10 wt.% rCF specimens, a printing temperature of 270 °C and a bed temperature of 110 °C were used. A printing speed of 40 mm/s, 100% infill percentage, an aligned rectilinear pattern, and a layer height of 0.2 mm were set for all three families. Before printing, filament spools were dehumidified in a drying oven for 24 h at 80 °C.

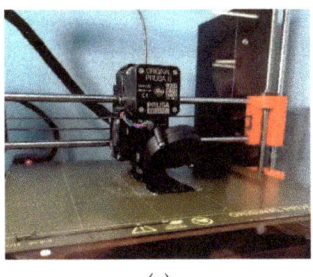 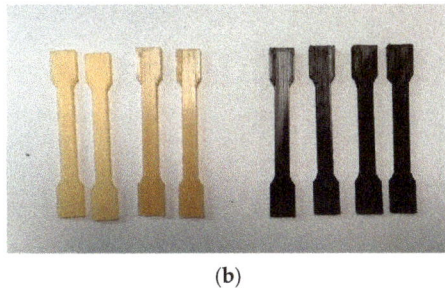

(**a**)   (**b**)

Figure 5. (**a**) PRUSA i3 3D printer and (**b**) 3D-printed dumbbell-shaped samples.

2.2. Methods

2.2.1. SEM Analysis and Sample Preparation (Carbon Sputtering Technique and Cryogenic Fracture)

As the first step, fracture surfaces of the three types of filaments used were analyzed, after brittle fracture in liquid nitrogen, using SEM, namely the neat filament of PA6,6 only and the pair of composite filaments with 5 and 10 wt.% reinforcement in recycled CFs. Prior to the SEM analysis, the specimens were sputter-coated with carbon to make the material conductive for the analysis. This pre-treatment was performed using an EM SCD005 vacuum sputter coater (Leica, Wetzlar, Germany).

2.2.2. Density Measurements

Regarding the characterization methods, the effective percentage of fiber by volume in the manufactured composite filaments was first evaluated by means of density measurements, assuming that the dispersion of fibers was homogeneous within the composite material. The density of rCFs was determined using an AccuPyc II 1340 helium pycnometer (MICROMERIT-ICS, Norcross, GA, USA), where the volume of a sample is estimated from a pressure change gradient upon expanding of a fixed amount of helium to a reference chamber of a known volume based on ideal gas law (ASTM UOP851). The system is accurate to within 0.03% of reading values [21]. The density of the composite filaments was measured in accordance with the Archimedean principle (buoyancy method in water) [22] using a commercial density determination kit of a 0.1 mg resolution analytical balance Mettler Toledo ME54 (Mettler Toledo, Worthington, OH, USA). The mass of the specimens was weighed in air and distilled water, and density (ρ) was computed according to Equation (1):

$$\rho = \frac{M_a \times \rho_w}{M_a - M_w} \tag{1}$$

where ρ is the density of the sample, M_a is the mass of the sample in air, M_w is the mass of the sample in water, and ρ_w is the density of water at the measured temperature. For each composite, 20 filament specimens (~3 cm length) were tested. The results obtained were then compared with those obtained with chemical digestion, carried out following the ASTM D3171 standard. The procedure was carried out using 95–97% sulfuric acid (Honeywell Fluka, Charlotte, NC, USA) and 30% hydrogen peroxide (J. T. Baker, Waltham, MA, USA).

2.2.3. Study of the Dimensional Distribution of the Fibers

To improve the filament production process with a view to future optimization, the effective dimensional distribution of the carbon microfibers was studied.

The dimensions of a fiber sample not yet subjected to any production process and fiber samples obtained following chemical dissolution of the matrix with formic acid on small portions of filaments (both at 5 and 10 wt.% of CFs) were analyzed. In both cases, portions of microfibers were isolated and placed on slides and the dimensional study was carried out using a Leica DMI5000 M optical microscope (Leica, Wetzlar, Germany) and Image J imaging software (version 2.3). A total of 20 images for each sample was captured, and a total of 90 measurements were made.

2.2.4. Tensile Testing of the Filaments

The filaments were tensile-tested with a Zwick/Roell universal testing machine (Z010, load cell: 10 kN, preload: 10 Mpa, Ulm, Germany) with a crosshead speed of 20 mm/min and a gauge length of 30 mm. Before carrying out the test, filament specimens (length of 150 mm) were conditioned in an oven at 80 °C for at least 24 h to avoid problems related to the moisture absorption tendency of the PA6,6 matrix. For each formulation (neat PA6,6, PA6,6 + 5% rCF, and PA6,6 + 10% rCF), 5 filament samples were tested.

The success of these tests is appreciable since no filament, during the execution of the tests, experienced failure or damage at the clamps but rather underwent necking, as in the best of desirable cases (Figure 6).

Figure 6. Tensile test of a filament.

2.2.5. Study of Fiber Homogeneity within the Filaments

To verify the uniformity of fiber distribution throughout the entire filament, the density of each filament type was measured with a scale that uses the principle of hydrostatic buoyancy. Samples about 3 cm in length were prepared by cutting the filament in different areas. Table 3 shows the average value of the calculated densities for each family (20 test specimens for each material) and their standard deviations.

Table 3. Standard deviation of fiber dispersion in the specimens.

Sample	Mean ± Std. Dev.
PA6,6 neat	$1.129 \pm 0.026 \text{ g/cm}^3$
PA6,6 + 5 wt.% rCF	$1.148 \pm 0.019 \text{ g/cm}^3$
PA6,6 + 10 wt.% rCF	$1.164 \pm 0.014 \text{ g/cm}^3$

The small values of the standard deviation are indicative of the homogeneity of fiber content along the filaments.

2.2.6. Tensile Tests of the 3D-Printed Samples

Four specimens from each family were subjected to tensile tests with the equipment already used for testing single filaments (Zwick/Roell Z010 universal testing machine). The test was carried out in accordance with the ISO 527-4 standard. The test parameters were a crosshead speed of 5 mm/min, a gauge length of 20 mm, and a preload of 1 MPa. Strain was measured with a contacting extensometer. As for the printed filaments, the tensile test was preceded by an oven-drying treatment of the specimens at 80 °C for 24 h. For each material, the average values of tensile strength and Young's modulus were obtained by testing on 4 specimens.

2.2.7. Microstructural Analysis

After mechanical characterization, the microstructural analysis of the cross sections of some 3D-printed samples from each family was performed.

As already described for filaments, in this case also, liquid nitrogen was used to image with SEM the fracture surfaces of the specimens.

3. Results

3.1. Preliminary Analysis of PA6,6 Neat and Composite Filaments for 3D Printing

Before studying the mechanical properties of the 3D-printed specimens, the morphology and properties of the extruded filaments were investigated.

SEM analysis was carried out to study the surface of the filaments and check for any defects; tensile tests were carried out, and the actual dimensions of the carbon microfibers were studied before and after extrusion.

3.1.1. SEM Analysis of Filament Fracture Surfaces

From the observation of the cross section of the filament of neat PA6,6 (Figure 7), no particular features emerged. The surface was typical of a brittle fracture mechanism (caused by cryogenic fracture), and defects, such as porosity or cavities, could not be detected. The filament showed smooth and round edges, with no air bubbles, indicating the successful removal of humidity during the oven-drying pre-treatment. The diameter was priorly measured with the digital caliper (20 measurements made along the filament spool) and was found to be 1.72 ± 0.10 mm.

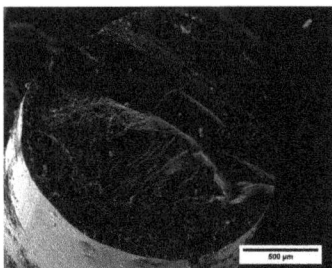

Figure 7. SEM micrograph of the neat PA6,6 filament (magnification 250×).

Also, the filaments of PA6,6 with 5 wt.% and 10 wt.% of recycled carbon fibers displayed proper circularity and were free of a porous surface (Figure 8a,b). In these cases, at higher magnifications, carbon fibers were found on the surface, which were perfectly incorporated within the polymeric matrix, as visible in Figure 8c. This result is indicative of a proper selection of extrusion parameters for manufacturing filaments implementing single-step extrusion processing. Indeed, one of the crucial challenges in the fabrication of fiber-reinforced polymeric filaments for 3D printing is to ensure a homogeneous microstructure free of scattered porosity and fiber agglomerates. Generally, a double-extrusion cycle is required to achieve a uniform microstructure and adequate mixing/distribution of fibers within the matrix [23]. However, this additional stage can be time-consuming, inducing further degradation mechanisms (thermal and mechanical degradation) in the constituent materials [24].

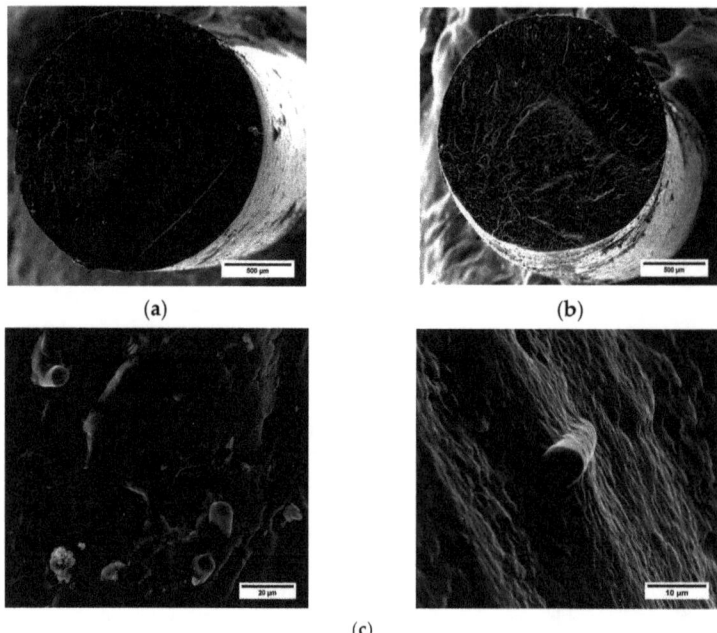

Figure 8. SEM micrograph of (**a**) PA6,6 + 5 wt.% rCF filament (magnification 250×) and (**b**) PA6,6 + 10 wt.% rCF filament (magnification 250×). (**c**) Details of rCFs embedded into the matrices (magnifications 5000× (left) and 8750× (right)).

3.1.2. Evaluation of the Effective Percentage of Reinforcement

The extrusion of polymer composites faces two crucial aspects: (1) obtaining a homogeneous polymer–fiber compound, with the aim of fabricating filaments with reinforcements well dispersed within the matrix, and (2) evaluating the real percentage of fibers integrating into the final composite, considering possible losses of materials that occur during processing into the extrusion line [25]. In order to obtain a material as homogeneous as possible within the filament, powdered PA6,6 was chosen as the raw material to make easier the mixing of the two phases. If pelleted PA6,6 had been used, it would have been difficult to homogenize the microfibers in the extruder.

To estimate the actual amount of CFs present in the manufactured 3D printing filaments, two parallel studies were carried out: the first using the densities of the raw materials involved and the composites produced to estimate the effective percentage of reinforcement by volume and the second, following what is described by the ASTM D3171 standard, carrying out an acid attack on the filament sections and measuring the quantity of residual fibers after filtration and drying processes.

For the first method, the densities of both raw materials used and the compounds themselves were taken into consideration, calculated using a balance that exploits the principle of hydrostatic thrust, while for the case of recycled carbon microfibers, the value was calculated using a helium pycnometer.

The density values and the percentage of fibers obtained at a theoretical level using the rule of mixtures are shown in Table 4.

The study of the actual percentage of carbon fibers present inside the composite filaments was performed using chemical digestion, as mentioned, following what is described by the ASTM D3171 standard, whose application is also illustrated in the article by Bowman et al. [26].

Table 4. Density values of CFs, neat PA6,6, and PA6,6 composites and estimated theoretical values of the percentage of reinforcement in composite filaments.

Material	Method	Value
rCFs	Helium pycnometer	1.917 g/cm^3
PA6,6 neat filament	Buoyancy balance	1.129 g/cm^3
PA6,6 + 5 wt.% rCF filament	Buoyancy balance	1.148 g/cm^3
PA6,6 + 10 wt.% rCF filament	Buoyancy balance	1.164 g/cm^3
	Percentage of rCF reinforcement	
PA6,6 + 5 wt.% rCF filament	Rule of mixtures	2.41% v/v
PA6,6 + 10 wt.% rCF filament		4.44% v/v

At the end of the various chemical etching phases, by applying the formulas given in the standard itself, it is possible to obtain the reinforcement content as a percentage by weight and by volume, the matrix content by weight and by volume, and the void volume fraction. In this specific case, the aim was to calculate the value of the reinforcement content as a volume percentage to compare the results with those obtained with the previously described density method. The values obtained are shown in Table 5.

Table 5. Effective percentage of reinforcement in composite filaments obtained with an acid attack.

Material	Value
PA6,6 + 5 wt.% rCF filament	3.40% v/v
PA6,6 + 10 wt.% rCF filament	6.79% v/v

The estimate of the volumetric percentage of CFs using the rule of mixtures (Table 4) differed from the values obtained using chemical digestion (Table 5). Primarily, the divergence can be attributed to the density measurements on the composite, which include the presence of voids in the samples. In contrast, chemical digestion treatment makes the measurement independent of the material's porosity, providing an almost real measure of the amount of reinforcement incorporated into the composites. Converting the designed weight fractions of 5 wt.% and 10 wt.% to the corresponding volumetric percentages resulted in volume contents of about 3.03% v/v and 6.20% v/v, respectively. Next, net of porosity, the filaments were loaded with an amount of reinforcement close to the target values. This finding indicates that the manufacturing process implemented in this study allowed proper control of the rCF content in the matrix.

3.1.3. Dimensional Study of Carbon Microfibers

With the aim to optimize the filament production process, the actual dimensional distribution of the carbon microfibers involved in this study was investigated.

The size of a sample of fibers deriving from the ball milling process and the samples obtained following matrix dissolution were analyzed and compared. Figure 9 shows the acquired micrograph of CFs for the assessment of fiber size distribution.

Figure 10 plots the fiber size distribution for as-received rCFs and fibers from the two composite samples. As expected, the average fiber length reduced following extrusion, due to the breakage that rCFs experienced during processing. Moreover, the greatest dimensional reduction occurred in the sample loaded with a higher rCF volume fraction. The mean fiber length was reduced by 23% and 37% in PA6,6 composites incorporating 5 wt.% and 10 wt.% of CFs, respectively. This trend was predictable and is justifiable due to the brittle behavior of carbon fibers. During the extrusion process, the melting of the polymer in twin-screw extrusion requires a large energy input, generates high local stresses and strains, which are detrimental to the fiber's dimensions [27]. In addition, a higher fiber content leads to a higher damage to the fiber size. The increased deterioration of fiber length for a higher rCF volume fraction is mainly attributed to the higher fiber–fiber interaction [28]. As the percentage of fibers added inside the dosing hopper increases, the stresses due to the interaction with the neighboring fibers also increases, enhancing breakage phenomena.

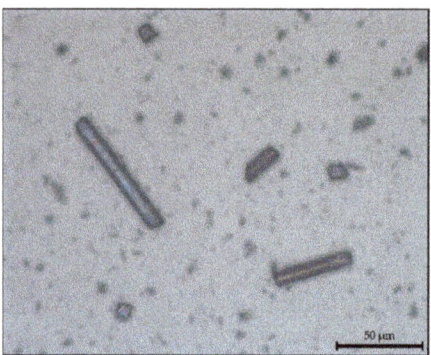

Figure 9. Carbon microfibers under the optical microscope (magnification 50×).

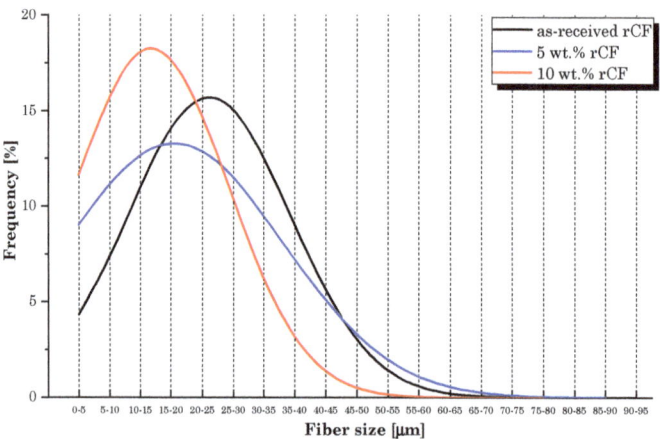

Figure 10. Fiber size distribution.

3.1.4. Tensile Tests of 3D Printing Filaments

The tensile test results are shown in Figure 11. As expected, the addition of rCFs resulted in an increase in Young's modulus and tensile strength values. Compared to the neat PA6,6 filament, the best performance was obtained for the composite specimens incorporating the highest fiber content (PA6,6 + 10 wt.% rCF), resulting in an increase up to 25% and 11% in the elastic modulus and strength, respectively. It is widely known that implementing reinforcing materials (such as CFs) for designing 3D-printable polymer composites overcomes the strength limitations of FDM-fabricated pure thermoplastic parts, while improving the functionality and applicability of the material [16]. The filaments produced in this work show promising mechanical performance when compared with similar commercial products obtained from virgin microfibers. For instance, the study conducted by Al-Mazrouei et al. [29] involved the use of a commercial 3D-printable nylon/CF filament (20 wt.% CF) having a declared tensile strength and Young's modulus of 66.3 MPa and 2.76 GPa, respectively.

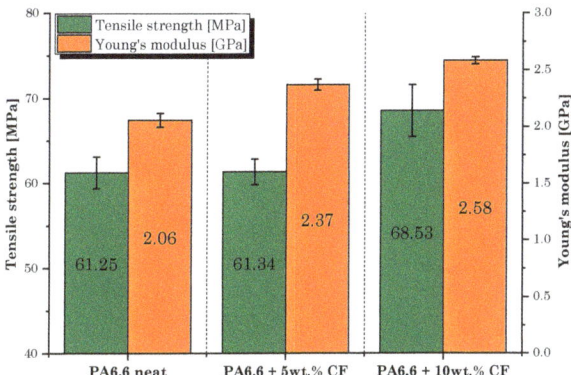

Figure 11. Tensile test results: tensile strength and Young's modulus values of each manufactured filament.

3.2. Characterization of Tensile Properties of Neat PA6,6 and Resulting Composites

3.2.1. Tensile Tests of 3D-Printed Specimens

Following the in-depth study of the properties and characteristics of the filaments for 3D printing produced by extrusion, the manufacturing and mechanical characterization of the printed dumbbell-shaped samples was performed. Figure 12 shows the average values obtained during the tensile tests.

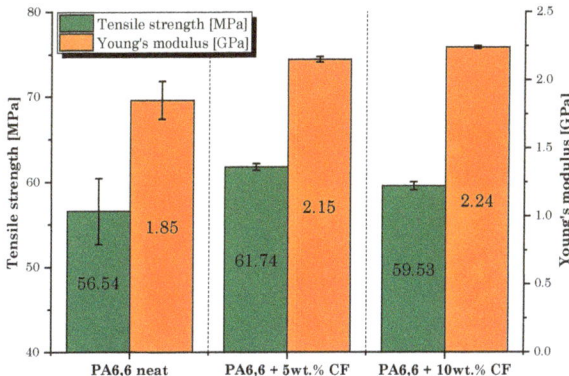

Figure 12. Tensile test results: tensile strength and Young's modulus values of 3D-printed dumbbell-shaped samples.

Tensile tests revealed that the role of microfibers in strengthening the PA6,6 matrix is similar to that detected during the filament characterization but clearly involves lower strength and stiffness values. Compared to bulk specimens, the layering effect in 3D-printed parts introduces anisotropy and reduces the capacity to resist tensile and shear load [30]. In addition, the improvement in strength was not regular with increasing rCF volume fraction. The sample filled with 10 wt.% of rCF, although with the maximum increase in the elastic modulus compared to the neat matrix (+21%), provided lower mechanical strength performance than the PA6,6 + 5 wt.% rCF sample. Indeed, the lowest dosage of rCFs induced the greatest improvement in tensile strength (+9% over the neat sample). The reason behind this behavior must be attributed to the presence of a large dusty fraction inside the microfiber feedstock used in this research and to the significant fiber breakage that occurs during compounding by extrusion and printing. This is in accordance with what was found in rCF size distribution analysis (Figure 10). It is well documented that for enhanced mechanical performance of fiber-reinforced composites, in general, a higher residual-length-to-diameter (l/d) ratio is better. As claimed by Lewicki et al. [31], the use of high-l/d-ratio (l/d > 50) CFs (~350 μm length) is necessary to achieve high-performance 3D-printable composites, while retaining a high degree of feature resolution. A proper choice of CF size can help increase the strength and stiffness of the thermoplastic polymer matrix significantly. Zhang et al. [32] found an increment of more than 40% and 140% in the tensile strength and elastic modulus of the acrylonitrile butadiene styrene (ABS) matrix, respectively, by integrating 15 wt.% of short CFs. Liao et al. [33] investigated the effect of CFs on PA12's performance, detecting an increase in the tensile strength and stiffness of up to 102% and 266%, respectively, following the addition of 10 wt.% of carbon fibers content reinforcement. Dul et al. [34] studied the effect of short CFs (15 wt.%) on the mechanical properties of a 3D-printed PA composite, revealing a maximum tensile strength of 96 MPa and a modulus of 7.9 GPa, with an increment of +34 and +147%, respectively, when compared to the neat PA sample. Therefore, the poor reinforcement effect was primarily attributable to the inadequate l/d ratio (l/d~4) of microfibers used in this work, which tends to undergo further decrease with extrusion processing. This evidence should be crucially taken into consideration for future process optimization.

By displaying the results of the tensile tests, it is worth noting that the scattering of the results in the composite samples is significantly lower than that of the plain matrix. A possible explanation can be traced back to the influence of the thermal conductivity of the carbon microfiller. A conductive filler would promote more favorable diffusion and adhesion conditions between filaments during printing, reducing the effect of interlayer defects on the mechanical response of the material [35]. In this context, SEM analysis will clarify the influence of rCFs on the microstructure of printed samples.

In any case, the excellent behavior of all the specimens during the execution of the tests should be underlined, which, as shown in the Figure 13, underwent elongation and necking mechanisms typical of a mono-material system. The deformation and failure mode of the dumbbell-shaped specimens indicated an adequate synergy between the constituents, as well as highlighting a minimal influence of the filaments debonding on the mechanical performance of the printed material.

(a)   **(b)**

Figure 13. 3D-printed composite specimens during tensile tests: (**a**) PA6,6 neat and (**b**) PA6,6 composite.

The stress–strain curves (Figure 14) of the neat PA6,6 sample and the PA6,6-CF composites clearly elucidate that the addition of carbon microfibers preserved the ductile behavior of the neat matrix. The stiffening induced by the rCF addition slightly reduced the elongation at break of the composites, ranging between 150% (10 wt.%. rCF) and 170% (5 wt.% rCF).

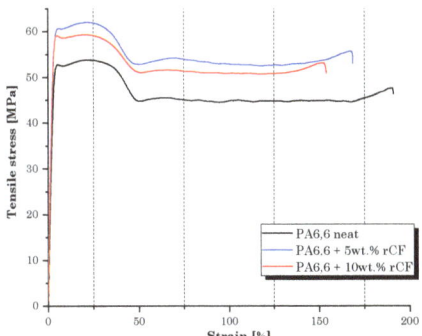

Figure 14. Stress–strain behavior of 3D-printed dumbbell-shaped samples.

3.2.2. Microstructural Analysis

Following tensile tests, microstructural analysis of the cross sections of the 3D-printed samples from each family was carried out. Regarding the surface of the neat PA6,6 specimen (Figure 15), it can be said that it was almost homogeneous; the printing filaments showed good mutual adhesion, to the point of being almost completely fused to each other, as can be seen from the presence of small inter-filament cavities.

 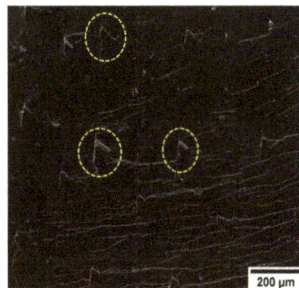

Figure 15. SEM micrographs of the neat PA6,6 specimen. The dotted circles highlight inter-filament voids in the matrix (magnifications 200× (**left**) and 500× (**right**)).

Even the surface of the PA6,6 specimens with 5% and 10% by weight of rCFs was quite homogeneous (Figure 16a,b), with the printing filaments showing good mutual adhesion. At the same magnification as in the previous case, the triangular-shaped voids formed between the beads in the printed parts were barely distinguishable.

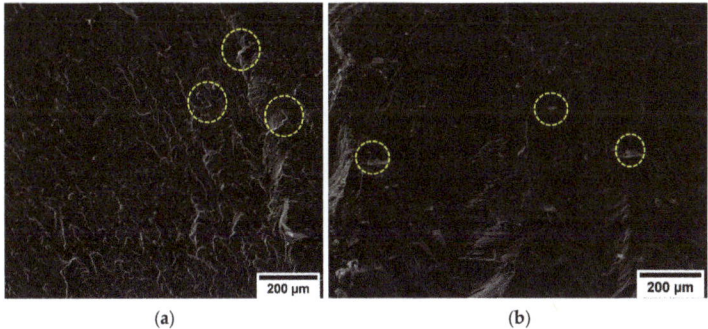

(**a**) (**b**)

Figure 16. SEM micrographs (magnification 500×) of (**a**) PA6,6 + 5 wt.% rCF and (**b**) PA6,6 + 10 wt.% rCF specimens. The dotted circles highlight inter-filament voids in the matrix.

From a first visual analysis, it was clear that the inter-filament void fraction of the specimens loaded with carbon microfibers was lower and less widespread than that of the pure material. By processing the SEM micrographs with ImageJ software, the percentage of voids in the matrices was estimated. The void fraction in the neat PA6,6 sample was around 2.20%. In the composite specimens, the porosity rate decreased to 0.35% and 0.30% for 5 wt.% rCF and 10 wt.% rCF, respectively. A similar effect was observed by Tekinalp et al. [36], who investigated the void formation in 3D-printed polymer composites incorporating short CFs. This result would consolidate the hypothesis, discussed in Section 3.2.1, concerning the improvement effect that conductive fibers would confer on the print quality and inter-layer densification of the material. While a potential microstructural quality improvement was observed, the CF-reinforced composite samples showed a limited increment in mechanical performance. These aspects provide a clear indication that fiber length optimization is the driving factor to achieve effective enhancement in strength behavior. Figure 17a shows the fractography of the PA6,6 + 5 wt.% rCF sample generated from the tensile tests. The fracture surface appeared completely free of any type of defect associated with the layer-by-layer production of 3D-printed parts. Figure 17b displays the fiber distribution in the matrix. In addition to the l/d ratio of the fiber, the fiber–matrix interface is one of the most influencing parameters for the final properties of the composite material, which are strongly interlinked to the ability to transfer across the interfacial zone [37]. The surface of rCFs looked smooth, and the matrix did not completely wrap around the fiber, which points to a low/moderate adhesion between polymer and carbon inclusions. Moreover, some holes were observed on the fracture surface, indicating that the fiber pull-out is the predominant mechanism of fracture rather than the inter-filament debonding (common occurrence in 3D-printed specimens). Fiber desizing due to pyrolysis treatment can affect the interfacial interaction experienced by the fibers with the polyamide matrix. However, to date, the mechanisms of adhesion between thermoplastics and reinforcing fibers are not fully understood. Some studies state that the thermal desizing process may facilitate several favorable fiber–matrix interactions (mechanical interlocking, non-covalent interactions, wetting) [38]. This is because the commercially available sizing agents for CFs are primarily tailored for thermosetting resins. Therefore, when applied to thermoplastic polymers, the sizing mechanisms cause adverse effects on the fiber–matrix interface properties. Conversely, other studies [39] have demonstrated that sizing removal is an ineffective method to enhance the CF–matrix interfacial bond (specifically for polyamide matrices). The non-optimal adhesion of the fibers to the matrix would suggest that the desizing induced by thermal recycling has no significant contribution to the interface properties. More detailed investigations are required to quantify the effect of CF–matrix compatibility on the final performance of the composite over the problem related to the l/d ratio of the reinforcement.

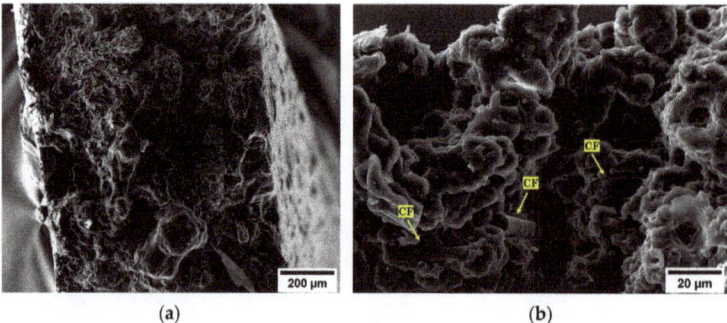

(a)   (b)

Figure 17. SEM micrographs of the fracture surface of the PA6,6 + 5 wt.% rCF specimen: (**a**) general view of the surface (magnification 500×) and (**b**) distribution of rCFs in the matrix (magnification 5 k×).

4. Optimization Hypothesis

In the light of what has been investigated, it can be said that the reinforcing action of rCFs was not significant due to the implementation of a filler with an unsuitable l/d ratio. For instance, the values of the Young's modulus and the tensile strength in our case for the composite loaded with 10 wt.% of rCFs were, respectively, 2.24 GPa and 59.53 MPa on average, which correspond to an improvement in mechanical behavior of 21% and 5%, respectively, compared to the neat PA6,6 specimens. These values are markedly lower than results available in the literature (reviewed in Section 3.2.1) implementing CFs in 3D-printed composites.

The main cause that led to such limited strength performance is certainly the unsuitable average size of the CFs used. As already mentioned, indeed, in the reinforcement used, there is a preponderant dusty fraction, and the longer fibers are then further crushed during the various phases of extrusion of the filament and subsequent printing. As mentioned, the fibers used as reinforcement are obtained by grinding, in a ball mill, inextricable agglomerates of recycled fibers. A possibility of optimization is therefore to act on the grinding phase, studying in detail an operation that allows grinding of the agglomerates, untangling the fibers, but obtaining fibers of adequate length to improve the mechanical properties of rCF-reinforced composites.

In this regard, it is possible to obtain an ideal length measurement through studies on the mechanical behavior of short-fiber composites. First, it is necessary to understand the effect of discontinuous fibers in a polymeric matrix by studying the mechanism of reinforcement of the fibers. The fibers exert their effect by limiting the deformation of the matrix, and in this way, the applied external load is transferred to the fibers by shear at the fiber–matrix interface. In short fibers, the tensile stress increases from zero at the ends up to a value σ_{max}, which would occur if the fibers were continuous. σ_{max} can be determined from Equation (2):

$$\frac{\sigma_{max}}{E_f} = \frac{\sigma_c}{E_l} \tag{2}$$

where σ_c is the stress applied to the composite and E_l can be determined via the rule of mixtures.

Therefore, there is a minimum fiber length that will allow the fiber to reach its full loading potential. The minimum fiber length at which maximum fiber stress can be achieved is called the load transfer length (l_t), and the value can be determined from a force balance (Equation (3)):

$$l_t = \frac{\sigma_{max} \times d}{2 \times \tau} \tag{3}$$

where τ is the shear strength of the fiber–matrix interface and d is the diameter of the fiber [40]. Note that l_t is also a function of the stress applied to the composite.

Let us consider the case of the composite material with 10 wt.% of rCFs. To carry out the calculation and estimate an attempt value for l_t for our specific case, we need to extrapolate the values of σ_{max}, τ, and Young's modulus of the CFs only (E_f).

Regarding σ_{max}, as a first approximation, it can be assumed that the specimen breaks during the tensile test for pull-out, as observed from the SEM analysis. Regarding the τ value, reference was made to the experimental value obtained for a PA matrix composite reinforced with desized CFs by Kim et al. [39]. The value of τ was set equal to 24 MPa. Finally, the value of Young's modulus for carbon fibers alone was considered equal to 230 GPa [41].

Applying the rule of mixtures (Equation (4)):

$$E_l = E_f V_f + E_m V_m \approx 17.4 \, \text{GPa} \tag{4}$$

with E_f and E_m being the modulus of the fibers and the matrix, respectively, and V_f and V_m the volume fraction of the fibers and the matrix, respectively.

Using Equation (2):

$$\sigma_{max} = E_f \left(\frac{\sigma_c}{E_l} \right) \approx 790 \text{ MPa}$$

Using Equation (3):

$$l_t = \frac{\sigma_{max} \times d}{2 \times \tau} = 0.115 \text{ mm} = 115 \text{ µm}$$

The fiber length estimated in the calculation widely exceeds the average size of carbon fillers integrated in the composite with 10 wt% of rCFs (~19 µm), supporting the poor performance improvement of the composite.

However, it should be noted that the ideal length of 115 µm must be the average length of the fibers already incorporated into the composite and not after the ball mill processing. It was verified that between the milled rCFs and the effective production of the composite material, the fibers undergo breaking, which leads to a maximum loss of about 37% in length (for PA6,6 + 10 wt.% rCF). To compensate for the damaging effect of extrusion, preserving an adequate l/d ratio for obtaining effectiveness in mechanical performance and ensuring adequate printability of the material, an average dimension ranging between 200 and 300 µm can be proposed. These values clearly consider the average size of the input waste carbon agglomerates, which are subjected to ball milling to obtain the filler implemented in this research.

5. Conclusions

The purpose of the work was to explore the possibility of implementing recycled microfibers, deriving from a secondary shredding of recycled carbon fibers, as reinforcement in the production of thermoplastic matrix composites optimized for 3D printing technology. The main results of the study are as follows:

- The extrusion parameters were successfully optimized to obtain composites filaments suitable for 3D printing processing.
- The mechanical characterization of the printed filaments revealed that rCFs increase the Young's modulus and tensile strength of the composites by up to 25% and 11% (10 wt.% of rCFs), in comparison to the performance of the neat sample, respectively.
- Tensile tests of printed specimens highlighted a similar increment in strength performance (+16% in Young's modulus and +9% in tensile strength for 5 wt.% rCFs and + 21% in Young's modulus and +5% in tensile strength for 10 wt.% rCFs). SEM analysis showed microstructures not affected by the common defects induced by the layer-by-layer deposition of additive fabrication. This demonstrates the achievement of well-selected printing parameters for the processing of the composites developed in this work.
- The average length of the microfibers used in this research was estimated at 30 µm, too short for consistently improving strength. In addition, a complication factor for this case is the gradual reduction in size that the fibers undergo following extrusion. The dimensional optimization of the output rCF fraction from the ball milling process is undoubtedly a challenge to be faced in order to maximize the mechanical performance of composites.

The field of additive manufacturing is currently in continuous expansion, and this, combined with the increasing interest in the field of eco-sustainable and circular composite materials, places this research in a strategic position particularly suitable for future studies and improvements. In addition to defining a more suitable processing for rCFs to ensure greater improvements in terms of strength performance, future studies will be based on the thermal characterization of the developed material. Measuring and analyzing thermal properties are crucial in the field of 3D printing, both in terms of optimization of the printing parameters and in terms of the end use of the product. With a comprehensive know-how of optimized material characteristics and manufacturing parameters, composite users can

Materials **2023**, *16*, 5436

achieve technical findings regarding the potential use of rCFs in additive fabrication. The cheaper price and lower carbon footprint of rCFs compared to virgin reinforcement are favorable for the eco-design of smart and high-performance composite parts for different industries, including robotics and automotive.

Author Contributions: Conceptualization, M.V., F.S. and J.T.; methodology, M.V. and M.S.; validation, M.V., M.S., F.S. and J.T.; formal analysis, M.S., I.R., C.S. and S.A.; investigation, S.A., I.R., M.S. and C.S.; data curation, S.A., C.S. and M.S.; writing—original draft preparation, S.A. and I.R.; writing—review and editing, M.V. and M.S.; supervision, M.V., F.S. and J.T.; funding acquisition, M.V. All authors have read and agreed to the published version of the manuscript.

Funding: This research received no external funding.

Informed Consent Statement: Informed consent was obtained from all subjects involved in the study.

Data Availability Statement: Not applicable.

Acknowledgments: Some of the researchers working on this study have a project carried out within the Made in Italy—Circular and Sustainable (MICS) Extended Partnership and received funding from the European Union Next-GenerationEU (Piano Nazionale di Ripresa e Resilienza (PNRR)—Missione 4 Componente 2, Investimento 1.3—D.D. 1551.11-10-2022, PE00000004). The authors would like to thank Christian Scopinich (Carbon Task Srl) for providing recycled carbon fibers implemented in the research activity. A heartfelt acknowledgement also goes to Luciano Fattore and Riccardo Martufi (Centro Saperi&Co Fablab, Sapienza University of Rome) for their valuable technical support in the 3D printing of the samples.

Conflicts of Interest: The authors declare no conflict of interest. This manuscript reflects only the authors' views and opinions, and neither the European Union nor the European Commission can be considered responsible for them.

References

1. Zhang, J.; Chevali, V.S.; Wang, H.; Wang, C.H. Current status of carbon fibre and carbon fibre composites recycling. *Compos. Part B Eng.* **2020**, *193*, 108053. [CrossRef]
2. Giorgini, L.; Benelli, T.; Brancolini, G.; Mazzocchetti, L. Recycling of carbon fiber reinforced composite waste to close their life cycle in a cradle-to-cradle approach. *Curr. Opin. Green Sustain. Chem.* **2020**, *26*, 100368. [CrossRef]
3. Naqvi, S.R.; Prabhakara, H.M.; Bramer, E.A.; Dierkes, W.; Akkerman, R.; Brem, G. A critical review on recycling of end-of-life carbon fibre/glass fibre reinforced composites waste using pyrolysis towards a circular economy. *Resour. Conserv. Recycl.* **2018**, *136*, 118–129. [CrossRef]
4. Valente, M.; Rossitti, I.; Biblioteca, I.; Sambucci, M. Thermoplastic Composite Materials Approach for More Circular Components: From Monomer to In Situ Polymerization, a Review. *J. Compos. Sci.* **2022**, *6*, 132. [CrossRef]
5. Nunes, A.O.; Viana, L.R.; Guineheuc, P.M.; da Silva Moris, V.A.; de Paiva, J.M.F.; Barna, R.; Soudais, Y. Life cycle assessment of a steam thermolysis process to recover carbon fibers from carbon fiber-reinforced polymer waste. *Int. J. Life Cycle Assess.* **2018**, *23*, 1825–1838. [CrossRef]
6. Anane-Fenin, K.; Akinlabi, E.T.; Akinlabi, E.T. Recycling of fibre reinforced composites: A review of current technologies. In Proceedings of the DII-2017 Conference on Infrastructure Development and Investment Strategies for Africa, Livingstone, Zambia, 30 August–1 September 2017.
7. Pimenta, S.; Pinho, S.T. Recycling carbon fibre reinforced polymers for structural applications: Technology review and market outlook. *Waste Manag.* **2011**, *31*, 378–392. [CrossRef]
8. Mazzocchetti, L.; Benelli, T.; D'Angelo, E.; Leonardi, C.; Zattini, G.; Giorgini, L. Validation of carbon fibers recycling by pyro-gasification: The influence of oxidation conditions to obtain clean fibers and promote fiber/matrix adhesion in epoxy composites. *Compos. Part A Appl. Sci. Manuf.* **2018**, *112*, 504–514. [CrossRef]
9. Giorgini, L.; Benelli, T.; Mazzocchetti, L.; Leonardi, C.; Zattini, G.; Minak, G.; Dolcini, E.; Tosi, C. Recovery of carbon fibers from cured and uncured carbon fiber reinforced composites wastes and their use as feedstock for a new composite production. *Polym. Compos.* **2015**, *36*, 1084–1095. [CrossRef]
10. Sayam, A.; Rahman, A.M.; Rahman, M.S.; Smriti, S.A.; Ahmed, F.; Rabbi, M.F.; Hossain, M.; Faruque, M.O. A review on carbon fiber-reinforced hierarchical composites: Mechanical performance, manufacturing process, structural applications and allied challenges. *Carbon Lett.* **2022**, *32*, 1173–1205. [CrossRef]
11. Fu, S.Y.; Lauke, B.; Mäder, E.; Yue, C.Y.; Hu, X.; Mai, Y.W. Hybrid effects on tensile properties of hybrid short-glass-fiber-and short-carbon-fiber-reinforced polypropylene composites. *J. Mater. Sci.* **2001**, *36*, 1243–1251. [CrossRef]
12. Dickson, A.N.; Abourayana, H.M.; Dowling, D.P. 3D printing of fibre-reinforced thermoplastic composites using fused filament fabrication—A review. *Polymers* **2020**, *12*, 2188. [CrossRef] [PubMed]

13. Blok, L.G.; Woods, B.K.S. 3D printed composites Benchmarking the state-of-the-art. In Proceedings of the 21st International Conference on Composite Materials, Xi'an, China, 20–25 August 2017. Available online: http://www.iccm21.org/index.php?m= content&c=index&a=lists&catid=5 (accessed on 1 January 2020).

14. Ferreira, R.T.L.; Amatte, I.C.; Dutra, T.A.; Bürger, D. Experimental characterization and micrography of 3D printed PLA and PLA reinforced with short carbon fibers. *Compos. Part B Eng.* **2017**, *124*, 88–100. [CrossRef]

15. Blok, L.G.; Longana, M.L.; Yu, H.; Woods, B.K. An investigation into 3D printing of fibre reinforced thermoplastic composites. *Addit. Manuf.* **2018**, *22*, 176–186. [CrossRef]

16. Ning, F.; Cong, W.; Hu, Y.; Wang, H. Additive manufacturing of carbon fiber-reinforced plastic composites using fused deposition modeling: Effects of process parameters on tensile properties. *J. Compos. Mater.* **2017**, *51*, 451–462. [CrossRef]

17. Wang, P.; Zou, B.; Ding, S.; Huang, C.; Shi, Z.; Ma, Y.; Yao, P. Preparation of short CF/GF reinforced PEEK composite filaments and their comprehensive properties evaluation for FDM-3D printing. *Compos. Part B Eng.* **2020**, *198*, 108175. [CrossRef]

18. Giani, N.; Mazzocchetti, L.; Benelli, T.; Picchioni, F.; Giorgini, L. Towards sustainability in 3D printing of thermoplastic composites: Evaluation of recycled carbon fibers as reinforcing agent for FDM filament production and 3D printing. *Compos. Part A Appl. Sci. Manuf.* **2022**, *159*, 107002. [CrossRef]

19. Cardona, C.; Curdes, A.H.; Isaacs, A.J. Effects of filament diameter tolerances in fused filament fabrication. *IU J. Undergrad. Res.* **2016**, *2*, 44–47. [CrossRef]

20. Haq, R.H.A.; Marwah, O.F.; Rahman, M.A.; Haw, H.F.; Abdullah, H.; Ahmad, S. 3D Printer parameters analysis for PCL/PLA filament wire using Design of Experiment (DOE). *Mater. Sci. Eng.* **2019**, *607*, 012001. [CrossRef]

21. Gabaude, C.M.; Guillot, M.; Gautier, J.C.; Saudemon, P.; Chulia, D. Effects of true density, compacted mass, compression speed, and punch deformation on the mean yield pressure. *J. Pharm. Sci.* **1999**, *88*, 725–730. [CrossRef]

22. Chutinan, S.; Platt, J.A.; Cochran, M.A.; Moore, B.K. Volumetric dimensional change of six direct core materials. *Dent. Mater.* **2004**, *20*, 345–351. [CrossRef]

23. Polline, M.; Mutua, J.M.; Mbuya, T.O.; Ernest, K. Recipe development and mechanical characterization of carbon fibre reinforced recycled polypropylene 3D printing filament. *Open J. Compos. Mater.* **2021**, *11*, 47–61. [CrossRef]

24. Capone, C.; Di Landro, L.; Inzoli, F.; Penco, M.; Sartore, L. Thermal and mechanical degradation during polymer extrusion processing. *Polym. Eng. Sci.* **2007**, *47*, 1813–1819. [CrossRef]

25. Yasim-Anuar, T.A.T.; Ariffin, H.; Norrrahim, M.N.F.; Hassan, M.A.; Andou, Y.; Tsukegi, T.; Nishida, H. Well-Dispersed Cellulose Nanofiber in Low Density Polyethylene Nanocomposite by Liquid-Assisted Extrusion. *Polymers* **2020**, *12*, 927. [CrossRef] [PubMed]

26. Bowman, S.; Jiang, Q.; Memon, H.; Qiu, Y.; Liu, W.; Wei, Y. Effects of styrene-acrylic sizing on the mechanical properties of carbon fiber thermoplastic towpregs and their composites. *Molecules* **2018**, *23*, 547. [CrossRef]

27. Berzin, F.; Beaugrand, J.; Dobosz, S.; Budtova, T.; Vergnes, B. Lignocellulosic fiber breakage in a molten polymer. Part 3. Modeling of the dimensional change of the fibers during compounding by twin screw extrusion. *Compos. Part A Appl. Sci. Manuf.* **2017**, *101*, 422–431. [CrossRef]

28. Fu, S.Y.; Lauke, B.; Mäder, E.; Yue, C.Y.; Hu, X.J. Tensile properties of short-glass-fiber-and short-carbon-fiber-reinforced polypropylene composites. *Compos. Part A Appl. Sci. Manuf.* **2000**, *31*, 1117–1125. [CrossRef]

29. Al-Mazrouei, N.; Al-Marzouqi, A.H.; Ahmed, W. Characterization and Sustainability Potential of Recycling 3D-Printed Nylon Composite Wastes. *Sustainability* **2022**, *14*, 10458. [CrossRef]

30. Wang, L.; Ma, G.; Liu, T.; Buswell, R.; Li, Z. Interlayer reinforcement of 3D printed concrete by the in-process deposition of U-nails. *Cem. Concr. Res.* **2021**, *148*, 106535. [CrossRef]

31. Lewicki, J.P.; Rodriguez, J.N.; Zhu, C.; Worsley, M.A.; Wu, A.S.; Kanarska, Y.; Horn, J.D.; King, M.J. 3D-printing of meso-structurally ordered carbon fiber/polymer composites with unprecedented orthotropic physical properties. *Sci. Rep.* **2017**, *7*, 43401. [CrossRef]

32. Zhang, W.; Cotton, C.; Sun, J.; Heider, D.; Gu, B.; Sun, B.; Chou, T.W. Interfacial bonding strength of short carbon fiber/acrylonitrile-butadiene-styrene composites fabricated by fused deposition modeling. *Compos. Part B Eng.* **2018**, *137*, 51–59. [CrossRef]

33. Liao, G.; Li, Z.; Cheng, Y.; Xu, D.; Zhu, D.; Jiang, S.; Guo, J.; Zhu, Y. Properties of oriented carbon fiber/polyamide 12 composite parts fabricated by fused deposition modeling. *Mater. Des.* **2018**, *139*, 283–292. [CrossRef]

34. Dul, S.; Fambri, L.; Pegoretti, A. High-performance polyamide/carbon fiber composites for fused filament fabrication: Mechanical and functional performances. *J. Mater. Eng. Perform.* **2021**, *30*, 5066–5085. [CrossRef]

35. Blanco, I.; Cicala, G.; Recca, G.; Tosto, C. Specific Heat Capacity and Thermal Conductivity Measurements of PLA-Based 3D-Printed Parts with Milled Carbon Fiber Reinforcement. *Entropy* **2022**, *24*, 654. [CrossRef] [PubMed]

36. Tekinalp, H.L.; Kunc, V.; Velez-Garcia, G.M.; Duty, C.E.; Love, L.J.; Naskar, A.K.; Blue, C.A.; Ozcan, S. Highly oriented carbon fiber–polymer composites via additive manufacturing. *Compos. Sci. Technol.* **2014**, *105*, 144–150. [CrossRef]

37. Sang, L.; Wang, C.; Wang, Y.; Wei, Z. Thermo-oxidative ageing effect on mechanical properties and morphology of short fibre reinforced polyamide composites–comparison of carbon and glass fibres. *RSC Adv.* **2017**, *7*, 43334–43344. [CrossRef]

38. Kiss, P.; Glinz, J.; Stadlbauer, W.; Burgstaller, C.; Archodoulaki, V.M. The effect of thermally desized carbon fibre reinforcement on the flexural and impact properties of PA6, PPS and PEEK composite laminates: A comparative study. *Compos. Part B Eng.* **2021**, *215*, 108844. [CrossRef]

39. Kim, D.K.; Kang, S.H.; Han, W.; Kim, K.W.; Kim, B.J. Facile method to enhance the mechanical interfacial strength between carbon fibers and polyamide 6 using modified silane coupling agents. *Carbon Lett.* **2022**, *32*, 1463–1472. [CrossRef]
40. Crawford, R.J. Mechanical Behaviour of Composites. In *Plastics Engineering*, 3rd ed.; Crawford, R.J., Ed.; Butterworth-Heinemann: Oxford, UK, 1998; pp. 168–244.
41. Tanaka, K.; Mizuno, S.; Honda, H.; Katayama, T.; Enoki, S. Effect of water absorption on the mechanical properties of carbon fiber/polyamide composites. *J. Solid Mech. Mater. Eng.* **2013**, *7*, 520–529. [CrossRef]

Article

Preliminary Study of Preheated Decarburized Activated Coal Gangue-Based Cemented Paste Backfill Material

Renlong Tang [1,2,*], Bingchao Zhao [1,2,*], Chuang Tian [1], Baowa Xu [1], Longqing Li [1,2], Xiaoping Shao [1,2] and Wuang Ren [3]

[1] Energy School, Xi'an University of Science and Technology, Xi'an 710054, China
[2] Key Laboratory of Western Mines and Hazards Prevention, Ministry of Education of China, Xi'an 710054, China
[3] School of Architecture and Civil Engineering, Xi'an University of Science and Technology, Xi'an 710054, China
* Correspondence: longrt@xust.edu.cn (R.T.); zhaobc913@163.com (B.Z.)

Abstract: This study proposes a novel idea of the use of coal gangue (CG) activation and preheated decarburized activated coal CG-based cemented paste backfill material (PCCPB) to realize green mining. PCCPB was prepared with preheated decarburized coal CG (PCG), FA, activator, low-dose cement, and water. This idea realized scale disposal and resource utilization of coal CG solid waste. Decarbonization and activation of CG crushed the material to less than 8 mm by preheated combustion technology at a combustion temperature of 900 °C and a decarbonization activation time of 4 min. The mechanism of the effect of different Na_2SO_4 dosages on the performance of PCCPB was investigated using comprehensive tests (including mechanical property tests, microscopic tests, and leaching toxicity tests). The results show that the uniaxial compressive strength (UCS) of C-S2, C-S3, and C-S4 can meet the requirements of backfill mining, among which the UCS of C-S3 with a curing time of 3 d and 28 d were 0.545 MPa and 4.312 MPa, respectively. Na_2SO_4 excites PCCPB at different curing time, and the UCS of PCCPB increases and then decreases with the increase in Na_2SO_4 dosage, and 3% of Na_2SO_4 had the best excitation effect on the late strength (28 d) of PCCPB. All groups' (control and CS1-CS4 groups) leachate heavy metal ions met the requirements of groundwater class III standard, and PCCPB had a positive effect on the stabilization/coagulation of heavy metal ions (Mn, Zn, As, Cd, Hg, Pb, Cr, Ba, Se, Mo, and Co). Finally, the microstructure of PCCPB was analyzed using FTIR, TG/DTG, XRD, and SEM. The research is of great significance to promote the resource utilization of coal CG residual carbon and realize the sustainable consumption of coal CG activation on a large scale.

Keywords: preheated decarburized; backfill; coal gangue; compressive strength; microstructure; leaching

Citation: Tang, R.; Zhao, B.; Tian, C.; Xu, B.; Li, L.; Shao, X.; Ren, W. Preliminary Study of Preheated Decarburized Activated Coal Gangue-Based Cemented Paste Backfill Material. *Materials* **2023**, *16*, 2354. https://doi.org/10.3390/ma16062354

Academic Editor: Daniela Fico

Received: 31 January 2023
Revised: 8 March 2023
Accepted: 13 March 2023
Published: 15 March 2023

1. Introduction

One of China's major strategies is sustainable development, and for the coal industry to develop sustainably, reasonable disposal of coal gangue solid waste is necessary. Coal is the ballast stone of China's energy security, but the coal mining process also produces a series of environmental problems, such as surface damage, coal gangue (CG) emissions, etc. [1,2]. CG accounts for 10–25% of the total coal mining [3], and over the years, China has accumulated more than 7 billion tons of CG solid waste and a growth rate of 150 million t/a [4]. CG stockpiles occupy a large number of land resources and also have potential environmental hazards such as spontaneous combustion and the leaching of heavy metal ions [5–7]. Backfill mining as a green mining method can reduce surface damage and effectively dispose of solid waste such as CG [8]. CG has poor cementation properties and is mainly used as aggregate in backfill mining [9,10]. However, the high carbon content of CG leads to poor interfacial bonding of the filler, which affects the performance of the cemented paste backfill (CPB) [11,12]. Borrowing the technical idea of metal mine tailings activation for clinker preparation [13–15], decarbonization and activation treatment of CG can fully use CG's heat and stimulate its activity, which is essential to realize the resourceization and sustainable utilization of CG.

Thermal activation is considered to be the most promising method for destroying the crystal structure of CG and improving its reaction activity [16]. Much research has been carried out to find the best activation process to achieve high chemical reactivity of CG. Organic components and carbon in CG can be burned off after calcination, and kaolinite in CG can also be gradually converted to meta kaolinite in the temperature range of 500–800 °C, thus improving the activity of volcanic ash [17–19]. Hao [20] investigated the effect of thermal activation conditions on the mineral phase and structural changes of CG minerals from the Junggar coalfield, Inner Mongolia, China, and showed that the highest volcanic ash activity was reached when calcined at 800 °C for 2 h. Song [21,22] calcined the Xuzhou CG and the results showed that the optimum thermal activation process for CG was 700 °C kept warm for 2 h when it showed the highest reactivity. Frías et al. [23] calcined 5 different samples of CG at different temperatures (500–900 °C) with a holding time of 2 h and used them in cement mortar to replace 12% of cement calcined at 800 °C. The cement mortar with coal CG had 18% higher compressive strength at 28 days than the control group. The results show that the volcanic ash produced by calcination at 600~800 °C has the best activity. However, the required calcination time is 1–2 h, and the long time required for thermal activation seriously restricts the scale of CG thermal activation.

Preheated combustion is an effective way to achieve short-time combustion of low-volatile fuels [24]. A new technology of preheated combustion of pulverized coal was first proposed by the All-Russian Institute of Thermal Engineering [25], where pulverized coal was preheated to about 816 °C and then burned with a minimum fluidization velocity of 0.4 m/s, which ensured pulverized coal burnout. The Institute of Engineering Thermophysics, Chinese Academy of Sciences, proposed a circulating fluidized bed preheating coupled with pulverized coal furnace combustion technology [26–29], where the circulating fluidized bed ignites and preheats the fuel in the first stage, and stable and efficient combustion of low-volatile-content fuel can be achieved in the pulverized coal furnace in the second stage with a minimum fluidization velocity of 0.67 m/s. Lyu and Wang Shuai et al. [30,31] used a two-stage descending tube furnace to achieve preheated combustion, studied the effects of preheating temperature, residence time, and combustion temperature on combustion exhaustion, and concluded that the pulverized coal could be fully combusted at a preheating time of 0.4 s. Wang Xuebin [32] developed the technology of constant temperature preheat decarburization of low-volatile-matter and low-calorific-value fuels. The 1st generation plant has been in stable operation for 2 years in Ganquanbao, Xinjiang, with an annual treatment of 600,000 tons of gasification slag, a combustion temperature of 900 °C, and a combustion time of 3.7–5.6 min. It has been proved that the preheat combustion technology can burn out the residual carbon of low-volatile-matter fuels in a short time. However, there are few studies on preheat decarbonization of CG, the activity of the CG residue after preheating decarbonization is unknown, and relevant research on the preparation of paste filling material as a gelling material has not been carried out. Therefore, preheat decarbonization activation of coal CG and carrying out preheat decarbonization CG-based CPB material research to promote the resource utilization of CG residue carbon to achieve sustainable consumption of CG activation scale has essential significance.

In order to fill the relevant gap in the literature, this paper prepares a cemented paste backfill with preheated decarburized activated CG (PCG), FA, activator, low-dose cement (most backfill materials use more than 10% cement dosage [33–36]), and water based on short-time preheated decarburization activation of CG. The innovations of this study are mainly in three aspects: (1) to propose a new idea of CG decarbonization activation and a new PCCPB using the described method; (2) to research the effect law of various Na_2SO_4 dosages on the performance of novel PCCPB; and (3) to reveal the micromechanism of novel PCCPB by means of FTIR, TG/DTG, XRD, and SEM. The research content provides new methods for decarbonizing and activating CG and large-scale utilization, promoting green and sustainable coal mining.

2. Materials and Methods

2.1. Materials

The raw materials for this experiment are PCG, fly ash (FA), aeolian sand (AS), cement (OPC), activator, and water. PCG and FA served as auxiliary materials, cement as cementitious materials, AS as aggregate, municipal water as mixing water, and Na_2SO_4 as an activator.

2.1.1. Preheated Decarburized Activated CG

The PCG used in the experiment was prepared using a preheated decarburization 108 device in Ganquanbao, Xinjiang, China. Its preparation process is shown in Figure 1. The device is based on the thermostatic preheated decarburization technology developed by Wang Xuebin [32], and a CG combustion temperature of 900 °C and combustion time of 4.0 min were used in this experiment. The PCG was ground for 10 min in the laboratory using a ball mill (SM-500, Wuxi, China) as this experimental material. Next, PCG's particle size distribution (PSD) after grinding was detected based on a laser scanner (Malvern 2000; Malvern, UK), as shown in Figure 2. The PSD curves indicated that the experimental coarse particles had relatively low PCG content. Moreover, the chemical properties of the prepared PCG were assessed, and the PCG was tested using an X-ray fluorescence spectrometer (XRF) (ARL spectrometer, Waltham, MA, USA). The chemical composition of PCG was tested (Table 1) and the main chemical constituents were SiO_2 (53.55%), Al_2O_3 (20.41%), and Fe_2O_3 (8.99%). In addition, PCG's microscopic morphology and mineral composition were tested by XRD (D8 Advance, Karlsruhe, Germany) and SEM (JSM-7610F, Akishima-shi, Japan). A SEM image of PCG with 2000 times magnification is shown in Figure 3, which indicates that the microscopic morphology of PCG is irregular with an angular shape. The mineralogical composition of PCG is shown in Figure 4. PCG is mainly composed of quartz and mullite. Moreover, the specific surface area of PCG was tested based on a specific surface area meter (BSD-660, Beijing, China) with the result of 3.17 m^2/g.

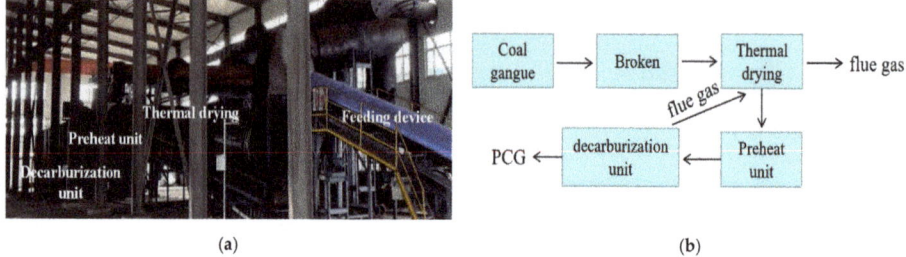

| (a) | (b) |

Figure 1. Equipment preparation process. (a) Site operation drawing of equipment; (b) flow chart of preheating decarburization.

Table 1. Main chemical compositions of PCG, FA, AS, and OPC.

Chemical Composition	OPC	AS	PCG	FA
Al_2O_3	5.53%	10.30%	20.41%	17.14%
SiO_2	22.36%	67.80%	53.55%	41.01%
CaO	65.08%	5.30%	3.76%	14.03%
Fe_2O_3	3.46%	5.80%	8.99%	14.47%
K_2O	0.62%	7.50%	1.62%	5.36%
Mg_2O	1.27%	1.78%	3.65%	4.31%
TiO_2	0.59%	0.35%	0.73%	1.60%
SO_3	/	/	1.27%	0.711%
Others	1.09%	1.17%	5.62%	0.83%

Materials **2023**, *16*, 2354

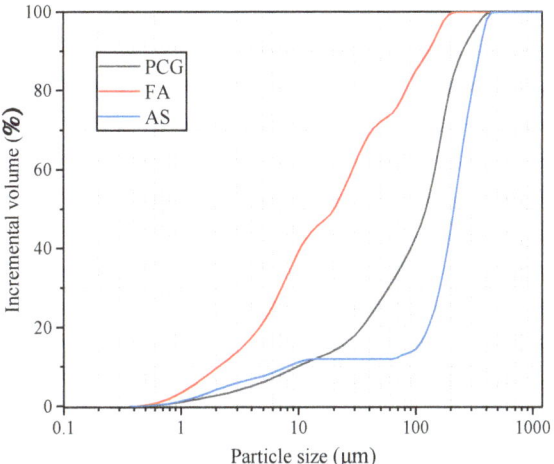

Figure 2. Particle size distribution of PCG, FA, and AS.

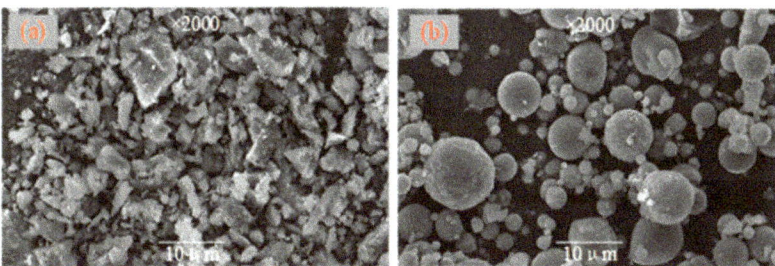

Figure 3. SEM result. (**a**) PCG; (**b**) FA.

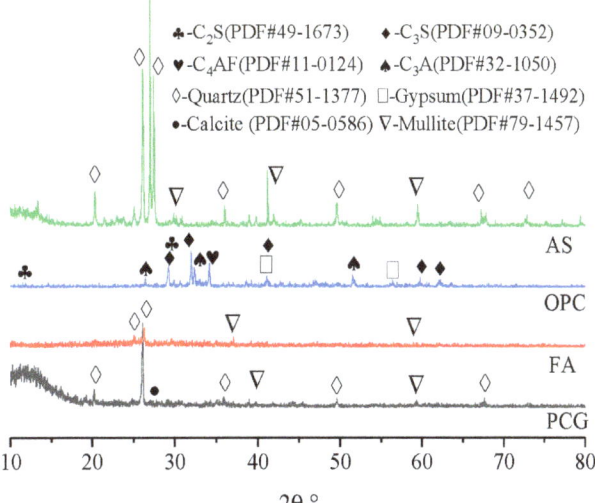

Figure 4. XRD result of raw materials.

2.1.2. Fly Ash

The FA was bought at the pineapple power plant in Yulin, Shaanxi Province, and the same analytical method as PCG was used for FA. Figure 2 and Table 1 show the d_{10}, $d_{60,}$ and d_{90} values of FA as 2.24 μm, 33.0 μm, and 135.3 μm, respectively. The PSD curves show that the content of the fine particles is shallow. The results of FA constituent testing are exhibited in Table 1. The main components are Al_2O_3, SiO_2, and Fe_2O_3 accounting for 86.65%. The microscopic morphology of FA is small and granular. The XRD detecting results are shown in Figure 4. The main minerals of fly ash are aluminosilicate (50–85%), sponge-like vitreous (10–30%), quartz (1–10%), iron oxide (3–25%), carbon particles (1–20%), and sulfate (1–4%).

2.1.3. Aeolian Sand

The AS used in the experiments was taken from Yuyang District, Yulin City. The main light minerals of AS are quartz, feldspar, and calcite, accounting for more than 90%. Heavy minerals are amphibole. Mica and epidote account for more than 5%. Figure 2 and Table 1 show the d_{10}, $d_{60,}$ and d_{90} values of AS as 8.27 μm, 256.03 μm, and 357.69 μm, respectively. The PSD curves show that the experiment's acceptable particle content is shallow, and the corresponding gradation is relatively discontinuous. The composition test results of AS can be seen in Table 1. The main components are SiO_2 and Al_2O_3, accounting for 78.1%.

2.1.4. Cement, Activator, and Water

Cement is a cementitious material, and its main mineralogical components are C_2S (20%), C_3S (50%), C_3A (7–15%), C_4AF (10–18%), etc. The main chemical composition is shown in Table 1. The main hydration products are calcium hydroxide (CH), calcium silicate hydrate (C-S-H), calcium aluminate hydrate (C-A-H), and calcium alumino-ferrite hydrate (C-A-F-H). The cement type used is ordinary Portland cement (OPC) 42.5, and the basicity of the cement is not more than 0.6%, in line with the Chinese national standard GB175-2007. Na_2SO_4 was selected as the activator. It was produced based on the standard of GB/T 9853-2008. The mixing water was mixed with municipal tap water.

2.2. Sample Preparation

According to the preliminary experimental study, it was determined that the PCCPB solid mass concentration of this experiment was 78%, the cement dosage was 3% of the total solid mass, the ash–sand ratio was fixed at 0.5, and the PCG, FA, cement, AS, and activator were formulated as shown in Table 2. The numbers in the table indicate different levels.

Table 2. Mixing proportions of PCCPB.

Group Number	Concentration (CO) [a], %	Cement Dosage (CD) [b], wt.%	AS Dosage (AD) [d], wt.%	FA/PCG	FA+PCG Content (FC) [c], wt.%	Activator Content (AC) [e], wt.%	Activator Type
Control					47	/	/
C-S1 [f]					46	1	
C-S2	78%	3%	50	7/3	45	2	Na_2SO_4
C-S3					44	3	
C-S4					43	4	

[a] CO: $\left(\frac{M_{AD}+M_{CD}+M_{FC}+M_{AC}}{M_{AD}+M_{CD}+M_{FC}+M_{AC}+M_{water}} \times 100\% \right)$ [b] CD: $\frac{M_{CD}}{M_{AD}+M_{CD}+M_{FC}+M_{AC}} \times 100\%$. [c] FC: $\frac{M_{FA}+M_{PCG}}{M_{AD}+M_{CD}+M_{FA}+M_{PCG}+M_{AC}} \times 100\%$. [d] AD: $\frac{M_{AD}}{M_{AD}+M_{CD}+M_{FC}+M_{AC}} \times 100\%$. [e] AC: $\frac{M_{AC}}{M_{AD}+M_{CD}+M_{FC}+M_{AC}} \times 100\%$. [f] C-S1: represents a concentration of 78%, an AS dosage of 50%, a cement dosage of 3%, a Na_2SO_4 content of 1%, an FA and PCG content of 46%, and an FA/PCG ratio of 7/3.

2.3. Experimental Setup and Method

2.3.1. Mechanical Property Test

Under the condition that PCCPB samples reached curing time (3 d, 7 d, 14 d, and 28 d), the UCS tests were conducted with a condition of 1.0 mm/min based on a DNS100 system (SinoTest, Changchun, China) in terms of the standard of GB/T 50081-2019. All experiments were performed three times, and the mean value of UCS was taken.

2.3.2. Microstructure Test

Fourier-transform infrared spectroscopy (FTIR) tests were performed on PCCPB samples that reached the curing time (3 d, 7 d, 14 d, and 28 d). FTIR tests were performed using a Nicolet iN10 Fourier transform micro-infrared spectrometer (Thermo Fisher, Waltham, MA, USA).

The change in weight loss with temperature was tested for cured 3 d and 28 d PCCPB samples using a TGA5500 device. Nitrogen was applied to block carbonization during the heating process. The heating rate in this experiment was 20 °C/min to 900 °C.

The Bruker X-ray diffractometer was used to test the PCCPB samples with a curing time of 3 d and 28 d. Then, the mineral composition of the samples was determined by comparing JADE6.0 software with powder diffraction file (PDF) cards. After the completion of the UCS test, a sample from the central part of the sample was selected and soaked for 48 h to terminate the hydration process. A scan rate of 5°/min and a spectral range of 5° to 80° were used for the tests.

The microstructure of the curing times of 3 d and 28 d specimens was observed using a JEOL JSM-6460LV SEM device. After the UCS test, the central part of the sample was selected and cut into 2 mm slices, which were put on the sample stage for testing.

2.3.3. Leaching Toxicity Test

The samples were crushed so that all sample particles passed a 3 mm sieve. The "Leaching Toxicity Leaching Method for Solid Wastes Inverted Shaking Method" (HJ557-2010) [37], using pure water as a leaching agent, was used. This standard assesses the likelihood that inorganic contaminants would leach from solid wastes and other solid materials into the surface or groundwater [38]. Figure 5 illustrates the test procedure.

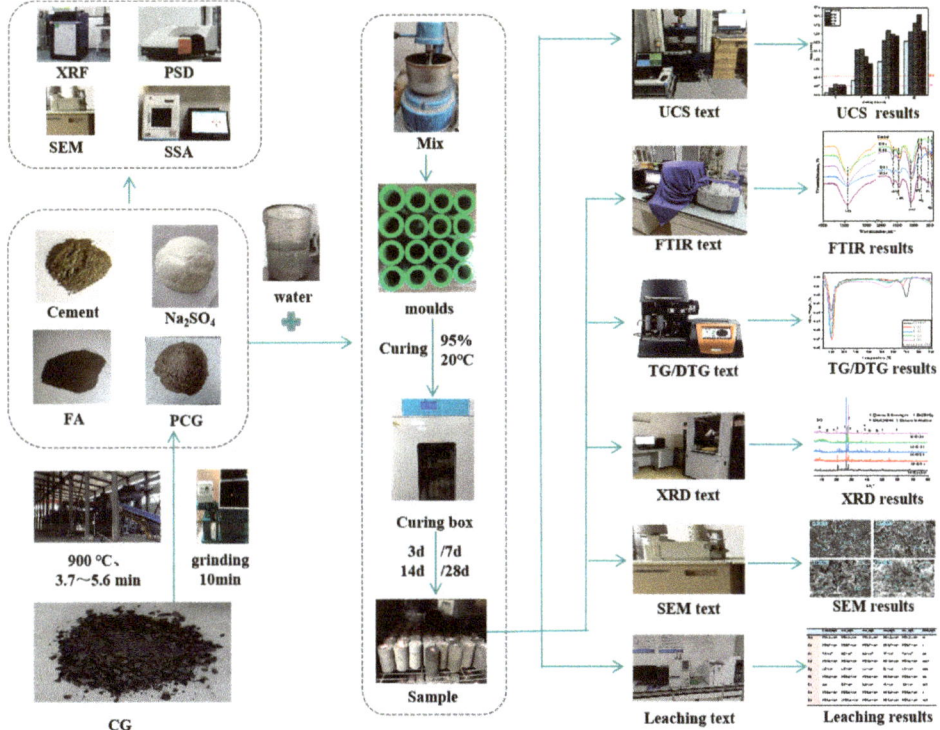

Figure 5. Experimental flowchart.

3. Results and Discussion

3.1. Mechanical Performance of PCCPB

3.1.1. Analysis of UCS of PCCPB

The mechanical properties of the PCCPB are critical indicators to evaluate its stability, which can be expressed by its UCS, coefficient of variance (CV) of UCS, and strong growth rate. Table 3 shows the results of the fundamental statistical analysis of the UCS of PCCPB, and the results demonstrate the mean ($\overline{X}$), standard deviation (SD), and coefficient of variance (CV). SD and CV measure the degree of dispersion of the data, where CV can be expressed by Equation (1) [39]. The SD of all PCCPB samples was kept below 0.13, and the CV was below 15% [40], indicating that the test data are available.

$$CV = \frac{SD}{\overline{X}} \times 100\% \tag{1}$$

Table 3. The statistical analysis of UCS results.

Sample	UCS, MPa 3 d No. of Sample (n): 3			R, %	UCS, MPa 7 d No. of Sample (n): 3			R,%	UCS, MPa 14 d No. of Sample (n): 3			R,%	UCS, MPa 28d No. of Sample (n): 3			R, %
	Mean	SD	CV		Mean	SD	CV		Mean	SD	CV		Mean	SD	CV	
Control	0.119	0.005	4.12	/	0.799	0.037	4.68	/	1.8	0.086	4.79	/	2.891	0.085	2.96	/
C-S1	0.376	0.015	3.99	215.97	2.438	0.052	2.13	205.13	2.924	0.109	3.72	62.44	3.4	0.075	2.22	17.61
C-S2	0.592	0.031	5.28	397.48	2.471	0.064	2.61	209.26	3.458	0.036	1.05	92.11	3.863	0.127	3.30	33.62
C-S3	0.545	0.008	1.44	357.98	2.057	0.038	1.83	157.45	3.284	0.066	2.02	82.44	4.312	0.095	2.20	49.15
C-S4	0.527	0.013	2.56	342.86	1.68	0.095	5.67	110.26	3.171	0.089	2.80	76.17	3.437	0.073	2.13	18.89

The variation of the UCS of PCCPB samples with the curing time and Na_2SO_4 dosage is exhibited in Figure 6. The UCS of PCCPB samples is positively related to curing time, which has been shown by other studies [41,42]. As can be seen from Figure 6, the UCS of C-S3 enhanced from 0.545 MPa (3 d) to 2.057 MPa (7 d), 3.284 MPa (14 d), and 4.312 MPa (28 d) after curing 3 d, 7 d, 14 d, and 28 d, respectively. The increase in UCS is caused by a large number of hydration products (C-(A)-S-H, calcium alumina, CH, and silicate, etc.) filling the internal pores, which leads to the higher compressive strength of the material, a conclusion confirmed in Section 3.2.4, the research of Liu et al. also confirmed this conclusion [43,44].

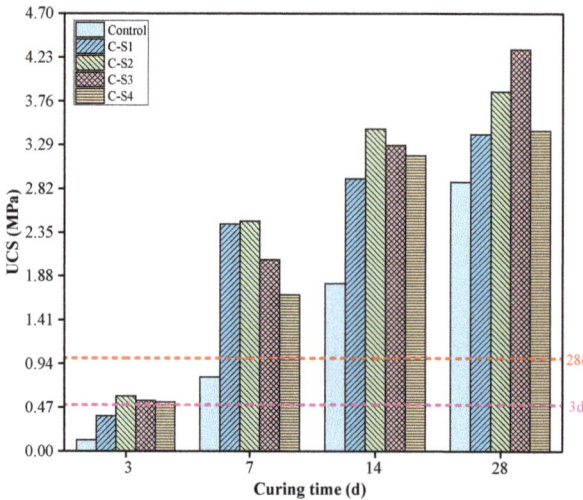

Figure 6. Effect of Na_2SO_4 dosage on the UCS of PCCPB.

From Figure 6, it can be seen that Na_2SO_4 has an excitation effect on PCCPB samples at different curing times, and the UCS of PCCPB increases and subsequently decreases during the process of increasing Na_2SO_4 dosage. Comparing the UCS of PCCPB at 3 d, 7 d, and 14 d, the highest UCS was found in the C-S2 group, indicating that adding 2% Na_2SO_4 had an apparent enhancement influence on the early strength of PCCPB [45], which indicated that more Na_2SO_4 in the early reaction process would inhibit the hydration reaction. For the curing time of PCCPB for 28 d, the UCS of C-S1, C-S2, C-S3, and C-S4 enhanced by 17.61%, 33.62%, 49.15%, and 18.89%, respectively, compared with the control group (the rates of increase in UCS of other groups are listed in the following table), and the UCS corresponding to the C-S3 group was the highest. When the addition of Na_2SO_4 dosage exceeded 3%, its UCS showed a decreasing trend, probably because the Na_2SO_4 that did not participate in the reaction remained in the pores of the PCCPB samples and produced sulfate erosion of the hydration products later [46].

According to the mechanical requirements of the CPB (3 d $\geq$ 0.5 MPa and 28 d $\geq$ 1.0 MPa) [47,48], the mechanical properties of C-S2, C-S3, and C-S4 meet the mechanical requirements of the CPB, and the UCS data of 3 d and 28 d were combined to determine CS3 as the best ratio.

3.1.2. Analysis of Elastic Modulus of PCCPB

The elastic modulus (EM) is a physical and mechanical parameter that directly affects the performance of the PCCPB and is also a necessary parameter for stability analysis, optimization of the structural parameters of the quarry, and numerical simulation. The EM can reflect the degree of bonding between aggregate particles in the PCCPB, and increasing the EM of the PCCPB can reduce the damage through the elastic buffering effect, thus improving the stability of the PCCPB [49].

Figure 7 shows the EM of PCCPB with different Na_2SO_4 dosages. The effect of Na_2SO_4 on the EM of PCCPB is similar to that of UCS, where the EM at any curing time increases with Na_2SO_4 and then decreases, with the highest EM corresponding to the C-S3 group and the most vigorous resistance to deformation of the PCCPB [50]. The EM of the C-S3 group increased by 80.1%, 85.9%, 29.5%, and 52.9% at 3 d, 7 d, 14 d, and 28 d of curing time, respectively, compared with the control group. Under the addition of Na_2SO_4 of more than 3%, the EM showed a decreasing trend.

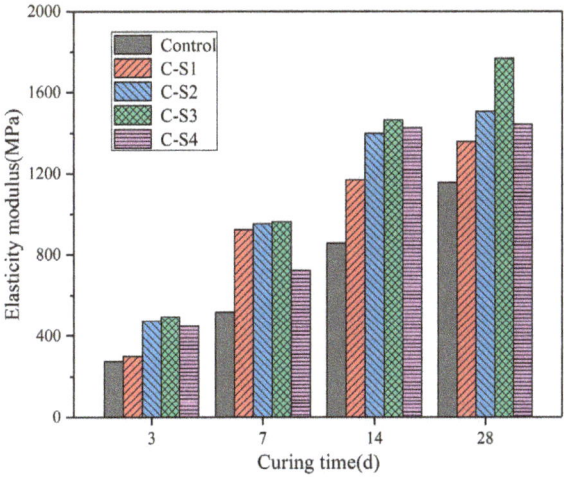

Figure 7. The EM of PCCPB with different Na_2SO_4 dosages.

Figure 8 reflects the variation of the EM of PCCPB with UCS, which shows that as the UCS increases, the EM increases, and vice versa. The relationship between UCS and EM of PCCPB was analyzed by the regression analysis method, including the square root of compressive strength, cube root of compressive strength, and quadratic function of UCS. The square root of UCS was found to be more suitable to describe the relationship between UCS and EM, and red mud material showed a similar relationship [51]. The final equation derived from the analysis is shown in Equation (2):

$$E = A \times UCS^{0.5} + B \tag{2}$$

where E represents the EM (MPa), and A and B are fitting parameters.

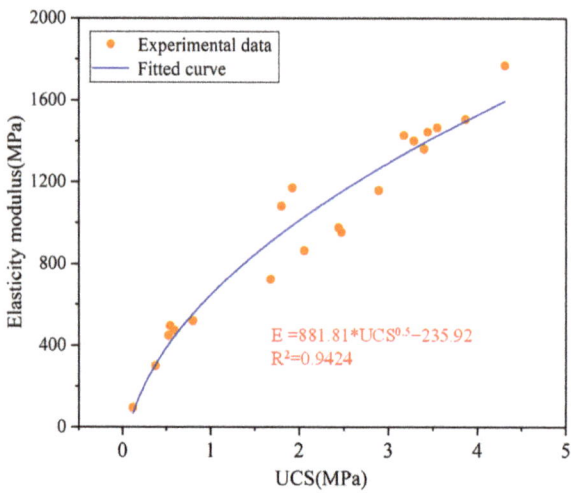

Figure 8. Fitting relationship between elastic modulus and UCS of PCCPB.

The fit coefficient $R^2 = 0.9424$ indicates that a quadratic function using the UCS's square root is consistent with the inspired relationship between the EM and the UCS of the PCCPB.

3.2. Microstructure of ARFGB

3.2.1. FTIR Results Analysis of PCCPB

The Fourier-transform infrared spectra (FTIR) of the PCCPB samples are shown in Figure 9. The bands corresponding to different Na_2SO_4 dosages and different curing times of the PCCPB samples shown in Figure 9 are around 3436 cm^{-1}, 1633 cm^{-1}, 1419 cm^{-1}, 1092 cm^{-1}, 871 cm^{-1}, 772 cm^{-1}, 612 cm^{-1}, and 463 cm^{-1}, respectively. The absorption peaks near 3436 cm^{-1} and 1633 cm^{-1} are O-H stretching vibration peaks and bending vibration peaks, respectively. These O-H characteristic absorption peaks originate from the structural water produced by the hydration reaction and a small part of the free water in the system [52]. The absorption peak near 468 cm^{-1} is due to the bending vibration of the Si-O bond [53]. This band represents quartz. The absorption peaks near 612 cm^{-1} are bending vibrations of Al-O, which correspond to the $[Si(OH)_5]^-$ monomeric structures in the 3D mesh of the gelling material [54]. The absorption peak near 871 cm^{-1} corresponds to the Si-OH band's bending vibration [55]. The band at 1419 cm^{-1} indicates that O-C-O bonds stretch in the presence of carbonates associated with calcite in the sample [56]. The peak near 772 cm^{-1} corresponds to the symmetric stretching vibration of Si-O-T dissolved in SiO_4 tetrahedra; thus, the dissolved Si-Al elements are involved in forming C-(A)-S-H gels; this is also confirmed by Li et al. [17]. The peak near 1092 cm^{-1} is the stretching vibration peak of asymmetric Si-O-Si or Si-O-Al. These absorption peaks show a structure of the Si-O-Al-O bond interconnection in the system. The $[Si(OH)_5]^-$ and $[Al(OH)_4]^-$ monomer structures in the system are connected by these bonds to form a polymer, which recombines into a three-dimensional network structure of the cementitious material [57].

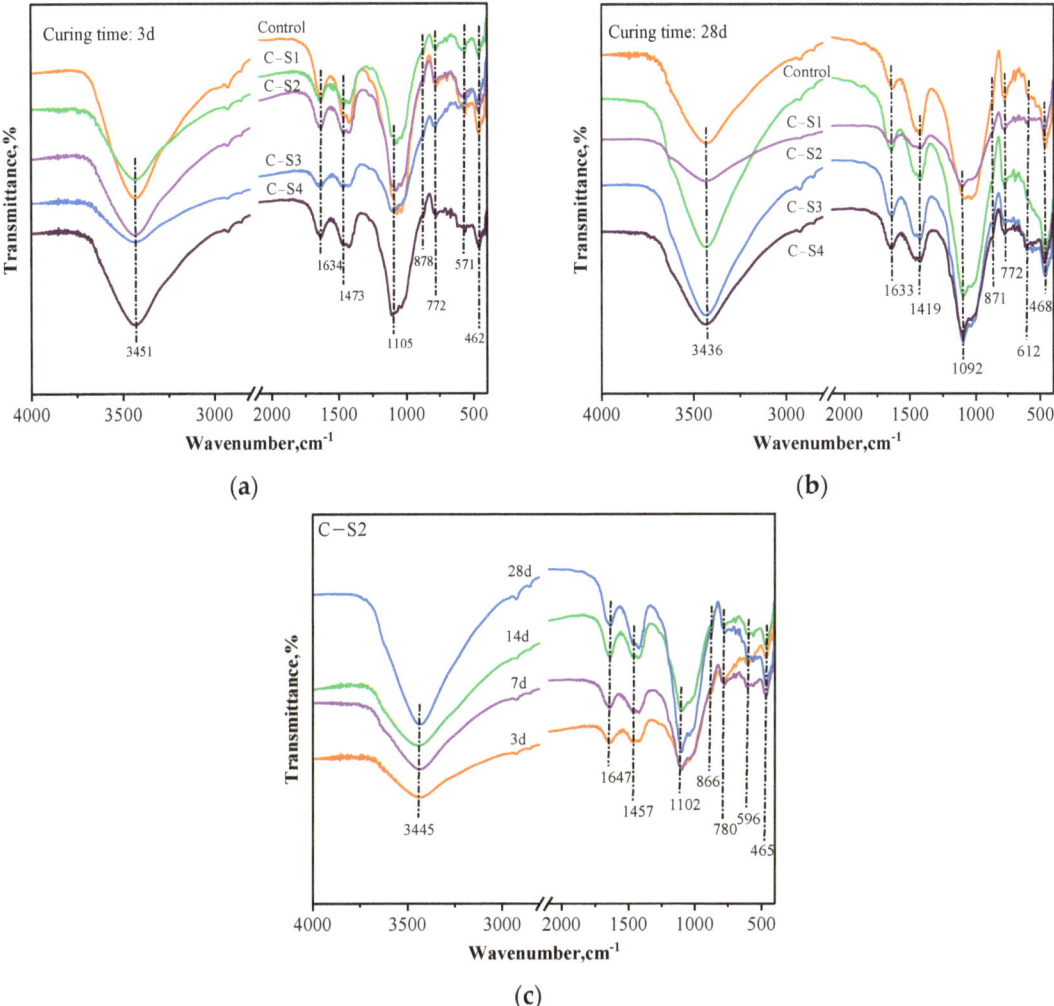

Figure 9. FTIR results of PCCPB samples. (**a**) 3 d; (**b**) 28 d; (**c**) C-S2.

3.2.2. TG-DTG Results Analysis of PCCPB

Figure 10 exhibits the TG–DTG data for different Na_2SO_4 dosages and curing times of PCCPB samples. By and large, all TG–DTG curves have similar characteristics with three prominent heat absorption peaks in the temperature ranges of 50–250 °C (zone I) and 400–450 °C (II). The heat absorption peak at 50–250 °C is associated with the evaporation of free water and dehydration of hydration products (C-(A)-S-H and AFt) [58]. The heat absorption peak at 400–450 °C is assigned to CH dehydration, and the decomposition of calcium carbonate and CH occurs at 600–750 °C [59]. In addition, a weaker heat absorption peak in zone IV above 750 °C is associated with the dehydroxylation of silicate minerals in the sample [60,61]; the weight loss reactions in zones I, II, and III also reveal the hydration reaction products of PCCPB, and the results are consistent with Section 3.2.3.

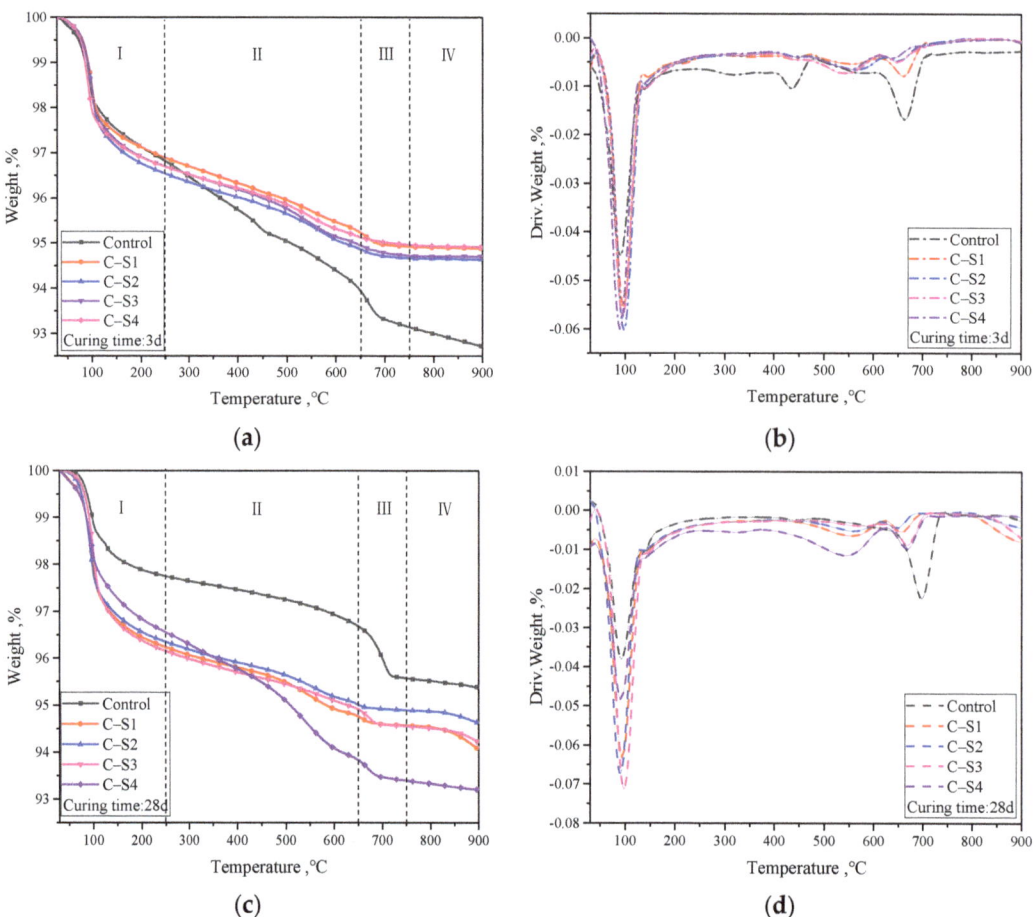

Figure 10. TG-DTG results of PCCPB samples. (**a**) 3d sample mass and temperature relationship (**b**) 3d sample mass loss rate of change and temperature (**c**) 28d sample mass and temperature relationship (**d**) 28d sample mass loss rate of change and temperature Table 4 reflects the weight loss rates of PCCPB with various Na$_2$SO$_4$ dosages, and the magnitude of the weight loss rate indirectly reflects the number of hydration products in a given temperature interval [62]. In the I zone, the weight loss of free water, C-(A)-S-H, and AFt in PCCPB was the first to increase and then to decrease with the increase in Na$_2$SO$_4$ dosage. The most significant weight loss in the PCCPB samples for curing time of 3 d was C-S2, which was 3.45% (zone I), of which the free water weight loss represented a relatively large ratio. The most significant weight loss in the PCCPB samples for a curing time of 28 d was C-S3, which was 3.84% (zone I), with C-(A)-S-H and AFt accounting for a relatively large proportion of the weight loss. This is consistent with the findings in Sections 3.1 and 3.2.4. In addition, mass loss in zone I was found to be higher for all samples at 28 d than at 3 d, indicating that PCCPB continued hydration with increasing age. This is consistent with the studies of Zhang et al. [63] and Feng et al. [64].

Table 4. Weight loss ratio of PCCPB with different Na_2SO_4 dosages.

Temperature Range, °C	Weight Change (3 d, 28 d), %				
	Control	C-S1	C-S2	C-S3	C-S4
I (30–250)	−2.25, −3.19	−3.11, −3.76	−3.45, −3.64	−3.3, −3.84	−3.31, −3.43
II (250–650)	−2.87, −1.06	−1.66, −1.47	−1.68, −1.34	−1.75, −1.24	−1.57, −2.72
III (650–750)	−0.80, −1.11	−0.31, −0.20	−0.19, −0.12	−0.22, −0.35	−0.16, −0.45
IV (750–900)	−0.42, −0.20	−0.04, −0.51	−0.03, −0.27	−0.03, −0.35	−0.06, −0.12

3.2.3. XRD Results Analysis of PCCPB

Figure 11 shows the XRD results of PCCPB. The diffraction peaks of quartz (PDF # 51–1377), ettringite (PDF # 37–1476), calcium hydroxide (PDF # 84–1263), C-(A)-S-H gel (PDF # 34–0002), calcite (PDF # 41–1475), and mullite (PDF # 83–1881) are observed. This is consistent with the results of Mota et al.'s [65] study on cement hydration products in pure cement and sodium sulfate environments. Calcium alumina, calcium hydroxide, and C-(A)-S-H gel are the hydration products generated in the specimens. The primary source of both is from both generation channels. The hydration of cement generates 1, see Equation (3)—the reaction of reactive SiO_2 generates 4 [66,67] and the other and Al_2O_3 in FA and PCG with CH in the hydration reaction of silicate cement, see Equations (5)–(7) [68,69]. The second channel, calcium alumina, is formed by the reaction of Al_2O_3 with CH to form C-A-H and then with gypsum. The formula for C-A-S-H is given in Equation (8). According to the XRD pattern of the original material, it contains quartz and mullite minerals and does not participate in the hydration reaction [70]. The calcite is derived from both the raw material and the specimen. Calcite is derived from the specimen's raw material and surface charring.

$$3CaO \cdot Al_2O_3(C_3A) + 3CaSO_4 \cdot 2H_2O + 30H_2O \rightarrow CaO \cdot Al_2O_3 \cdot 3CaSO_4 \cdot 32H_2O(AFt) \tag{3}$$

$$3Ca \cdot SiO_2 + xH_2O \rightarrow xCaO \cdot SiO_2 \cdot yH_2O(C-S-H) + (3-x)Ca(OH)_2 \tag{4}$$

$$Al_2O_3 + Ca(OH)_2 + xH_2O \rightarrow CaO \cdot Al_2O_3 \cdot xH_2O(C-A-H) \tag{5}$$

$$3CaO \cdot Al_2O_3 \cdot 6H_2O + 3CaSO_4 + 26H_2O \rightarrow 3CaO \cdot Al_2O_3 \cdot 3CaSO_4 \cdot 32H_2O(AFt) \tag{6}$$

$$SiO_2 + Ca(OH)_2 + xH_2O \rightarrow CaO \cdot SiO_2 \cdot xH_2O(C-S-H) \tag{7}$$

$$Al_2O_3 + Ca(OH)_2 + 2SiO_2 + xH_2O \rightarrow CaO \cdot Al_2O_3 \cdot 2SiO_2 \cdot xH_2O(C-A-S-H) \tag{8}$$

From the result in Figure 11a, quartz, calcite, and mullite diffraction peaks do not change significantly in all plots as the Na_2SO_4 dosage increases from 0 wt.% to 4 wt.%. The variation is more significant for calcium alumina and C-(A)-S-H gels, where the intensity of the diffraction peaks of these two hydration products first becomes stronger from a weak diffraction peak and then changes to a weak diffraction peak. Specifically, the dosage of Na_2SO_4 increased from 0 wt.% to 2 wt.%, and the diffraction peak of chalcocite at 2θ was enhanced at 9°. However, the dosage of the Na_2SO_4 was increased to 4 wt.%, and the diffraction peak of chalcocite in the C-S4 sample was decreased. In addition, a similar phenomenon was observed for the C-(A)-S-H gel diffraction peak at 2θ of 39°. This implies that the diffraction peaks of the hydration products in the C-S2 specimens are the strongest at the curing time of 3 d. This is consistent with the UCS of all specimens at 3 d. This suggests that Na_2SO_4 as an activator facilitates the generation of calcium alumina and improves early strength development, which agrees with Razali et al. [71] and Zhao et al. [72]. However, with excess Na_2SO_4, the slurry viscosity increases, leading to the precipitation of C-(A)-S-H gel in the early stage, which inhibits the positive proceeding of hydration [73,74]. In addition, comparing Figure 11a,b, it can be seen that the diffraction peaks of calcite and C-(A)-S-H gels in the curing time of 28 d of PCCPB were significantly enhanced, especially for calcite with 2θ of 16°. The diffraction peaks of calcite with 2θ of 49° were significantly enhanced compared with those in the 3 d samples, especially for the C-S4 sample. Notably, the diffraction peaks of calcium hydroxide with 2θ of 18° and 34° do not change significantly in all samples.

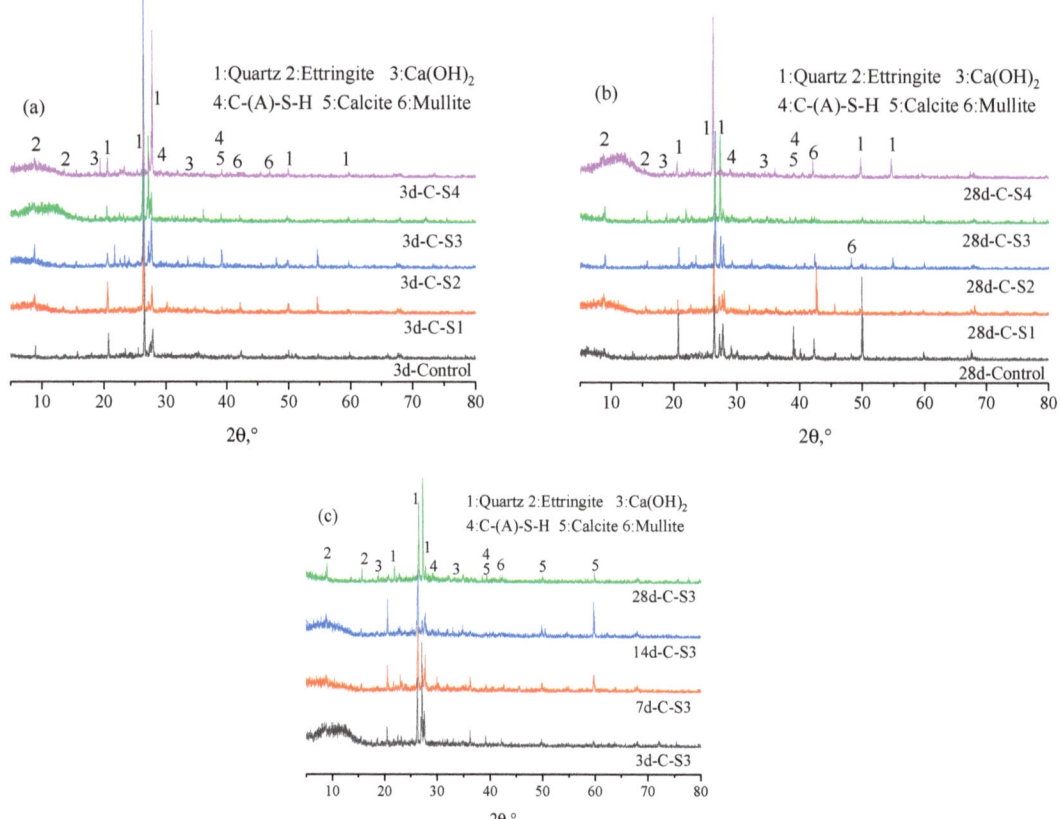

Figure 11. XRD results of PCCPB. (**a**) 3 d; (**b**) 28 d; (**c**) C-S3.

On the one hand, it may be due to the small cement content used in this case of only 3 wt.% of the solid mass, which generates less calcium hydroxide. On the other hand, it may be that the excitation of Na_2SO_4 promotes the hydrolysis of calcium oxide in FA, replenishing the calcium hydroxide consumed by hydration. In order to analyze the effect of curing time on the hydration products, C-S3 samples were selected for testing, and the curing time of the specimens was 3 d, 7 d, 14 d, and 28 d. The XRD results are shown in Figure 11c. The diffraction peaks of the hydration products (calcium alumina, C-(A)-S-H gel) in the samples changed from weak to firm as the curing time increased from 3 d to 28 d. This indicates that the content of hydration products increases with the curing time, increasing strength. Ouyang et al. [75] also obtained a similar conclusion and found that the diffraction peaks of the hydration products (AFt, calcium alumina, and C-(A)-S-H gel) of the sand-based cemented paste filling material were enhanced with the extension of curing age from 1 d to 7 d and 28 d.

3.2.4. SEM Results Analysis of PCCPB

Figure 12 shows the SEM results of PCCPB samples for the curing times of 3 d and 28 d. It was observed that the microscopic morphology of PCCPB samples showed significant differences depending on curing time and the ratio of Na_2SO_4. With the increase in Na_2SO_4, the microscopic morphology of all PCCPB samples showed a pattern from loose to dense and slightly loose. Under the condition of a curing time of 3 d, the control group was the loosest and most porous. When Na_2SO_4 was increased to 2 wt.%, the microscopic morphology of the C-S2 group changed more, from loose to relatively dense, and the matrix of small granular FA and PCG was reduced. After curing for 28 d, abundant needle-like calcarenite (whose length is greater than 1 μm) and irregularly shaped C-(A)-S-H gels were observed in microscopic images of all samples, which had relatively more hydration products and a more extensive distribution range compared with the 3 d samples. The intercalation of calcarenite between the pores of PCCPB increases the denseness of the specimens. However, the control group has fewer hydration products and a relatively loose structure. The microstructure of PCCPB is mainly composed of C-(A)-S-H gels, fine needle-like AFt crystals, hexagonal lamellar CH, unhydrated AS particles, and PCG. This differs from the hydration products of binders made from blast furnace slag [76,77], and the reported results for the hydration product of pure cement are consistent [78,79]. In addition, the formation of hydration products provides the basis for the UCS of C-(A)-S-H.

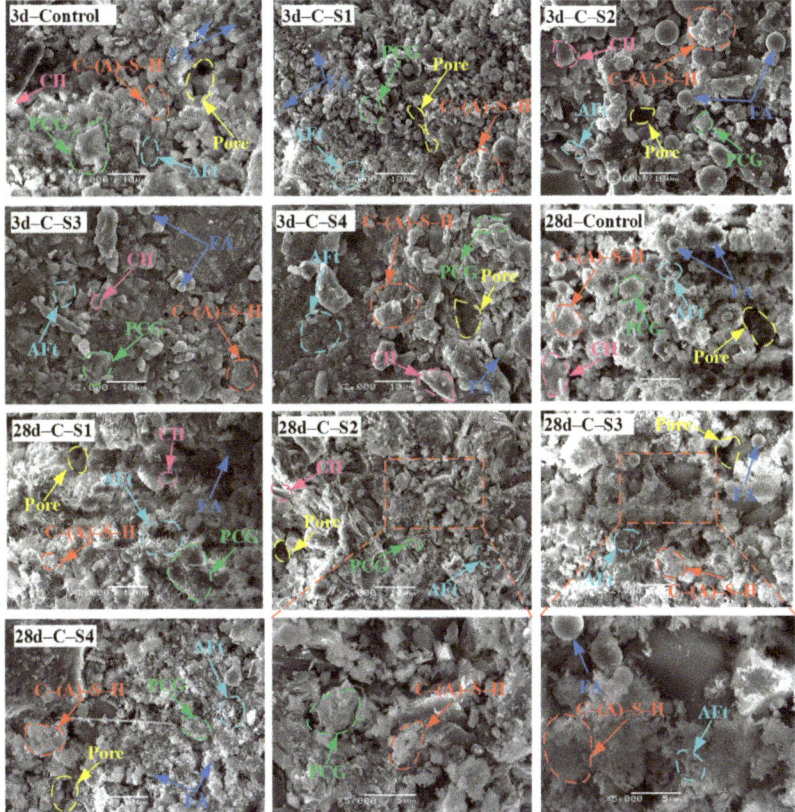

Figure 12. SEM results of PCCPB.

Figure 13 shows the SEM results of C-S3 group samples. From Figure 13, it can be seen that the microstructure of the samples becomes dense as the age increases from 3 d to 28 d with the gradual increase in calcium alumina (AFt) and C-(A)-S-H gel. It was also found that the number of unreacted FA and PCG particles decreased. This is because the silica–oxygen tetrahedral monomer and aluminum–oxygen tetrahedral monomer in FA and PCG are involved in the generation of C-(A)-S-H gels and are connected, gradually forming a monolithic structure [80]. The hydrated calcium silica-aluminate structure has a specific skeletal effect [81,82], thus increasing the UCS of the samples with curing time. This is consistent with the findings of the previous UCS and XRD studies. Zhang and Tang et al. [55] reached similar conclusions. For example, Tang et al. [83] studied fly ash–aeolian sand filling materials and found that AFt and C-(A)-S-H gels in the samples increased with the extension of curing age, and the micro-morphology of the samples changed from loose to dense.

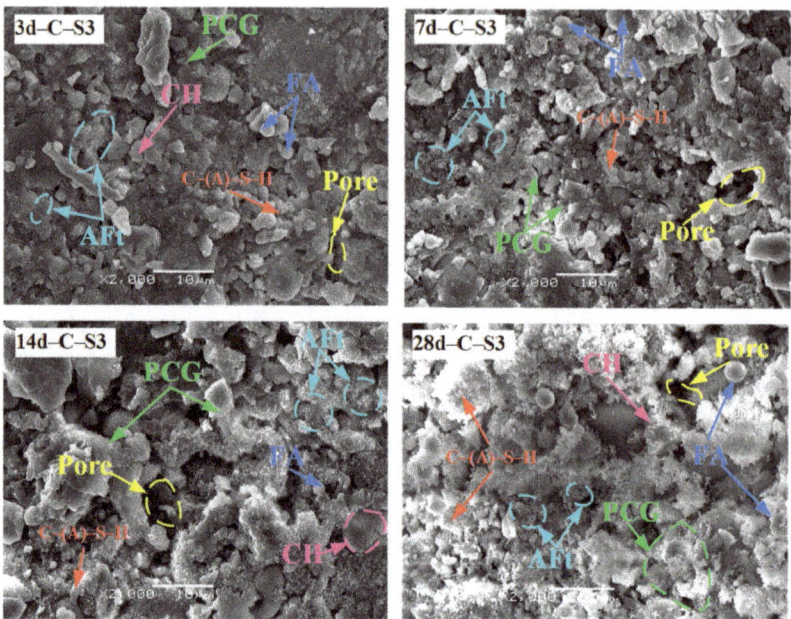

Figure 13. SEM results of C-S3 group samples.

3.3. Leaching Toxicity Results Analysis of PCCPB

When using PCCPB as a backfill material, it is necessary to evaluate the impact of PCCPB weight metal ion leaching on groundwater [84]. Table 5 exhibits the leaching data of heavy metal ions from PCCPB in 28 days of curing. The standard limits are taken from HJ/T 300-2007, which show that the heavy metal ions of leachate in the control group and CS1-CS4 group meet the requirements of groundwater class III standard, which indicates that PCCPB meets the environmental safety requirements and PCCPB is a promising technology for heavy metal curing and stabilization. The CS1-CS4 group is lower than the control group, and the CS3 group is the lowest for three main reasons: (1) C-(A)-S-H and CH are formed by PCCPB hydration, as shown in Sections 3.2.2 and 3.2.3. C-(A)-S-H and CH can immobilize heavy metals on the gelling particles through chemical bonding and physical adsorption/wrapping microstructure [79,85,86]. The lowest CS-4 value is attributed more to C-(A)-S-H and CH products (Section 3.2.3), confirmed by its macroscopic uniaxial compressive strength. (2) Calcareous aluminate neutralizes heavy metal ions within the needles by chemical replacement. There is ample evidence that calcarenite can trap heavy metal elements such as Cd and Cu [87]. (3) The solidification/stabilization of Cr by PCCPB is due to the physical adsorption of hydration products such as C-(A)-S-H gels and sodalite. In addition, hydration refines the pore structure and blocks the transport channels of Cr ions, facilitating solidification/stabilization [88]. Calcite, calcium silicate, calcium hydroxide, and calcium silicate may play a key role in immobilizing As species and heavy metals in PCCPB [89].

Table 5. Leaching results of ARFGB.

	Control, mg/L	C-S1, mg/L	C-S2, mg/L	C-S3, mg/L	C-S4, mg/L	Limit, mg/L
Mn	ND 1.2×10^{-4}	ND 1.2×10^{-4}	ND 1.2×10^{-4}	ND 1.2×10^{-4}	ND 1.2×10^{-4}	0.1
Zn	ND 6.7×10^{-4}	ND 6.7×10^{-4}	ND 6.7×10^{-4}	ND 6.7×10^{-4}	ND 6.7×10^{-4}	1
As	7.5×10^{-3}	6.8×10^{-3}	6.2×10^{-3}	4.7×10^{-3}	7.0×10^{-3}	0.01
Cd	ND 5.0×10^{-5}	ND 5.0×10^{-5}	ND 5.0×10^{-5}	ND 5.0×10^{-5}	ND 5.0×10^{-5}	0.005
Hg	1.8×10^{-4}	1.4×10^{-4}	1.1×10^{-4}	8.1×10^{-5}	1.3×10^{-4}	0.001
Pb	ND 9.0×10^{-5}	ND 9.0×10^{-5}	ND 9.0×10^{-5}	ND 9.0×10^{-5}	ND 9.0×10^{-5}	0.01
Cr	0.01	8.4×10^{-3}	6.2×10^{-3}	4.3×10^{-3}	5.2×10^{-3}	0.05
Cu	ND 8.0×10^{-5}	ND 8.0×10^{-5}	ND 8.0×10^{-5}	ND 8.0×10^{-5}	ND 8.0×10^{-5}	1
Ba	ND 2.0×10^{-4}	ND 2.0×10^{-4}	ND 2.0×10^{-4}	ND 2.0×10^{-4}	ND 2.0×10^{-4}	0.7
Ni	ND 6.0×10^{-5}	ND 6.0×10^{-5}	ND 6.0×10^{-5}	ND 6.0×10^{-5}	ND 6.0×10^{-5}	0.02
Ag	ND 4.0×10^{-5}	ND 4.0×10^{-5}	ND 4.0×10^{-5}	ND 4.0×10^{-5}	ND 4.0×10^{-5}	0.05
Se	2.3×10^{-3}	1.8×10^{-3}	1.5×10^{-3}	1.1×10^{-3}	1.6×10^{-3}	0.01
Mo	0.02	0.017	0.014	0.012	0.013	0.07
Sb	1.9×10^{-3}	1.6×10^{-3}	1.4×10^{-3}	1.1×10^{-3}	1.3×10^{-3}	0.005
Co	ND 3.0×10^{-5}	1ND 3.0×10^{-5}	ND 3.0×10^{-5}	ND 3.0×10^{-5}	ND 3.0×10^{-5}	0.05

4. Conclusions

This study comprehensively evaluated the properties of a novel PCCPB prepared from PCG, FA, activator, low-dose cement, and water. The effects of different activator dosages on the mechanical properties, microstructure, and leaching risk of PCCPB were investigated. Based on the experimental results, the following main conclusions can be drawn:

(1) The UCS of C-S2, C-S3, and C-S4 can meet the requirements of backfill mining, among which the UCS of C-S3 is 0.545 and 4.312 MPa at 3 d and 28 d, respectively. Na_2SO_4 has an excitation effect on PCCPB at different curing times, and the UCS of PCCPB increases and then decreases with the increase in Na_2SO_4. The 3% Na_2SO_4 has the best excitation effect on the later strength (28 d) of PCCPB.

(2) The main hydration products of PCCPB are C-(A)-S-H gel and calcium alumina (AFt), and the effect of different Na_2SO_4 dosages on the content and micromorphology of the gelling products of PCCPB is the main reason for changing its mechanical properties and leaching risk.

(3) All groups' (control group and CS1–CS4 group) leachate heavy metal ions meet the groundwater class III standard requirements, and PCCPB meets the environmental safety requirements. Based on the mechanical properties and leaching results, C-S3 was the best ratio.

(4) Preheated decarburization (combustion temperature 900 °C, time 4 min) is an effective method of CG activation, and with PCG, FA, activator, low-dose cement, and water, preparation of the novel PCCPB is feasible.

Author Contributions: Conceptualization, B.Z.; data curation, R.T.; formal analysis, R.T. and C.T.; investigation, L.L., X.S. and W.R.; methodology, R.T.; project administration, B.Z.; resources, L.L. and X.S.; software, B.X., C.T. and R.T.; supervision, R.T.; validation, B.Z. and X.S.; visualization, B.X.; writing—original draft, R.T.; writing—review and editing, C.T. and B.X. All authors have read and agreed to the published version of the manuscript.

Funding: This research has the support of the Natural Science Basic Research Project of Shaanxi Province (Grant No. 2023-JC-QN-0399) and the National Natural Science Foundation of China (Grant No. 52074208).

Informed Consent Statement: Informed consent was obtained from all subjects involved in the study.

Data Availability Statement: The data used to support the findings of this study are included in the article.

Conflicts of Interest: The authors declare no conflict of interest.

References

1. Qian, M.; Miao, X.; Xu, J. Green Mining of Coal Resources Harmonizing with Environment. *J. China Coal Soc.* **2007**, *1*, 1–7.
2. Xiu, Z.; Wang, S.; Ji, Y.; Wang, F.; Ren, F.; Nguyen, V.T. Loading Rate Effect On the Uniaxial Compressive Strength (UCS) Behavior of Cemented Paste Backfill (CPB). *Constr. Build. Mater.* **2021**, *271*, 121526. [CrossRef]
3. Huang, G.; Ji, Y.; Li, J.; Hou, Z.; Dong, Z. Improving Strength of Calcinated Coal Gangue Geopolymer Mortars Via Increasing Calcium Content. *Constr. Build. Mater.* **2018**, *166*, 760–768. [CrossRef]
4. Ma, H.; Zhu, H.; Wu, C.; Chen, H.; Sun, J.; Liu, J. Study On Compressive Strength and Durability of Alkali-Activated Coal Gangue-Slag Concrete and its Mechanism. *Powder Technol.* **2020**, *368*, 112–124. [CrossRef]
5. Qin, L.; Gao, X. Properties of Coal Gangue-Portland Cement Mixture with Carbonation. *Fuel* **2019**, *245*, 1–12. [CrossRef]
6. Wu, Y.; Yu, X.; Hu, S.; Shao, H.; Liao, Q.; Fan, Y. Experimental Study of the Effects of Stacking Modes On the Spontaneous Combustion of Coal Gangue. *Process Saf. Environ. Protect.* **2018**, *123*, 39–47. [CrossRef]
7. Guo, S.; Zhang, J.; Li, M.; Zhou, N.; Song, W.; Wang, Z.; Qi, S. A Preliminary Study of Solid-Waste Coal Gangue Based Biomineralization as Eco-Friendly Underground Backfill Material: Material Preparation and Macro-Micro Analyses. *Sci. Total Environ.* **2021**, *770*, 145241. [CrossRef]
8. Guo, Z.; Qiu, J.; Jiang, H.; Xing, J.; Sun, X.; Ma, Z. Flowability of Ultrafine-Tailings Cemented Paste Backfill Incorporating Superplasticizer: Insight From Water Film Thickness Theory. *Powder Technol.* **2021**, *381*, 509–517. [CrossRef]
9. Zhang, J.; Zhang, Q.; Spearing, A.J.S.; Miao, X.; Guo, S.; Sun, Q. Green Coal Mining Technique Integrating Mining-Dressing-Gas Draining-Backfilling-Mining. *Int. J. Min. Sci. Technol.* **2017**, *27*, 17–27. [CrossRef]
10. Zhang, Q.; Wang, Z.; Zhang, J.; Jiang, H.; Wang, Y.; Yang, K.; Tian, X.; Yuan, L. Integrated Green Mining Technology of "Coal Mining-Gangue Washing-Backfilling-Strata Control-System Monitoring"-Taking Tangshan Mine as a Case Study. *Environ. Sci. Pollut. Res. Int.* **2021**, *29*, 5798–5811. [CrossRef]
11. Liu, C.; Deng, X.; Liu, J.; Hui, D. Mechanical Properties and Microstructures of Hypergolic and Calcined Coal Gangue Based Geopolymer Recycled Concrete. *Constr. Build. Mater.* **2019**, *221*, 691–708. [CrossRef]
12. Bai, G.; Zhu, C.; Liu, C.; Liu, B. An Evaluation of the Recycled Aggregate Characteristics and the Recycled Aggregate Concrete Mechanical Properties. *Constr. Build. Mater.* **2020**, *240*, 117978. [CrossRef]
13. Khudyakova, T.M.; Kolesnikova, O.G.; Zhanikulov, N.N.; Botabaev, N.E.; Kenzhibaeva, G.S.; Iztleuov, G.M.; Suigenbaeva, A.Z.; Kutzhanova, A.N.; Ashirbaev, H.A.; Kolesnikova, V.A. Low-Basicity Cement, Problems and Advantages of its Utilization. *Refract. Ind. Ceram.* **2021**, *62*, 369–374. [CrossRef]
14. Kolesnikova, O.; Syrlybekkyzy, S.; Fediuk, R.; Yerzhanov, A.; Nadirov, R.; Utelbayeva, A.; Agabekova, A.; Latypova, M.; Chepelyan, L.; Volokitina, I.; et al. Thermodynamic Simulation of Environmental and Population Protection by Utilization of Technogenic Tailings of Enrichment. *Materials* **2022**, *15*, 6980. [CrossRef]
15. Kolesnikova, O.; Vasilyeva, N.; Kolesnikov, A.; Zolkin, A. Optimization of Raw Mix Using Technogenic Waste to Produce Cement Clinker. *Min. Inf. Anal. Bull.* **2022**, *60*, 103–115. [CrossRef]
16. Zhang, Y.; Ling, T. Reactivity Activation of Waste Coal Gangue and its Impact On the Properties of Cement-Based Materials—A Review. *Constr. Build. Mater.* **2020**, *234*, 117424. [CrossRef]
17. Li, C.; Wan, J.; Sun, H.; Li, L. Investigation On the Activation of Coal Gangue by a New Compound Method. *J. Hazard. Mater.* **2010**, *179*, 515–520. [CrossRef]
18. Zhang, Y.; Zhang, Z.; Zhu, M.; Cheng, F.; Zhang, D. Decomposition of Key Minerals in Coal Gangues During Combustion in O2/N2 and O2/Co2 Atmospheres. *Appl. Therm. Eng.* **2018**, *148*, 977–983. [CrossRef]
19. Xu, B.; Liu, Q.; Ai, B.; Ding, S.; Frost, R.L. Thermal Decomposition of Selected Coal Gangue. *J. Therm. Anal. Calorim.* **2018**, *131*, 1413–1422. [CrossRef]
20. Hao, R.; Li, X.; Xu, P.; Liu, Q. Thermal Activation and Structural Transformation Mechanism of Kaolinitic Coal Gangue From Jungar Coalfield, Inner Mongolia, China. *Appl. Clay Sci.* **2022**, *223*, 106508. [CrossRef]
21. Li, L.; Zhang, Y.; Zhang, Y.; Sun, J.; Hao, Z. The Thermal Activation Process of Coal Gangue Selected From Zhungeer in China. *J. Therm. Anal. Calorim.* **2016**, *126*, 1559–1566. [CrossRef]
22. Song, X.; Gong, C.; Li, D. Study On Structural Characteristic and Mechanical Property of Coal Gangue in Activation Process. *J. Chin. Ceram. Soc.* **2004**, *3*, 358–363.
23. Frías, M.; de Rojas, M.I.S.; García, R.; Valdés, A.J.; Medina, C. Effect of Activated Coal Mining Wastes On the Properties of Blended Cement. *Cem. Concr. Compos.* **2012**, *34*, 678–683. [CrossRef]
24. Rabovitser, J.; Bryan, B.; Knight, R.; Nester, S.; Ake, T. Development and Testing of a Novel Coal Preheating Technology for Nox Reduction From Pulverized Coal-Fired Boilers. *Gas* **2003**, *1*, 4.
25. Man, C.; Zhu, J.; Ouyang, Z.; Liu, J.; Lyu, Q. Experimental Study On Combustion Characteristics of Pulverized Coal Preheated in a Circulating Fluidized Bed. *Fuel Process. Technol.* **2018**, *172*, 72–78. [CrossRef]
26. Pan, F.; Zhu, J.; Liu, J. Experimental Study and Numerical Simulation of Preheating Combustion in Circulating Fluidized Bed. *Clean Coal Technol.* **2021**, *27*, 180–188.
27. Lyu, Q.; Wang, J.; Zhu, J. Experimental Study On Combustion Characteristics of Anthracite Pulverized Coal Preheated by Circulation Fluidized Bed. *Boil. Technol.* **2011**, *42*, 23–27.
28. Zhu, J.; Ouyang, Z.; Lu, Q. An Experimental Study On Nox Emissions in Combustion of Pulverized Coal Preheated in a Circulating Fluidized Bed. *Energy Fuels* **2013**, *27*, 7724–7729. [CrossRef]

29. Zhu, S.; Lyu, Q.; Zhu, J.; Liang, C. Experimental Study On No X Emissions of Pulverized Bituminous Coal Combustion Preheated by a Circulating Fluidized Bed. *J. Energy Inst.* **2018**, *92*, 247–256. [CrossRef]

30. Lv, Z.; Xiong, X.; Yu, S.; Tan, H.; Xiang, B.; Huang, J.; Peng, J.; Li, P. Experimental Investigation On No Emission of Semi-Coke Under High Temperature Preheating Combustion Technology. *Fuel* **2021**, *283*, 119293. [CrossRef]

31. Wang, S.; Gong, Y.; Niu, Y.; Hui, S. Study On No Formation During Preheating-Combustion Coupling of Pulverized Coal. *Proc. CSEE* **2020**, *40*, 2951–2959.

32. Shi, Z.; Wang, G.; Wang, X.; Chen, Y.; Yu, W.; Miao, Y.; Peng, Y.; Tan, H. Study On Combustion Characteristics of Preheating Decarburization Process for Fine Slag in Coal Gasification. *Clean Coal Technol.* **2021**, *27*, 105–110.

33. Shao, X.; Tian, C.; Li, C.; Fang, Z.; Zhao, B.; Xu, B.; Ning, J.; Li, L.; Tang, R. The Experimental Investigation On Mechanics and Damage Characteristics of the Aeolian Sand Paste-Like Backfill Materials Based On Acoustic Emission. *Materials* **2022**, *15*, 7235. [CrossRef] [PubMed]

34. Zhou, N.; Dong, C.; Ouyang, S.; Deng, X.; Du, E. Feasibility Study and Performance Optimization of Sand-Based Cemented Paste Backfill Materials. *J. Clean Prod.* **2020**, *259*, 120798. [CrossRef]

35. Deng, X.; Zhang, J.; Klein, B.; Zhou, N.; De Wit, B. Experimental Characterization of the Influence of Solid Components On the Rheological and Mechanical Properties of Cemented Paste Backfill. *Int. J. Miner. Proc.* **2017**, *168*, 116–125. [CrossRef]

36. Shao, X.; Sun, J.; Xin, J.; Zhao, B.; Sun, W.; Li, L.; Tang, R.; Tian, C.; Xu, B. Experimental Study On Mechanical Properties, Hydration Kinetics, and Hydration Product Characteristics of Aeolian Sand Paste-Like Materials. *Constr. Build. Mater.* **2021**, *303*, 124601. [CrossRef]

37. *HJ 557-2010*; Institute of Solid Waste Pollution Control Technology, Solid Waste-Extraction Procedure for Leaching Toxicity-Horizontal Vibration Method. China Environmental Science Press: Beijing, China, 2010.

38. Huang, Z.; Su, X.; Zhang, J.; Luo, D.; Chen, Y.; Li, H. Study On Leaching of Heavy Metals From Red Mud-Phosphogypsum Composite Materials. *Inorg. Chem. Ind.* **2022**, *54*, 133–140.

39. Bhattacharyya, G.K.; Johnson, R.A. *Statistical Concepts and Methods*; Wiley & Sons: Hoboken, NJ, USA, 1977.

40. Behera, S.K.; Ghosh, C.N.; Mishra, D.P.; Singh, P.; Mishra, K.; Buragohain, J.; Mandal, P.K. Strength Development and Microstructural Investigation of Lead-Zinc Mill Tailings Based Paste Backfill with Fly Ash as Alternative Binder. *Cem. Concr. Compos.* **2020**, *109*, 103553. [CrossRef]

41. Qi, T.; Feng, G.; Guo, Y.; Zhang, Y.; Ren, A.; Kang, L.; Guo, J. Experimental Study On the Changes of Coal Paste Backfilling Material Performance During Hydration Process. *J. Min. Saf. Eng.* **2015**, *32*, 42–48.

42. Shao, X.; Wang, L.; Li, X.; Fang, Z.; Zhao, B.; Tao, Y.; Liu, L.; Sun, W.; Sun, J. Study On Rheological and Mechanical Properties of Aeolian Sand-Fly Ash-Based Filling Slurry. *Energies* **2020**, *13*, 1266. [CrossRef]

43. Liu, L.; Fang, Z.; Qi, C.; Zhang, B.; Guo, L.; Song, K. Experimental Investigation On the Relationship Between Pore Characteristics and Unconfined Compressive Strength of Cemented Paste Backfill. *Constr. Build. Mater.* **2018**, *179*, 254–264. [CrossRef]

44. Liu, L.; Xin, J.; Qi, C.; Jia, H.; Song, K. Experimental Investigation of Mechanical, Hydration, Microstructure and Electrical Properties of Cemented Paste Backfill. *Constr. Build. Mater.* **2020**, *263*, 120137. [CrossRef]

45. Phuong, T.B.; Yuko, O.; Kenji, K. Effect of Sodium Sulfate Activator On Compressive Strength and Hydration of Fly-Ash Cement Pastes. *J. Mater. Civ. Eng.* **2020**, *32*, 04020117.

46. Li, W.; Fall, M. Sulphate Effect On the Early Age Strength and Self-Desiccation of Cemented Paste Backfill. *Constr. Build. Mater.* **2016**, *106*, 296–304. [CrossRef]

47. Ercikdi, B.; Baki, H.; İzki, M. Effect of Desliming of Sulphide-Rich Mill Tailings On the Long-Term Strength of Cemented Paste Backfill. *J. Environ. Manag.* **2013**, *115*, 5–13. [CrossRef] [PubMed]

48. Liu, L.; Xin, J.; Feng, Y.; Zhang, B.; Song, K.I. Effect of the Cement–Tailing Ratio On the Hydration Products and Microstructure Characteristics of Cemented Paste Backfill. *Arab. J. Sci. Eng.* **2019**, *44*, 6547–6556. [CrossRef]

49. Rao, Y.; Shao, Y.; Huang, Y.; Sun, X.; Li, Y.; Li, J. Effect of Superplasticizer On the Elastic Modulus of Super Fine Tailings Filling Body. *Min. Res. Dev.* **2016**, *36*, 31–35.

50. Li, X.; Liu, C. Mechanical Properties and Damage Constitutive Model of High Water Material at Different Loading Rates. *Adv. Eng. Mater.* **2018**, *20*, 1701098. [CrossRef]

51. Choe, G.; Kang, S.; Kang, H. Mechanical Properties of Concrete Containing Liquefied Red Mud Subjected to Uniaxial Compression Loads. *Materials.* **2020**, *13*, 854. [CrossRef]

52. Andoni, A.; Delilaj, E.; Ylli, F.; Taraj, K.; Korpa, A.; Xhaxhiu, K.; Çomo, A. FTIR Spectroscopic Investigation of Alkali-Activated Fly Ash: Atest Study. *Zaštita Mater.* **2018**, *59*, 539–542. [CrossRef]

53. Criado, M.; Fernández-Jiménez, A.; Palomo, A. Alkali Activation of Fly Ash: Effect of the SiO_2/Na_2O Ratio Part I: FTIR Study. *Microporous Mesoporous Mat.* **2007**, *106*, 180–191. [CrossRef]

54. Nguyen, H.T. Evaluation On Formation of Aluminosilicate Network in Ternary-Blended Geopolymer Using Infrared Spectroscopy. *Solid State Phenom.* **2019**, *296*, 99–104. [CrossRef]

55. Zhang, N.; Li, H.; Zhao, Y.; Liu, X. Hydration Characteristics and Environmental Friendly Performance of a Cementitious Material Composed of Calcium Silicate Slag. *J. Hazard. Mater.* **2016**, *306*, 67–76. [CrossRef]

56. Xin, J.; Liu, L.; Xu, L.; Wang, J.; Yang, P.; Qu, H. A Preliminary Study of Aeolian Sand-Cement-Modified Gasification Slag-Paste Backfill: Fluidity, Microstructure, and Leaching Risks. *Sci. Total Environ.* **2022**, *830*, 154766. [CrossRef]

57. Lecomte, I.; Henrist, C.; Liégeois, M.; Maseri, F.; Rulmont, A.; Cloots, R. (Micro)-Structural Comparison Between Geopolymers, Alkali-Activated Slag Cement and Portland Cement. *J. Eur. Ceram. Soc.* **2006**, *26*, 3789–3797. [CrossRef]
58. Ouellet, S.; Bussière, B.; Aubertin, M.; Benzaazoua, M. Microstructural Evolution of Cemented Paste Backfill: Mercury Intrusion Porosimetry Test Results. *Cem. Concr. Res.* **2007**, *37*, 1654–1665. [CrossRef]
59. Abdul-Hussain, N.; Fall, M. Unsaturated Hydraulic Properties of Cemented Tailings Backfill that Contains Sodium Silicate. *Eng. Geol.* **2011**, *123*, 288–301. [CrossRef]
60. Liu, S.; Wang, L.; Li, Q.; Song, J. Hydration Properties of Portland Cement-Copper Tailing Powder Composite Binder. *Constr. Build. Mater.* **2020**, *251*, 118882. [CrossRef]
61. Scrivener, K.L.; John, V.M.; Gartner, E.M. Eco-Efficient Cements: Potential Economically Viable Solutions for a Low-CO_2 Cement-Based Materials Industry. *Cem. Concr. Res.* **2018**, *114*, 2–26. [CrossRef]
62. Quanji, Z. Thixotropic Behavior of Cement-Based Materials: Effect of Clay and Cement Types. Master's Thesis, Iowa State University, Ames, IA, USA, 2010.
63. Zhang, S.; Niu, D.; Luo, D. Enhanced Hydration and Mechanical Properties of Cement-Based Materials with Steel Slag Modified by Water Glass. *J. Mater. Res. Technol.* **2022**, *21*, 1830–1842. [CrossRef]
64. Feng, J.; Sun, J. A Comparison of the 10-Year Properties of Converter Steel Slag Activated by High Temperature and an Alkaline Activator. *Constr. Build. Mater.* **2020**, *234*, 116948. [CrossRef]
65. Mota, B.; Matschei, T.; Scrivener, K. Impact of NaOH and Na_2SO_4 On the Kinetics and Microstructural Development of White Cement Hydration. *Cem. Concr. Res.* **2018**, *108*, 172–185. [CrossRef]
66. Sun, Q.; Tian, S.; Sun, Q.; Li, B.; Cai, C.; Xia, Y.; Wei, X.; Mu, Q. Preparation and Microstructure of Fly Ash Geopolymer Paste Backfill Material. *J. Clean Prod.* **2019**, *225*, 376–390. [CrossRef]
67. Kong, X.; Lu, Z.; Zhang, C. Recent Development On Understanding Cement Hydration Mechanism and Effects of Chemical Admixtures On Cement Hydration. *J. Chin. Ceram. Soc.* **2017**, *45*, 274–281.
68. Justnes, H.; Sellevold, E.J.; Lundevall, G. High Strength Concrete Binders. Part a: Reactivity and Composition of Cement Pastes with and without Condensed Silica Fume. In Proceedings of the 4th International Conference on Fly Ash, Silica Fume, Slag and Natural Pozzolana in Concrete, Istanbul, Turkey, 3–8 May 1992.
69. Kipkemboi, B.; Zhao, T.; Miyazawa, S.; Sakai, E.; Hirao, H. Effect of C3S Content of Clinker On Properties of Fly Ash Cement Concrete. *Constr. Build. Mater.* **2020**, *240*, 117840. [CrossRef]
70. Liu, L.; Ruan, S.; Qi, C.; Zhang, B.; Tu, B.; Yang, Q.; Song, K.I.I.L. Co-Disposal of Magnesium Slag and High-Calcium Fly Ash as Cementitious Materials in Backfill. *J. Clean Prod.* **2021**, *279*, 123684. [CrossRef]
71. Razali, N.N.; Sukardi, M.A.; Sopyan, I.; Mel, M.; Salleh, H.M.; Rahman, M.M. The Effects of Excess Calcium On the Handling and Mechanical Properties of Hydrothermal Derived Calcium Phosphate Bone Cement. *IOP Conf. Ser. Mater. Sci. Eng.* **2018**, *290*, 012053. [CrossRef]
72. Zhao, Y.; Qiu, J.; Zhang, S.; Guo, Z.; Ma, Z.; Sun, X.; Xing, J. Effect of Sodium Sulfate On the Hydration and Mechanical Properties of Lime-Slag Based Eco-Friendly Binders. *Constr. Build. Mater.* **2020**, *250*, 118603. [CrossRef]
73. Somna, K.; Jaturapitakkul, C.; Kajitvichyanukul, P.; Chindaprasirt, P. Naoh-Activated Ground Fly Ash Geopolymer Cured at Ambient Temperature. *Fuel* **2011**, *90*, 2118–2124. [CrossRef]
74. Salih, M.A.; Ali, A.A.A.; Farzadnia, N. Characterization of Mechanical and Microstructural Properties of Palm Oil Fuel Ash Geopolymer Cement Paste. *Constr. Build. Mater.* **2014**, *6*, 592–603. [CrossRef]
75. Ouyang, S.; Huang, Y.; Zhou, N.; Li, J.; Gao, H.; Guo, Y. Experiment On Hydration Exothermic Characteristics and Hydration Mechanism of Sand-Based Cemented Paste Backfill Materials. *Constr. Build. Mater.* **2022**, *318*, 125870. [CrossRef]
76. Roy, D.M.; Ldorn, G.M. Hydration, Structure, and Properties of Blast Furnace Slag Cements, Mortars, and Concrete. *J. Proc.* **1982**, *79*, 444–457.
77. Rachid, H.S.; Walid, M.; Mahfoud, B.; Richard, L.; Keith, T.; Agnes, Z.; NorEdine, A. Mechanical Properties and Microstructure of Low Carbon Binders Manufactured From Calcined Canal Sediments and Ground Granulated Blast Furnace Slag (GGBS). *Sustainability* **2021**, *13*, 9057.
78. Poon, C.; Azhar, S.; Anson, M.; Wong, Y. Strength and Durability Recovery of Fire-Damaged Concrete After Post-Fire-Curing. *Cem. Concr. Res.* **2001**, *31*, 1307–1318. [CrossRef]
79. Tang, P.; Chen, W.; Xuan, D.; Cheng, H.; Poon, C.S.; Tsang, D.C.W. Immobilization of Hazardous Municipal Solid Waste Incineration Fly Ash by Novel Alternative Binders Derived From Cementitious Waste. *J. Hazard. Mater.* **2020**, *393*, 122386. [CrossRef]
80. Ding, Z.; Zhou, J.; Su, Q.; Wang, Q.; Sun, H. Mechanical Properties of Geopolymer Recycled Aggregate Concrete. *J. Shenyang Jianzhu Univ. Nat. Sci.* **2021**, *37*, 138–146.
81. Ding, Z.; Hong, X.; Zhu, J.; Tian, B.; Fang, Y. Alkali-Activated Red Mud-Slag Cementitious Materials. *J. Chin. Electron Microsc. Soc.* **2018**, *37*, 145–153.
82. Liu, J.; Li, Z.; Zhang, M.; Wang, S.; Hai, R. Mechanical Property and Polymerization Mechanism of Red Mud Geopolymer Cement. *J. Build. Mater.* **2022**, *25*, 178–183.
83. Tang, R.; Zhao, B.; Li, C.; Xin, J.; Xu, B.; Tian, C.; Ning, J.; Li, L.; Shao, X. Experimental Study On the Effect of Fly Ash with Ammonium Salt Content On the Properties of Cemented Paste Backfill. *Constr. Build. Mater.* **2023**, *369*, 130513. [CrossRef]

84. Türkel, S. Long-Term Compressive Strength and some Other Properties of Controlled Low Strength Materials Made with Pozzolanic Cement and Class C Fly Ash. *J. Hazard. Mater.* **2006**, *137*, 261–266. [CrossRef]

85. Chen, X.; Guo, Y.; Ding, S.; Zhang, H.; Xia, F.; Wang, J.; Zhou, M. Utilization of Red Mud in Geopolymer-Based Pervious Concrete with Function of Adsorption of Heavy Metal Ions. *J. Clean Prod.* **2019**, *207*, 789–800. [CrossRef]

86. Li, B.; Zhang, S.; Li, Q.; Li, N.; Yuan, B.; Chen, W.; Brouwers, H.J.H.; Yu, Q. Uptake of Heavy Metal Ions in Layered Double Hydroxides and Applications in Cementitious Materials: Experimental Evidence and First-Principle Study. *Constr. Build. Mater.* **2019**, *222*, 96–107. [CrossRef]

87. Peng, D.; Wang, Y.; Liu, X.; Tang, B.; Zhang, N. Preparation, Characterization, and Application of an Eco-Friendly Sand-Fixing Material Largely Utilizing Coal-Based Solid Waste. *J. Hazard. Mater.* **2019**, *373*, 294–302. [CrossRef] [PubMed]

88. Luo, Z.; Zhi, T.; Liu, L.; Mi, J.; Zhang, M.; Tian, C.; Si, Z.; Liu, X.; Mu, Y. Solidification/Stabilization of Chromium Slag in Red Mud-Based Geopolymer. *Constr. Build. Mater.* **2022**, *316*, 125813. [CrossRef]

89. Bah, A.; Jin, J.; Ramos, A.O.; Bao, Y.; Ma, M.; Li, F. Arsenic(V) Immobilization in Fly Ash and Mine Tailing-Based Geopolymers: Performance and Mechanism Insight. *Chemosphere* **2022**, *306*, 135636. [CrossRef]

 *materials*

Article

The Effect of Process Conditions on Sulfuric Acid Leaching of Manganese Sludge

Jafar Safarian *, Ariel Skaug Eini, Markus Antonius Elinsønn Pedersen and Shokouh Haghdani

Department of Materials Science and Engineering, Norwegian University of Science and Technology, 7034 Trondheim, Norway; arielse@stud.ntnu.no (A.S.E.); mapeders@stud.ntnu.no (M.A.E.P.); shokouhhaghdani@gmail.com (S.H.)
* Correspondence: jafar.safarian@ntnu.no

Abstract: Manganese sludge, an industrial waste product in the ferroalloy industry, contains various components and holds significant importance for sustainable development through its valorization. This study focuses on characterizing a manganese sludge and investigating its behavior during sulfuric acid leaching. The influence of process conditions, including temperature, acid concentration, liquid to solid ratio, and leaching duration, was examined. The results revealed that Mn, Zn, and K are the main leachable components, and their overall leaching rates increase with increasing temperature, liquid to solid ratio, and time. However, the acid concentration requires optimization. High leaching rates of 90% for Mn, 90% for Zn, and 100% for K were achieved. Moreover, it was found that Pb in the sludge is converted to sulfate during the leaching, which yields a sulfate concentrate rich in $PbSO_4$. The leaching process for Mn and Zn species appears to follow a second or third order reaction, and the calculation of rate constants indicated that Mn leaching kinetics are two to five times higher than those for Zn. Thermodynamic calculations were employed to evaluate the main chemical reactions occurring during leaching.

Keywords: manganese; sludge; zinc; lead; potassium; sulfuric acid; ferromanganese; submerged arc furnace; kinetics; manganese carbonate

Citation: Safarian, J.; Eini, A.S.; Pedersen, M.A.E.; Haghdani, S. The Effect of Process Conditions on Sulfuric Acid Leaching of Manganese Sludge. *Materials* **2023**, *16*, 4591. https://doi.org/10.3390/ma16134591

Academic Editor: Daniela Fico

Received: 26 May 2023
Revised: 21 June 2023
Accepted: 22 June 2023
Published: 25 June 2023

1. Introduction

Manganese (Mn) is an important element widely employed in both ferrous and non-ferrous metallurgy, mainly in its metallic form, to produce various alloys. The compounds of Mn have numerous applications, and one particularly significant use is the incorporation of manganese IV oxide (MnO_2) in battery manufacturing. The main processing route for manganese ores involves their utilization in the production of manganese ferroalloys through a carbothermic reduction process in a submerged arc furnace (SAF) [1]. Previous studies have investigated the thermochemistry of the SAF process and the underlying reaction mechanisms involved in the production of Mn metal [2]. In the SAF process, the furnace charge mainly consists of Mn oxides, which are largely transformed into the Mn ferroalloy product after reduction. In addition, a smaller portion of Mn ends up being processed into slag as a byproduct [3]. Furthermore, the off-gas furnace carries fine particles (dust), which are separated from the gas using dedusting techniques [4,5]. When wet scrubbers are employed to capture the dust, the wet particles are separated from the gas through water spray, resulting in the formation of a sludge known as Mn sludge [5]. This Mn sludge contains significant amounts of valuable elements, including Mn, Zn, and Pb, making its valorization crucial for sustainable development and the circular economy. A study by Shen et al. [6] has characterized manganese furnace dust regarding the charge materials and indicated significant amounts of Mn and Zn in the dust. It is worth noting that the recycling of Mn sludge into the SAF process is difficult due to the presence of volatile components, such as Zn and K. Currently, this sludge is considered an industrial hazardous waste and is landfilled, emphasizing the importance of its valorization. Pyrometallurgical processing of SAF dust/sludge to recycle it into the SAF process has technical and economic limitations. However, there is growing interest in the application of hydrometallurgical processes, which offer potential profitability through the extraction of valuable metals [7].

Extensive research has been conducted on the utilization of hydrometallurgical processes for extracting Mn from both primary and secondary raw materials. The production of electrolytic Mn metal (EMM) and electrolytic Mn dioxide (EMD) through hydrometallurgical processing of Mn ores are mature processes. As mentioned earlier, the Mn yield in the SAF process is relatively low, with a portion of Mn lost mostly in the slag byproduct and dust. Therefore, there has been considerable interest in recovering Mn from these streams. Various studies in the literature have investigated the recovery of Mn from Mn-containing slags using different acids, such as H_2SO_4 [8–13]. Notably, high Mn recovery rates of 88% and 90% have been reported. Although the generation of Mn sludge is smaller compared to the slag byproduct in the SAF process, it represents a more complex material. The Mn sludge contains a significant amount of Mn, along with various highly volatile substances found in the SAF charge, such as Zn, Pb, K, Cd, etc. Sancho et al. [14] studied the extraction of electrolytic Mn metal from Mn sludge. Their process involved primary leaching using H_2SO_4, purification steps, and an electrolysis phase for electrolytic manganese production, resulting in the production of high-purity Mn metal. Similarly, Ghafarizadeh et al. [15] focused on the reductive leaching of SAF dust from the ferromanganese process using sulfuric acid, with additional reducing agents, such as oxalic acid, hydrogen peroxide, and glucose. They observed a Mn leaching rate of up to 40% using 0.5 to 3 M H_2SO_4 solutions at 70 °C within 90 min, with a liquid to solid ratio (L/S) of 10. Furthermore, they found that the presence of oxalic acid and hydrogen peroxide significantly increased the Mn leaching recovery (almost complete recovery), while glucose had a lesser effect. Sadeghi et al. [16] conducted the reductive acid leaching of a Mn residue using 1 M H_2SO_4. They employed glucose as the reducing agent, which resulted in high leaching rates of over 90% for Mn and Zn. Additionally, they observed that microwave-assisted leaching indicates a faster leaching rate compared to ultrasound-assisted and conventional leaching techniques.

In the present study, the leaching behavior of a Mn sludge using H_2SO_4 solutions is investigated through experimental work, aiming to evaluate the effect of process conditions. In hydrometallurgical processes used to produce Mn and Zn metals, sulfuric acid has been used as the leaching agent. Since the Mn sludge has a significant amount of these two metals, in the present study, sulfuric acid leaching was considered as the leaching agent to use. Various materials characterization techniques are employed to gain insights into the Mn sludge. Furthermore, thermodynamic calculations are performed to further understand the acid leaching process of Mn sludge.

2. Materials and Methods

2.1. Materials and Preparation

The manganese sludge used in this study was received from the ferromanganese industry and contained a significant amount of water. To minimize errors in mass measurement, the sludge was dried in an oven at approximately 90 °C for a period of two days. Sulfuric acid (H_2SO_4) solutions with concentrations of 0.7, 1.2, and 1.7 molar (M) were prepared using the standard methods of acid dilution by adding distilled water. Additionally, concentrated sulfuric acid (98.3% H_2SO_4) was diluted.

2.2. Leaching Experiments

The leaching experiments were performed in a beaker using a magnetic stirrer with a rotation speed of 350 rpm on a hot plate. Once the H_2SO_4 solution reached the target temperature, the Mn sludge powder was gradually added to the stirred solution. In all experiments, a precise amount of 10 ± 0.05 g of dried sludge powder was used. Different volumes of leaching solutions, namely 12.5, 25, 50, and 100 mL, were employed to achieve liquid to solid ratios (L/S) of 1.25, 2.5, 5, and 10 mL/g, respectively. To investigate the effect of temperature, six experiments were conducted at 30 °C and 50 °C, employing three different L/S ratios under fixed leaching conditions using a 1.7 M H_2SO_4 solution. The remaining experiments were carried out at 50 °C. To study the impact of leaching duration, leaching tests were performed for durations ranging from 10 to 120 min using 1.2 and 1.7 M H_2SO_4 solutions. After the leaching process, the leaching residue was separated by filtration using a Büchner funnel. The residue was then washed with distilled water, dried in an oven, and its mass was measured. The pH of the solution was measured for selected experiments both before and after the leaching process.

2.3. Characterization of Materials and Products

The chemical compositions of the dried sludge powder and leaching residues were measured using X-ray fluorescence (XRF) with a device from Thermo Fisher, Degerfors, Sweden. For XRF testing, sample preparation was conducted using the flux fusion method. Phase analysis of these materials was performed by X-ray diffraction (XRD) using the Bruker D8 A25 DaVinciTM, Karlsruhe, Germany, with CuK radiation (wavelength of 1.54 Å). The measurement range was set from 10 to 80° with a step size of 0.03. It is worth mentioning that XRD analysis was employed to study all the leaching residues. In addition, selected leaching solutions were analyzed using inductively coupled plasma-mass spectrometry (ICP-MS). Furthermore, the leaching residue samples were characterized by scanning electron microscope (SEM) (Zeiss Ultra FESEM, National Institute of Standards and Technology, Gaithersburg, MD, USA) equipped with an XFlash® 4010 Detector supplied by the Bruker Corporation (Billerica, MA, USA) for energy-dispersive X-ray spectroscopy (EDS).

3. Results

3.1. The Characteristics of Mn Sludge

Table 1 displays the XRF analysis for the Mn sludge, showing a significant amount of Mn. Moreover, Figure 1 illustrates the XRD spectrum for the Mn sludge, which indicates that the Mn exists mainly in the form of a carbonate compound as rhodochrosite ($MnCO_3$). Previous studies have also reported the presence of $MnCO_3$ in this type of sludge [5]. Therefore, in the XRF analysis presented in Table 1, this compound was specifically considered, while the most stable oxides were considered for the other metals. It is important to note that due to the high intensity of the diffracted X-ray signal from $MnCO_3$, the identification of other minor phases was not possible.

Table 1. Chemical composition of Mn sludge measured by XRF (wt%).

$MnCO_3$	Fe_2O_3	MgO	SiO_2	Al_2O_3	CaO	TiO_2	BaO	ZnO	PbO	Na_2O	K_2O	P_2O_5	CdO	SO_3
63.65	1.92	0.61	1.75	1.14	0.96	0.03	0.09	20.08	1.66	2.88	4.28	0.09	0.09	0.79

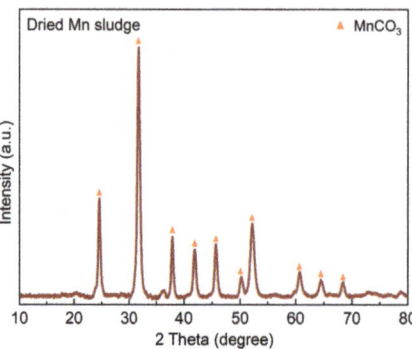

Figure 1. XRD spectrum of the dried Mn sludge with the identified phase.

3.2. Leachability of Mn Sludge

3.2.1. Effect of Temperature

Figure 2 presents the measured chemical compositions of the leaching solutions, obtained using 1.7 M H$_2$SO$_4$ solution for three L/S ratios at 30 °C and 50 °C. The analysis reveals that Mn, Zn, and K are the main elements being dissolved into the solution, followed by Na, Mg, etc. Clearly, higher L/S ratios result in lower concentrations of Mn, Zn, and K in the solutions, indicating that for the lowest L/S ratio, significant portions of these elements are leached. Moreover, increasing the leaching temperature from 30 °C to 50 °C generally leads to a higher leaching rate of the elements.

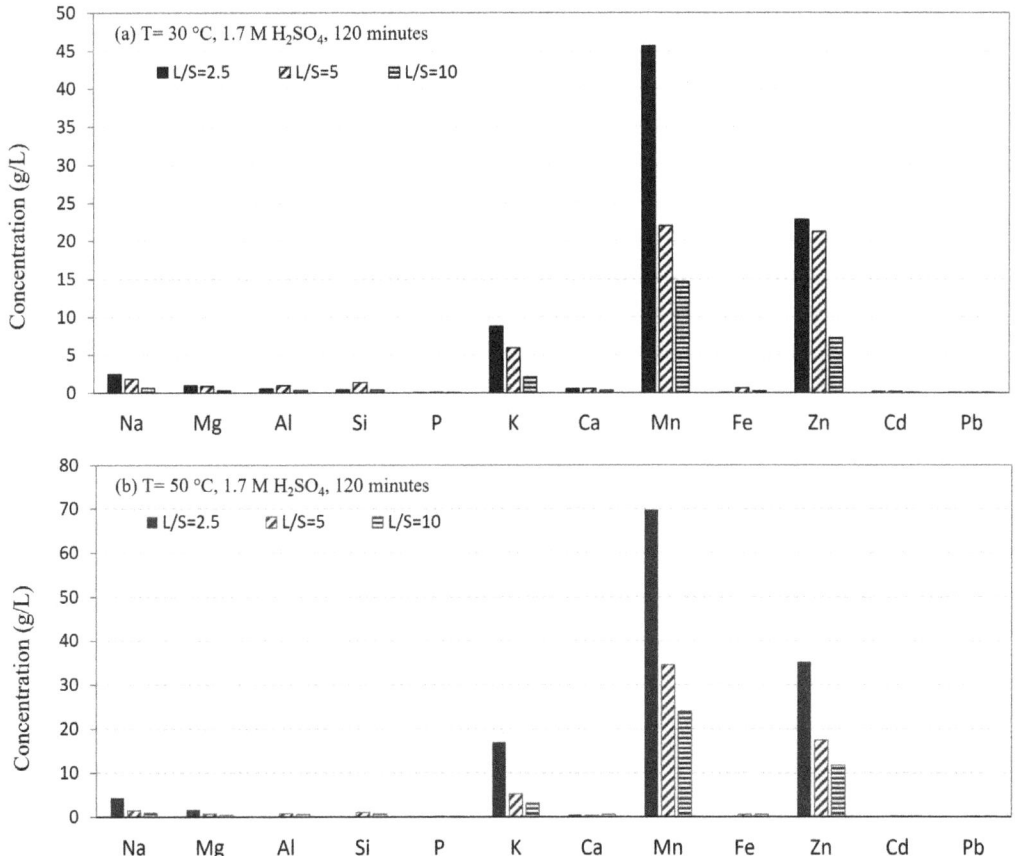

Figure 2. The concentrations of elements in solutions after leaching at different L/S ratios and temperatures using 1.7 M H$_2$SO$_4$ solution at (**a**) 30 °C and (**b**) 50 °C.

To gain a better understanding of the temperature effect, the recovery rates of Mn, Zn, and K were calculated for different L/S ratios, as illustrated in Figure 3. The results demonstrate that as the temperature rises from 30 °C to 50 °C, the recovery of metals into the leachates increases, considering the slightly lower accuracy observed for L/S = 5. Therefore, the temperature was fixed at 50 °C for all the other experiments.

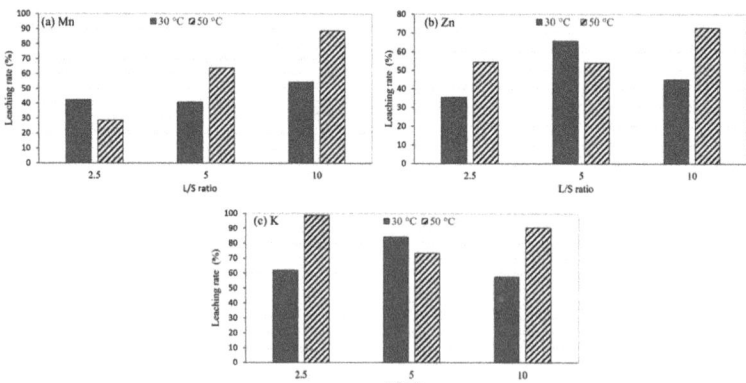

Figure 3. The leaching rates for different temperatures and L/S ratios of (**a**) Mn, (**b**) Zn, and (**c**) K based on XRF measurements.

3.2.2. Effect of Solution Volume and Concentration

The leaching behavior of the sludge is influenced by the L/S ratio and acid concentration, as displayed in Figure 4. It is worth mentioning that the L/S ratio during the leaching process was not constant due to gas release via the chemical reactions involved and the formation of bubbles. Hence, the L/S ratio values that are presented are the numbers at initial. It is evident that increasing the L/S ratio for a given acid concentration leads to a higher leaching rate, with Mn and Zn recoveries reaching approximately 90% and 70%, respectively. Figure 4a indicates that different acid concentrations result in the leaching of most of the Mn. Higher acid concentrations lead to increased Mn leaching rates at fixed leaching durations for each L/S ratio. However, for higher L/S ratios, the Mn leaching rate becomes less dependent on the acid concentration. A similar trend is observed for the leaching of Zn from the sludge (Figure 4b), where 66% to 77% of Zn is dissolved using 0.7 M to 1.7 M acid solutions. Figure 4c shows that the dissolution of K during the leaching process is consistently higher than that of Mn and Zn, following the same trend with changes in the L/S ratio. Moreover, the differences in the K leaching rates for different acid concentrations at a given L/S ratio are smaller compared to Mn and Zn leaching rates.

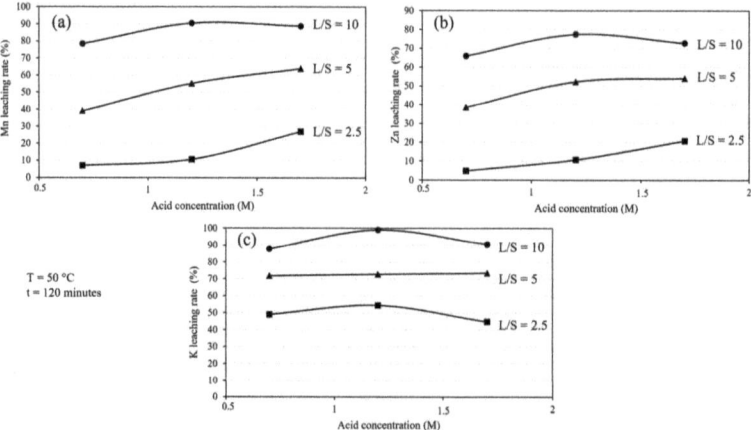

Figure 4. The effect of L/S ratio and acid concentration on the leaching rates of (**a**) Mn, (**b**) Zn, and (**c**) K at 50 °C and duration of 120 min, based on XRF measurements.

3.2.3. Effect of Leaching Duration

Figure 5 shows the effect of leaching duration on the dissolution of Mn (a), Zn (b), and K (c) from the sludge under fixed temperature and L/S ratio. The experimental data consistently indicate a rapid initial stage of leaching, followed by a much slower stage.

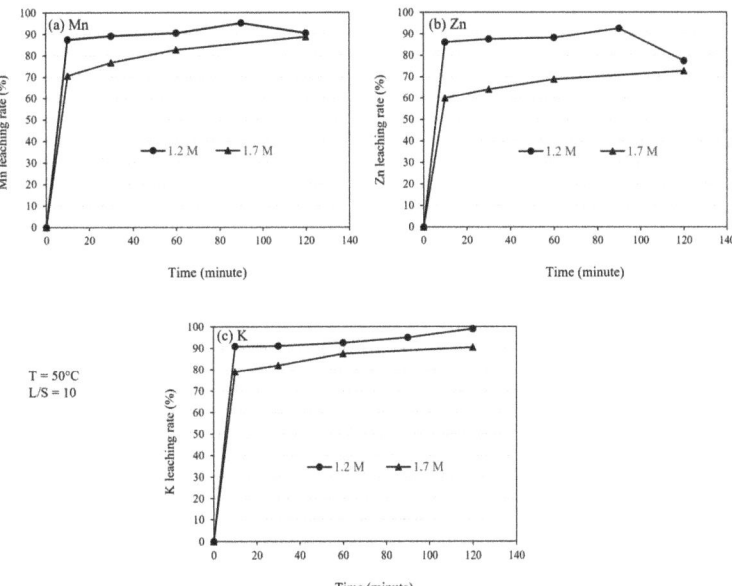

Figure 5. The leaching rates of (**a**) Mn, (**b**) Zn, and (**c**) K using different acid concentrations, namely 1.2 M and 1.7 M H_2SO_4 solutions at 50 °C for L/S ratio of 10.

As illustrated in Figure 5, a significant portion of Mn, Zn, and K is leached within the first 10 min. Furthermore, the initial leaching rate is high and not significantly dependent on the acid concentration. However, the leaching extent is higher for the 1.2 M solution compared to the 1.7 M solution. Nevertheless, this difference becomes insignificant after 120 min of leaching. Obtaining a lower leaching rate for Mn and Zn with 1.2 M solution in the sample leached for 120 min than that leached for 90 min is difficult to explain, and the measured concentrations may be outliers, thus further experimental work is needed to clarify this area.

The pH measurements reveal an increase in the solution pH during the leaching process, with greater pH changes observed when lower acid volumes are used. For example, when using a 1.7 M solution, the pH increases from approximately −0.23 to approximately 5 for L/S ratios of 1.25 and 2.5, while it reaches approximately 5 for an L/S ratio of 5. In several experiments using a 1.7 M solution at L/S = 10, the pH increases within a range from 0.16 to 0.44. The final pH for 0.7 and 1.2 M H_2SO_4 concentrations is also approximately five. However, since the pH changes were not continuously measured, they cannot be used to study the leaching rate.

3.3. The Phase Analysis of Residue

Figure 6 displays the XRD spectra of the leaching residues obtained from experiments conducted at different leaching durations using a 1.2 M H_2SO_4 solution and an L/S ratio of 10 in comparison to the Mn sludge. The results show that during the leaching process, $MnCO_3$ from the sludge is leached, while lead sulfate ($PbSO_4$) simultaneously forms. The XRD analysis data presented in Figure 6 confirm that the presence of other crystalline residue components is not significant compared to these two phases.

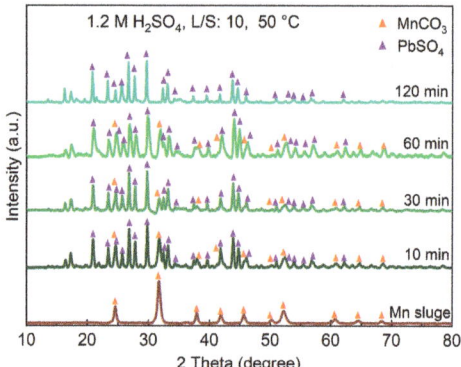

Figure 6. The XRD spectra of leaching residue samples at different leaching durations using 1.2 M H_2SO_4 solution and an L/S of 10 at 50 °C.

The identified phases in the leaching residues obtained using various acid concentrations and L/S ratios are shown in 7. When a 0.7 M acid solution is used, a significant amount of $MnCO_3$ remains unbleached, even at higher L/S ratios, which is consistent with the previously presented results regarding the recovery of Mn into the solutions. Comparing the XRD spectra in Figure 7 indicates that the $PbSO_4$ phase appears in the sludge leached at L/S = 2.5 when a 1.7 M acid solution is used, while for lower acid concentrations, $PbSO_4$ is observed at L/S = 5. Moreover, for this L/S ratio, more intense $PbSO_4$ peaks and correspondingly weaker $MnCO_3$ peaks are observed when a stronger acid is used. The XRD spectra of the residue samples obtained from leaching conditions with L/S = 10 and a 1.7 M acid solution were quite similar to the sample leached with a 1.2 M solution for L/S = 10, presented in Figure 6, and they all exhibit $PbSO_4$ as the main phase.

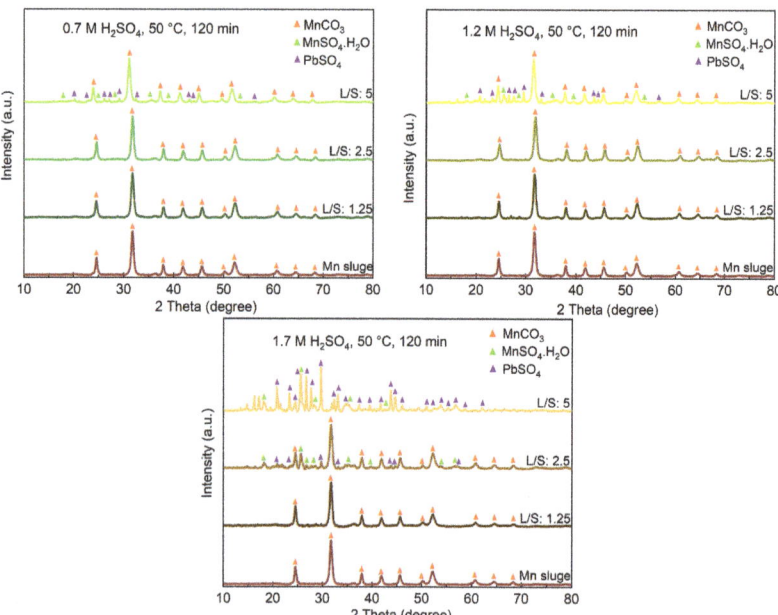

Figure 7. The XRD spectra of leaching residues obtained in leaching with different acid concentrations and L/S ratios.

3.4. The Microstructural Analysis of Residue

Figure 8 presents the elemental X-ray mapping of a representative leaching residue, demonstrating a high Mn recovery rate of approximately 89%. The figure illustrates that, during the leaching process, both small particles and relatively large agglomerates can be formed. The agglomerates exhibit a significant amount of Pb and S, which are closely associated, confirming the dominance of the $PbSO_4$ phase as the main compound in the residue. Additionally, a distinct layer of $PbSO_4$ is observed on the surface of the agglomerates. The distribution of Mn, Fe, Al, Si, and Ca within the sample is not uniform, whereas Zn, K, and Na are observed to be distributed throughout the sample.

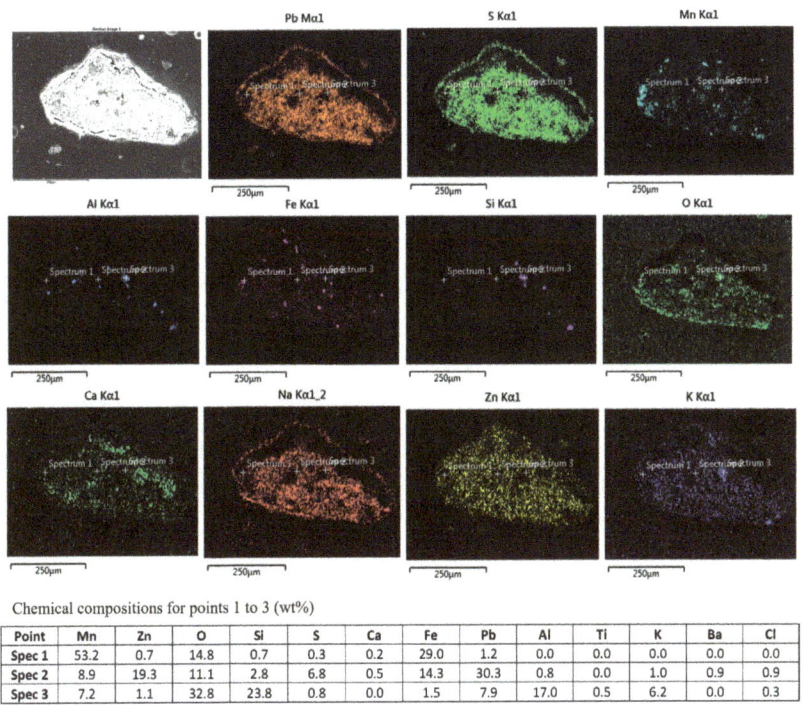

Chemical compositions for points 1 to 3 (wt%)

Point	Mn	Zn	O	Si	S	Ca	Fe	Pb	Al	Ti	K	Ba	Cl
Spec 1	53.2	0.7	14.8	0.7	0.3	0.2	29.0	1.2	0.0	0.0	0.0	0.0	0.0
Spec 2	8.9	19.3	11.1	2.8	6.8	0.5	14.3	30.3	0.8	0.0	1.0	0.9	0.9
Spec 3	7.2	1.1	32.8	23.8	0.8	0.0	1.5	7.9	17.0	0.5	6.2	0.0	0.3

Figure 8. The X-ray mapping of elements in an agglomerate of the leaching residue, and EDS point analysis for selected particles.

The distribution of Mn, Fe, and O as well as the analysis of point 1 in Figure 8 indicates that Fe, Mn, and O are accumulated in some particles. Similarly, the distribution of Al, Si, and O, as observed in point 3, demonstrates a similar pattern for these elements. On the other hand, the analysis of point 2 reveals the coexistence of all major elements, including Pb, Zn, Fe, Mn, S, and O, in certain areas of the sample.

4. Discussion

4.1. Mn Sludge Composition

Obtaining an overview of the composition and formation of Mn sludge components is important to understand the leaching behavior of the sludge. The minor oxides components of sludge, such as SiO_2, Al_2O_3, CaO, and MgO (Table 1) originated from the SAF charge via the transportation of fine particles (dust) by the furnace gas. However, Mn, Zn, and K components end up in the sludge via a different mechanism with chemical reactions involved. Mn in the sludge was found in the form of $MnCO_3$, as shown in Figure 1, whereas Mn ores used in the ferroalloy industry are typically in the form of oxides (e.g., MnO_2, Mn_2O_3) or silicates (e.g., $MnSiO_3$). The presence of $MnCO_3$ in the submerged arc off-gas

Materials **2023**, *16*, 4591

furnace can be attributed to the significant amounts of CO and CO_2 gases generated during the furnace operation. These gases are produced from the charged solid carbon reductant through the carbothermic reduction of MnO by C, the gaseous reduction of higher Mn oxides by CO gas, and the Boudouard reaction, as discussed previously [2]. In addition, carbonate compounds present in the furnace charge, such as $CaCO_3$, $MgCO_3$, and K_2CO_3 are thermally decomposed at elevated temperatures, resulting in the generation of CO_2 gas within the furnace burden. Given the significant amount of CO in the SAF gas outlet, it is expected that the Mn-containing compound leaving the furnace is mainly in the form of MnO at temperatures ranging from 200 to 400 °C. The decrease in gas temperature causes an increase in the CO_2 content, leading to the conversion of MnO to $MnCO_3$. This change in gas composition results from the conversion of CO to CO_2. Consequently, $MnCO_3$ is formed through the following reaction:

$$MnO + CO_2 = MnCO_3 \tag{1}$$

By calculating the changes in Gibbs energy with temperature for various CO_2 partial pressures for chemical reaction (1), it was found that the formation of $MnCO_3$ occurs at temperatures below 340 °C, 335 °C, 325 °C, 320 °C, and 300 °C for CO_2 partial pressures of 1 atm, 0.8 atm, 0.6 atm, 0.4 atm, and 0.2 atm, respectively. Clearly, the reaction temperature is not significantly affected by the CO_2 partial pressure, and the reaction proceeds at temperatures below 300 °C. Similar reactions may take place for some minor components present in the dust, such as K and Zn, leading to the formation of their carbonates. The presence of these phases was investigated previously [5]. The studied sludge contained a high amount of Zn, which could be present in the forms of ZnO and $ZnCO_3$. However, these compounds were not detected in the XRD analysis, possibly because ZnO exists in an amorphous form, resulting from its formation from the gas phase. The amount of K in the sludge is much lower than Mn and Zn, while higher leaching rates for K are observed in comparison with Mn and Zn (Figure 5). This may indicate that it has a higher reactivity with sulfuric acid and a large portion of it is dissolved rapidly. It is noted that this may confirm good contact between the solid fine particles with the solution during the leaching process.

4.2. The Kinetics of Leaching

Mn and Zn are the most valuable metals in the sludge due to their abundance and leachability. The leaching process of these elements from the sludge takes place in two stages: an initial fast stage followed by a slower secondary stage, as shown in Figure 5. To determine the reaction rate constant, the conversion fractions of $MnCO_3$ to Mn^{2+} (α) and ZnO to Zn^{2+} were calculated based on the leachability data for 1.2 and 1.7 M solutions at different durations. The α-values were tested for different chemical reaction orders, and it was found that the first-order reaction does not fit the leaching process for Mn and Zn. However, the second and third-order reactions, described by Equations (2) and (3), respectively, provide a better fit to the data, as demonstrated in Figure 9, with the exception of one outlier measurement for the 1.7 M solution. It is important to note that the vertical axis values in Figure 9 represent the α functions given by the following equations:

$$F_2(\alpha) = \frac{1}{1-\alpha} - 1 = k_2 t, \tag{2}$$

$$F_3(\alpha) = 0.5 \left(\frac{1}{(1-\alpha)^2} - 1 \right) = k_3 t \tag{3}$$

where k_2 and k_3 are the reaction rate constants, and t is the leaching time.

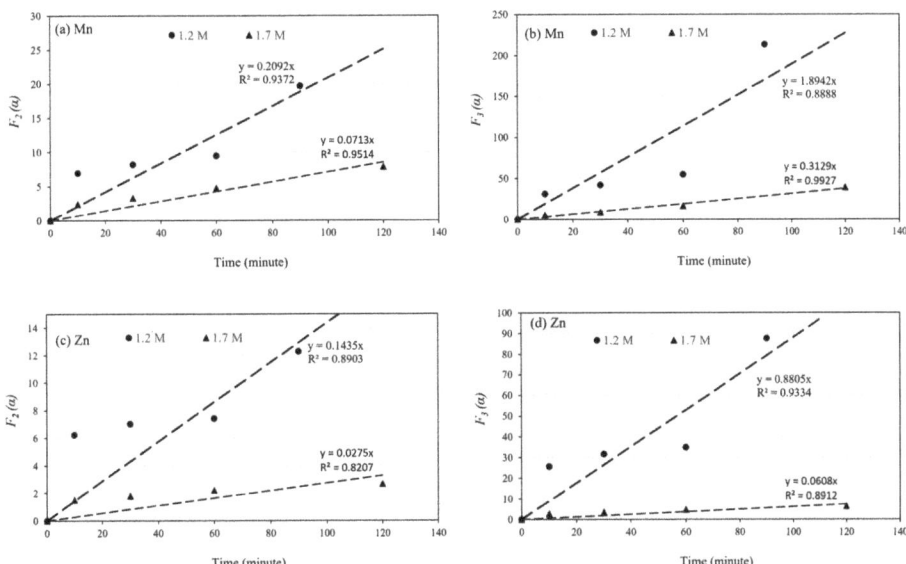

Figure 9. The relationship between F_2 (α) and F_3 (α) with leaching time for two different acid concentrations of 1.2 and 1.7 M. Mn: (**a**,**b**), Zn: (**c**,**d**).

Based on the above equations, the slope of the fitted lines in Figure 9 represents the reaction rate constants k_2 and k_3. These values were calculated and presented in Table 2. The obtained rate constant values show that the leaching rate of Mn is 3 to 6 times higher for the 1.2 M solution compared to the 1.7 M solution. Similarly, the leaching rate of Zn is 4 to 15 times higher for the 1.2 M solution compared to the 1.7 M solution. This observation may explain the reason for the higher leaching rate of Mn in the 1.2 M solution compared to the 1.7 M solution for an L/S ratio of 10, as shown in Figures 4 and 5. Furthermore, comparing the rate constant values for Mn and Zn leaching indicates that the leaching of Mn is 2 to 5 times faster than the leaching of Zn. However, the difference in leaching rates between Mn and Zn is greater for the 1.7 M solution compared to the 1.2 M solution.

Table 2. Rate constant values for Mn and Zn leaching from the sludge for different acid concentrations.

Solution Concentration	Rate Constant k_2 (s^{-1})		Rate Constant k_3 (s^{-1})	
	Mn	Zn	Mn	Zn
1.2 M	3.5×10^{-3}	2.4×10^{-3}	3.16×10^{-2}	1.47×10^{-2}
1.7 M	1.2×10^{-3}	5×10^{-4}	5.2×10^{-3}	1.0×10^{-3}

4.3. Thermochemistry of Reactions

The results showed a significant mass transport of Mn and Zn from the sludge into the solution during leaching. The leaching chemistry of these elements is discussed as follows.

4.3.1. Mn Leaching

It was observed that the majority of Mn in the sludge exists in the form of $MnCO_3$. During the leaching process, Mn is leached, resulting in an increase in pH and the consumption of acid. On the other hand, significant gas evolution occurs during leaching, suggesting the following reaction:

$$MnCO_3(s) + H_2SO_4(aq) = MnSO_4(aq) + H_2O(l) + CO_2(g). \tag{4}$$

Clearly, the generation of CO_2 gas within the system is the main mechanism for bubble formation. The formation of bubbles combined with the applied magnetics stirring may explain the observed rapid leaching rates of Mn, as shown in Figure 5 The Eh–pH diagram of the Mn–S–H_2O system, depicted in Figure 10 using FactSage thermodynamic software, illustrates that Mn^{2+} and $MnSO_4$(aq) are formed during leaching, considering the above reaction. The final pH of the solution was measured within the range of 0.16 to 5, indicating that the solution is within the Mn^{2+} and $MnSO_4$(aq) region.

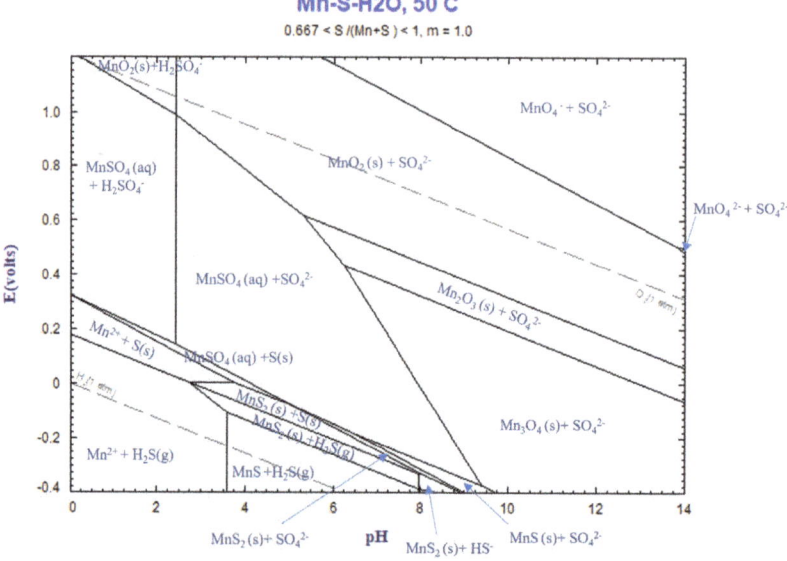

Figure 10. Eh—pH diagram for Mn–S–H_2O system at 50 °C, calculated by FactSage thermodynamic software version 8.2.

The highest obtained Mn leaching rate of about 90% (Figures 4 and 5), the shape of the leaching curves in Figure 5, and observing no $MnCO_3$ phase in the samples leached by 1.2 M and 1.7 M solutions at L/S = 10 collectively indicate that the majority of $MnCO_3$ has been leached. It is possible that the remaining Mn has not had proper contact with the acid due to the formation of $PbSO_4$ on sludge agglomerates (Figure 8) that causes a further reduction in the Mn leaching rate, a kind of passivation. The other explanation is that a small portion of Mn (less than 10 wt%) exists in other amorphous minor phases (not $MnCO_3$) that were not observed in XRD spectra and such phases may not be leachable.

4.3.2. Zn Leaching

As mentioned earlier, Zn in the sludge exists in the form of ZnO and $ZnCO_3$. In the presence of an H_2SO_4 solution, these compounds dissolve through the following reactions:

$$ZnO(s) + H_2SO_4(aq) = ZnSO_4(aq) + H_2O. \tag{5}$$

$$ZnCO_3(s) + H_2SO_4(aq) = ZnSO_4(aq) + H_2O(l) + CO_2(g). \tag{6}$$

The Eh–pH diagram for the Zn–S–H_2O system in Figure 11 illustrates that the aqueous $ZnSO_4$ remains in a stable phase within the range of final pH values measured between 0.16 and 5. Consequently, the chemical reaction (2) takes place, leading to the dissolution of Zn in the sludge. Like chemical reaction (2), the formation of CO_2 gas through chemical reaction (2) may contribute to the formation of gas bubbles during leaching.

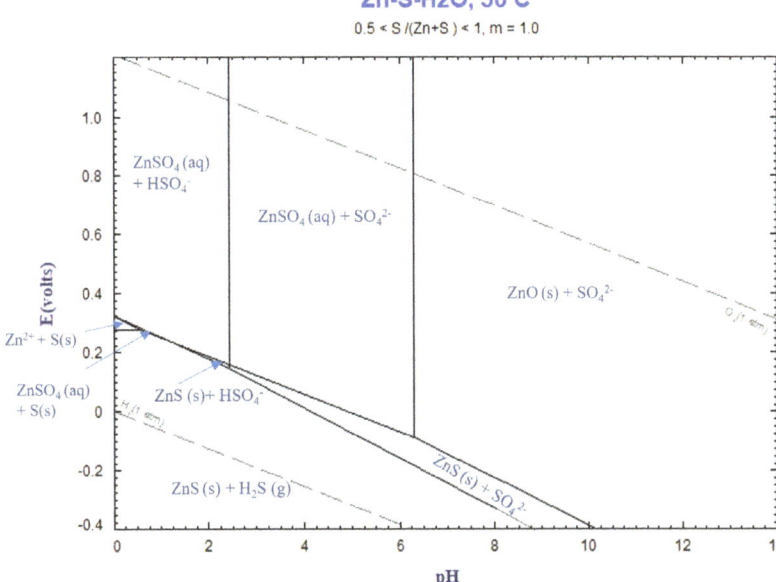

Figure 11. Eh–pH diagram for Zn–S–H$_2$O system at 50 °C, calculated by FactSage thermodynamic software version 8.2.

Chemical reactions (5) and (26) exhibit significant progress when using high concentrations of acids and higher L/S ratios. For example, when employing 1.2 M and 1.7 M acid solutions with an L/S ratio of 10, the acid content is sufficient to leach Mn and Zn significantly. However, when using a 1.2 M acid solution with L/S ratios of 1.25 and 2.5, it is theoretically possible to leach up to 30% and 60% of Mn, respectively. Nevertheless, due to the simultaneous leaching of Zn and K, the leaching of Mn is lower than the theoretical values. On the other hand, as the system approaches equilibrium, the leaching does not proceed further. This explains the existence of unbleached sludge at lower L/S ratios, as depicted in Figure 7. Based on the obtained results, the best leaching conditions can be achieved by using acid solutions with concentrations of 1.2 M and higher in combination with high L/S ratios.

Figure 5 indicates that the initial leaching rates of Mn and Zn with 1.2 M and 1.7 M solutions are similar for the L/S = 10. However, the extent of leaching in the initial stage and further leaching kinetics in the second slow stage are higher for the 1.2 M solution than the 1.7 M solution. These differences may indicate that the concentration of sulfuric acid reactants above 1.2 M solution is not important for the process kinetics. Hence, we may suggest that the leaching process is more dependent on the mass transport of the chemical reaction products away from the reaction surface. Obviously, the mass transport of MnSO$_4$(aq) and ZnSO$_4$(aq) is faster in the more dilute solution of 1.2 M solution than that for the 1.7 M solution, and this may explain the above observations.

4.3.3. Lead Sulfate Formation

In all the leaching experiments, the formation of solid PbSO$_4$ and the insignificant dissolution of Pb into the solution were observed. This can be attributed to the stability of solid PbSO$_4$ in the system, as predicted by the Eh–pH diagram for the Pb–S–H$_2$O system in Figure 12. The final pH values of all the experiments ranged from 0.16 to approximately 5, within which solid PbSO$_4$ remains in a stable phase.

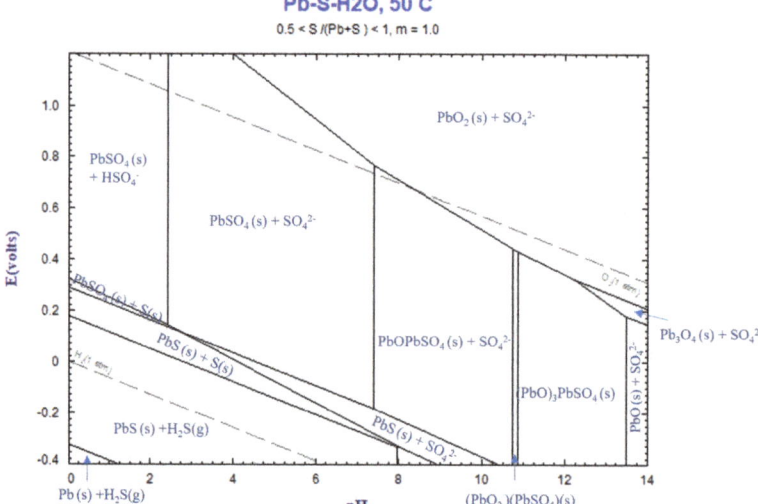

Figure 12. Eh–pH diagram for Pb–S–H$_2$O system at 50 °C, calculated by FactSage thermodynamic software version 8.2.

In the presence of sulfuric acid with the sludge, the formation of PbSO$_4$ can be explained by the following reaction:

$$PbO(s) + H_2SO_4(aq) = PbSO_4(s) + H_2O. \tag{7}$$

The complete conversion of PbO to PbSO$_4$ during leaching is evident from the intense peaks of PbSO$_4$ observed in the XRD spectra of several residue samples. Clearly, the other components of unbleached sludge are found in significantly lower amounts in samples with high recoveries of Mn and Zn. The extensive progress of reaction (2) can be attributed to the particle size of PbO in the sludge, which is expected to be very fine. These particles likely condense from the gas phase as Pb gas exits SAF and subsequently oxidizes to PbO (which has a high melting point of 888 °C) before condensation.

5. Conclusions

The leaching behavior of a manganese sludge using sulfuric acid solutions was investigated, and the main conclusions are as follows:

- The leaching rate of Mn sludge is fast, resulting in high leaching rates of 90% for both Mn and Zn.
- The leaching rates of Mn, Zn, and K increase with increasing temperature, liquid to solid ratio, and duration. Mn exhibits a higher maximum leaching rate compared to Zn.
- Comparing acid solutions, the leaching rates of Mn and Zn are slightly higher using a 1.2 M H$_2$SO$_4$ solution compared to a 1.7 M H$_2$SO$_4$ solution, while the leaching rates are much lower with a 0.7 M acid solution.
- Kinetic studies of leaching with 1.2 M and 1.7 M H$_2$SO$_4$ acid solutions reveal that the reaction rate constant for Mn dissolution exceeds 1.2×10^{-3} s^{-1}, while for Zn dissolution, it is higher than 5×10^{-4} s^{-1}.
- During the leaching process, aqueous sulfate solutions of MnSO$_4$ and ZnSO$_4$ are formed from the dissolution of solid MnCO$_3$ and ZnO (or ZnCO$_3$) present in the sludge. This process is accompanied by the generation of CO$_2$ gas and bubble formation.
- The Pb content in the sludge is fully converted to solid PbSO$_4$ during leaching, and samples with high Mn and Zn leaching rates yield a residue of lead sulfate.

Author Contributions: Conceptualization, J.S.; methodology, J.S., A.S.E., M.A.E.P. and S.H.; software, J.S., A.S.E., M.A.E.P. and S.H.; validation, J.S., A.S.E., M.A.E.P. and S.H.; formal analysis, J.S., A.S.E., M.A.E.P. and S.H.; investigation, J.S., A.S.E., M.A.E.P. and S.H.; resources, J.S.; data curation, J.S., A.S.E., M.A.E.P. and S.H.; writing—original draft preparation, J.S., A.S.E., M.A.E.P. and S.H.; writing—review and editing, J.S. and S.H.; visualization, J.S., A.S.E., M.A.E.P. and S.H.; supervision, J.S.; project administration, J.S.; funding acquisition, J.S. All authors have read and agreed to the published version of the manuscript.

Funding: This research was funded by the SFI Metal Production (Centre for Research-based Innovation, grant number 237738). The APC was also funded by the SFI Metal Production.

Institutional Review Board Statement: Not applicable.

Informed Consent Statement: Not applicable.

Data Availability Statement: Data is available in a publicly accessible repository that does not issue DOIs. Publicly available datasets were analyzed in this study. This data can be found at https://www.ntnu.edu/metpro (accessed on 20 May 2023).

Acknowledgments: The authors gratefully acknowledge the financial support from the Research Council of Norway and the partners of the SFI Metal Production.

Conflicts of Interest: The funders had no role in the design of the study; in the collection, analysis, or interpretation of data; in the writing of the manuscript; or in the decision to publish the results. On behalf of all the authors, the corresponding author states that there is no conflict of interest.

References

1. Olsen, S.E.; Tangstad, M.; Lindstad, T. *Production of Manganese Ferroalloys*; Sintef: Trondheim, Norway; Academic Press: New York, NY, USA, 2007.
2. Safarian, J. Duplex Process to Produce Ferromanganese and Direct Reduced Iron by Natural Gas. *ACS Sustain. Chem. Eng.* **2021**, *9*, 5010–5026. [CrossRef]
3. Safarian, J. Kinetics and Mechanisms of Reduction of MnOcontaining Silicate Slags by Selected forms of Carbonaceous Materials. Ph.D. Thesis, NTNU, Trondheim, Norway, 2007.
4. Gaal, S.; Tangstad, M.; Ravary, B. Recycling of waste materials from the production of FeMn and SiMn. *Twelfth Int. Ferroalloys Congr. Hels. Finl.* **2010**, 81–88.
5. Ravary, B.; Hunsbedt, L. Progress in recycling sludge from off-gas cleaning of manganese alloy furnaces. In Proceedings of the thirteenth International Ferroalloys Congress Efficient technologies in ferroalloy industry, Almaty, Kazakhstan, 9–13 June 2013; pp. 1023–1027.
6. Shen, R.; Zhang, G.; Dell'Amico, M.; Brown, P.; Ostrovski, O. Characterisation of Manganese Furnace Dust and Zinc Balance in Production of Manganese Alloys. *ISIJ Int.* **2005**, *45*, 1248–1254. [CrossRef]
7. Binnemans, K.; Jones, P.T.; Fernández, M.; Torres, V.M. Hydrometallurgical Processes for the Recovery of Metals from Steel Industry By-Products: A Critical Review. *J. Sustain. Met.* **2020**, *6*, 505–540. [CrossRef]
8. Mohanty, J.K.; Sahoo, P.K.; Nathsarma, K.C.; Panda, D.; Paramguru, R.K. Characterization and leaching of ferromanganese slag. *Mining, Met. Explor.* **1998**, *15*, 30–33. [CrossRef]
9. Baumgartner, S.J. Groot. The recovery of manganese products from ferromanganese slag using a hydrometallurgical route. *J. South. Afr. Inst. Min. Metall.* **2014**, *114*, 331–340.
10. Groot, D.; Kazadi, D.; Pollmann, H.; de Villiers, J.; Redtmann, T.; Steenkamp, J. The recovery of manganese and generation of a valuable residue from ferromanganese slags by a hydrometallurgical route. In Proceedings of the Thirteenth International Ferroalloys Congress Efficient Technologies in Ferroalloy Industry, Almaty, Kazakhstan, 9–13 June 2013; pp. 1051–1059.
11. Kazadi, D.; Groot, D.; Pöllmann, H.; de Villiers, J.; Redtmann, T. Utilization of Ferromanganese slags for manganese extraction and as a cement additive. In Proceedings of the Advances in Cement and Concrete Technology in Africa Conference, Emperor's Palace, Johannesburg, South Africa, 28–30 January 2013; pp. 983–995.
12. Yan, S.; Qiu, Y.-R. Preparation of electronic grade manganese sulfate from leaching solution of ferromanganese slag. *Trans. Nonferrous Met. Soc. China* **2014**, *24*, 3716–3721. [CrossRef]
13. Kazadi, D.; Groot, D.; Steenkamp, J.; Pöllmann, H. Control of silica polymerisation during ferromanganese slag sulphuric acid digestion and water leaching. *Hydrometallurgy* **2016**, *166*, 214–221. [CrossRef]
14. Sancho, J.; Fernández Pérez, B.; Ayala, J.N.; Recio, J.C.; Rodríguez, C.; Bernardo, J.L. Method of obtaining electrolytic manganese from ferroalloy production waste. In Proceedings of the 1st Spanish National Conference on Advances in Materials Recycling and Eco—Energy, Madrid, Spain, 12–13 November 2009; pp. 126–128.

15. Ghafarizadeh, B.; Rashchi, F. Separation of Manganese and Iron from Reductive Leaching Liquor of Electric Arc Furnace Dust of Ferromanganese Production Units by Solvent Extraction. *Miner. Eng.* **2011**, *24*, 174–176. [CrossRef]
16. Sadeghi, S.M.; Ferreira, C.M.; Soares, H.M. Evaluation of two-step processes for the selective recovery of Mn from a rich Mn residue. *Miner. Eng.* **2019**, *130*, 148–155. [CrossRef]

Article

Study on the Effect of Temperature on the Crystal Transformation of Microporous Calcium Silicate Synthesized of Extraction Silicon Solution from Fly Ash

Dong Kang [1,2,3,4], Zhijie Yang [1,2,3,4,*], De Zhang [1,2,3,4], Yang Jiao [1,2,3,4], Chenyang Fang [1,2,3,4] and Kaiyue Wang [1,2,3,4]

1   School of Mining and Technology, Inner Mongolia University of Technology, Hohhot 010051, China
2   Key Laboratory for Green Development of Mineral Resources, Inner Mongolia University of Technology, Hohhot 010051, China
3   Inner Mongolia Engineering Research Center of Geological Technology and Geotechnical Engineering, Inner Mongolia University of Technology, Hohhot 010051, China
4   Key Laboratory of Geological Hazards and Geotechnical Engineering Defense in Sandy and Drought Regions at Universities of Inner Mongolia Autonomous Region, Inner Mongolia University of Technology, Hohhot 010051, China
*   Correspondence: yangzj@imut.edu.cn; Tel.: +86-19975544835

Abstract: In this study, microporous calcium silicate was synthesized from a silicon solution of fly ash extracted by soaking in strong alkali as a silicon source. By means of XRD, TEM, FTIR, and thermodynamic calculations, the crystal evolution and growth process of microporous calcium silicate were studied under the synthesis temperature of 295~365 K. The results show that calcium silicate is a single-chain structure of the Si–O tetrahedron: Q1 type Si–O tetrahedron is located at both ends of the chain, and the middle is the $[SiO_4^{4-}]$ tetrahedron connected by $[O^{2-}]$ coplanar, and Ca^{2+} is embedded in the interlayer structure of calcium silicate. The formation rate and crystallization degree of calcium silicate hydrate were positively correlated with temperature. When the synthesis temperature was 295 K, its particle size was about 8 μm, and when the synthesis temperature was 330 K, a large number of amorphous microporous calcium silicate with a particle size of about 14 μm will be generated. When the temperature was above 350 K, the average particle size was about 17 μm. The microporous calcium silicate showed obvious crystalline characteristics, which indicate that the crystallization degree and particle size of microporous calcium silicate could be controlled by a reasonable synthesis temperature adjustment.

Keywords: microporous calcium silicate; extraction silicon solution; crystal transformation; fly ash

Citation: Kang, D.; Yang, Z.; Zhang, D.; Jiao, Y.; Fang, C.; Wang, K. Study on the Effect of Temperature on the Crystal Transformation of Microporous Calcium Silicate Synthesized of Extraction Silicon Solution from Fly Ash. *Materials* **2023**, *16*, 2154. https://doi.org/10.3390/ma16062154

Academic Editor: Daniela Fico

Received: 31 December 2022
Revised: 2 March 2023
Accepted: 4 March 2023
Published: 7 March 2023

1. Introduction

Calcium silicate hydrate (C–S–H) is a general term for various ternary compounds composed of CaO–SiO2–H2O through different molar ratios. In the process of the high-value utilization [1–6] of fly ash [7–10], the amorphous silicon element is extracted by the alkali leaching method, and the obtained high-alkali extraction silicon solution (ESS) contains a large amount of silicon element. Microporous calcium silicate [11–15] is a kind of C–S–H with a special structure and is dynamically synthesized using amorphous silicon in ESS as a silicon source and adding a calcium source. It can be used as building materials [16–18], thermal insulation materials [19,20], papermaking filler, adsorbent, etc., so has a high utilization value. Many scholars such as Fang Qi and Lothenbach have studied the synthesis of microporous calcium silicate at temperatures ranging from 20 °C to 80 °C. The results show that microporous calcium silicate with low crystallinity is easy to generate at low temperatures, and its particle size gradually increases with the increase in the synthesis temperature [21]. Zhijie Yang [22] found that with the increase in the C/S molar ratio, the average pore volume, average pore diameter, and average specific surface area of the synthesized calcium silicate minerals decreased first, then increased and then decreased again. Grangeon [23] believed that the increase in the synthesis temperature would promote the agglomeration behavior of microporous calcium silicate and the formation of more dimers, thus leading to the increase in the synthetic particle size. Sonja Haastrup [14] synthesized microporous calcium silicate at 60~95 °C, and the experimental results showed that the synthesis of microporous calcium silicate with a short-range order in the structure had the characteristics of a long-range disorder. The increase in the calcium silicon ratio will lead to the shortening of the silicate chain, and the metal cation will be attached to the negatively charged hydroxyl group, resulting in the reduction in an available hydroxyl group in the synthesis of microporous calcium silicate. The increase in temperature will promote the dissolution of amorphous silicon, thus accelerating the synthesis process of microporous calcium silicate. S. Shaw et al. found that when the synthesis temperature increased to 190~310 °C, the synthetic products evolved from poorly crystallized microporous calcium silicate into synthetic products with a higher crystallized degree such as tobermorite and xonotlite-type calcium silicate crystals [24]. At present, the structure of calcium silicate synthesized under similar synthetic conditions is very different, and the microstructure, growth process, and crystal transformation law of calcium silicate are still not fully understood.

Temperature is a crucial influencing factor for the synthesis of calcium silicate minerals by ESS. Studying the influence of different synthesis temperatures on the generated calcium silicate is an extremely important way to master its crystal development and transformation process. In this paper, ESS was used as the silicon source and lime milk as the calcium source. In the range of synthesis temperature from 295 K (22 °C) to 365 K (92 °C), different synthesis temperatures were set as reaction variables to synthesize microporous calcium silicate. The synthetic products were analyzed by XRD, TEM, FTTR, and other modern analysis and testing techniques, and the thermodynamic reaction formula was established. The Gibbs free energies of various calcium silicate minerals in the synthetic products were calculated. Combining the experimental results with thermodynamic calculation, the formation sequence and possibility of various calcium silicate minerals synthesized were judged. The purpose was to synthesize microporous calcium silicate from the amorphous silicon of solid waste-fly ash, and to clarify the influence of temperature on the crystal transformation and growth process of microporous calcium silicate synthesized from ESS. Thus, the transformation process and development rule of microporous calcium silicate were mastered, which provided a scientific basis for the synthesis of microporous calcium silicate by ESS.

2. Materials and Methods

2.1. Materials

The fly ash (C = chemical composition is shown in Table 1) used in this test came from Datang International Power Generation Co. Ltd. in Inner Mongolia, China. NaOH (Kermel, China) was the analytical reagent for ESS preparation, distilled water was used as the experimental water, and a CaO (Kermel, China) analytical reagent was used as the calcium source for the synthesis of microporous calcium silicate.

Table 1. Chemical composition of fly ash.

Constituents	Content (g/L)
SiO_2	47.9
Al_2O_3	40.1
CaO	4.09
MgO	0.65
Fe_2O_3	3.43
K_2O	0.84
Na_2O	0.32
SO_3	0.52

2.2. Methods

2.2.1. Preparation of ESS

ESS was prepared in a constant temperature water bath (SN-HWS-260, Shangyi, China). Under the condition of reaction temperature of 363 K (90 °C), the fly ash was soaked with 10% sodium hydroxide solution, the liquid–solid ratio was 4:1, and the reaction time was 90 min. Amorphous silicon reacts with NaOH in the fly ash to form a sodium silicate solution. In the form of $Na_2O \cdot xSiO_2$ in the solution, and through circulation pump (SHZ-D Yuying, China) filtration, ESS can be obtained after solid and liquid separation, where the residual solid was used to extract aluminum and other valuable elements. The ESS components obtained in this experiment are shown in Table 2.

Table 2. Chemical composition of ESS.

Constituents	Content (g/L)
SiO_2	40~60
NaOH	60~80
Al^{3+}	0.5~2
Fe^{3+}	0.01~0.05

2.2.2. Preparation of Microporous Calcium Silicate

Dissolve CaO in water to form a $Ca(OH)_2$ solution with a concentration of 150 g/L, and let stand for 30 min. After mixing ESS with lime milk at a silicon calcium molar ratio of 1:1, and adding distilled water to a solid–liquid ratio of 1:10, the synthesis experiment was carried out in a constant temperature water bath (Shangyi, China, SN-HWS-260) with a reaction time of 2 h. The microporous calcium silicate was synthesized at the following temperatures: 295 K (22 °C), 310 K (37 °C), 330 K (57 °C), 350 K (77 °C), 360 K (87 °C), and 365 K (92 °C). The filter cake obtained from the reaction products after vacuum filtration was dried at a drying temperature of 350 K for 24 h. The cake was manually ground with an agate mortar and passed through a 200-mesh sieve. As determined by XRF, the main chemical components of the synthesized calcium silicate are shown in Table 3, and the experimental process is shown in Figure 1. Finally, the microscopic morphology of microporous calcium silicate was observed by SEM (Hitachi S-4800, Japan), the structure of microporous calcium silicate was analyzed by XRD (PANalytical, The Netherlands, X'Pert Powder 3, Cu target, 40 kv, scanning range of 10~100°, step size 0.02°), FTIR (SHIMADZU, IRTracer-100, Japan) as well as TG and DSC (NETZSCH, Germany STA-449–F5 Heating rate of 5 °C/min), and the particle size of the synthesized samples at different temperatures was analyzed by a laser particle size analyzer (WJL-606, Jingke, China, refractive index 1.0).

Table 3. The composition of microporous calcium silicate was synthesized at different temperatures.

Number	Temperature	Proportion (%)					
		CaO	SiO_2	Al_2O_3	Na_2O	MgO	Others
1	295 K	55.14	37.31	3.35	0.71	0.70	2.79
2	310 K	42.26	50.65	2.44	2.27	0.10	2.28
3	330 K	42.14	50.52	2.38	2.44	0.17	2.35
4	350 K	41.30	50.41	2.40	3.20	0.29	2.40
5	360 K	39.70	52.21	2.10	3.95	0.21	1.83
6	365 K	40.04	50.04	2.34	3.97	0.20	3.05

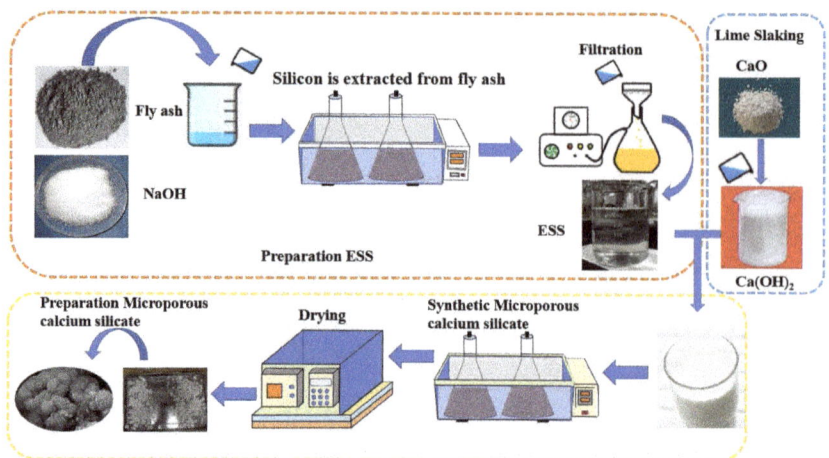

Figure 1. Preparation of microporous calcium silicate by the ESS form fly ash.

3. Results and Discussion

3.1. Thermodynamic Analysis

Synthesis of hydrated calcium silicate is a very complex process. At present, more than 30 kinds of stable calcium silicate hydrate have been found. Under the same conditions of reaction time, silica–calcium ratio, alkali concentration, and stirring speed, calcium silicate products with different silica–calcium ratios were formed continuously. The calcium silicate minerals formed in the early stage also evolve continuously, thus forming some single or multiple composite calcium silicate products. In this paper, thermodynamic formulas were established from the perspective of thermodynamics to analyze the formation sequence and the possibility of various calcium silicate minerals under specific conditions. Equations (1) and (2) were used as the thermodynamic calculation formulas for each reactant and product. Using Equations (3)–(5) as the formulas for calculating the Gibbs free energy of various calcium silicate-like products:

$$\triangle G_T = \triangle H_{298}^{\ominus} - T \triangle S_{298}^{\ominus} + \int_{298}^{T} \triangle C_p dT - T \int_{298}^{T} \frac{\triangle C_p}{T} dT \tag{1}$$

$$\triangle C_p = a + b * 10^{-3}T + c * 10^{-5} \, T^{-2} \tag{2}$$

$$\triangle H_{298}^{\ominus} = \left(\Sigma \triangle H_{298}^{\ominus}\right)_{resultant} - \left(\Sigma \triangle H_{298}^{\ominus}\right)_{reactant} \tag{3}$$

$$\triangle C_p = \left(\Sigma C_p\right)_{resultant} - \left(\Sigma \triangle C_p\right)_{reactant} \tag{4}$$

$$\triangle S_{298}^{\ominus} = \left(\Sigma \triangle S_{298}^{\ominus}\right)_{resultant} - \left(\Sigma \triangle S_{298}^{\ominus}\right)_{reactant} \tag{5}$$

The chemical reaction equations of various calcium silicate minerals in the synthetic products are shown as Equations (6)–(13):

$$3Ca(OH)_2(aq) + 2Na_2SiO_3(aq) + 2H_2O = 3CaO{\cdot}2SiO_2{\cdot}3H_2O(s) + 4NaOH(aq) \tag{6}$$

$$Ca(OH)_2(aq) + 2Na_2SiO_3(aq) + 3H_2O = CaO{\cdot}2SiO_2{\cdot}2H_2O(s) + 4NaOH(aq) \tag{7}$$

$$2Ca(OH)_2(aq) + Na_2SiO_3(aq) + 0.17H_2O = 2CaO{\cdot}SiO_2{\cdot}1.17H_2O(s) + 2NaOH(aq) \tag{8}$$

$$5Ca(OH)_2(aq) + 6Na_2SiO_3(aq) + 4H_2O = 5CaO{\cdot}6SiO_2{\cdot}3H_2O(s) + 12NaOH(aq) \tag{9}$$

$$5Ca(OH)_2(aq) + 6Na_2SiO_3(aq) + 6.5H_2O = 5CaO{\cdot}6SiO_2{\cdot}5.5H_2O(s) + 12NaOH(aq) \tag{10}$$

$$2Ca(OH)_2(aq) + 3Na_2SiO_3(aq) + 3.5H_2O = 2CaO{\cdot}3SiO_2{\cdot}2.5H_2O(s) + 6NaOH(aq) \tag{11}$$

$$Ca(OH)_2(aq) + Na_2SiO_3(aq) + H_2O = CaO \cdot SiO_2 \cdot H_2O(s) + 2NaOH(aq) \quad (12)$$

$$6Ca(OH)_2(aq) + 6Na_2SiO_3(aq) + H_2O = 6CaO \cdot 6SiO_2 \cdot H_2O(s) + 12NaOH(aq) \quad (13)$$

The related thermodynamic parameters of the reactants and products are shown in Table 4. Thermodynamic calculations were performed using HSC Chemistry (6.0), an integrated thermodynamic database software developed by the Outokumpu Research Center, Finland. The results were analyzed and plotted by Origin software (2021), as shown in Figure 2.

Table 4. The related thermodynamic parameters of the products and reactants.

Chemical Formula	$\triangle H_{298}^{\ominus}$ (kJ/mol)	$\triangle S_{298}^{\ominus}$ (kJ × K)	$a + b \times 10^{-3} + c \times 10^{-5} \, T^{-2}$		
			a	b	c
$3CaO \cdot 2SiO_2 \cdot 3H_2O$	−4782.312	312.126	341.163	188.698	−61.3790
$CaO \cdot 2SiO_2 \cdot 2H_2O$	−3138.837	171.126	187.485	78.241	−43.304
$2CaO \cdot SiO_2 \cdot 1.17H_2O$	−2665.208	160.666	171.711	93.722	−30.962
$5CaO \cdot 6SiO_2 \cdot 3H_2O$	−9935.745	513.168	600.613	312.545	−87.111
$5CaO \cdot 6SiO_2 \cdot 5.5H_2O$	−10,686.770	611.492	462.750	791.194	0.000
$2CaO \cdot 3SiO_2 \cdot 2.5H_2O$	−4920.384	271.542	332.503	151.879	−73.429
$CaO \cdot SiO_2 \cdot H_2O$	−1917.845	112.707	263.112	64.952	−36.401
$6CaO \cdot 6SiO_2 \cdot H_2O$	−10,024.864	507.519	553.334	272.797	−76.776
$Ca(OH)_2$	−1002.947	−74.517	89.248	33.150	−10.348
Na_2SiO_3	−1561.427	113.847	112.789	76.665	−19.708
H_2O	−285.830	69.950	186.884	−464.247	−19.565
$NaOH$	−469.863	44.769	12,683.218	−52,451.899	−2228.366

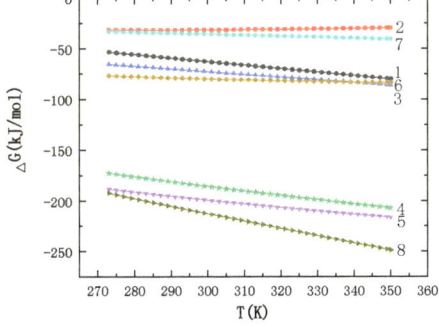

Figure 2. The relationship between the synthesis temperature and Gibbs free energy; 1—$3CaO \cdot 2SiO_2 \cdot 3H_2O$; 2—$CaO \cdot 2SiO_2 \cdot 2H_2O$; 3—$2CaO \cdot SiO_2 \cdot 1.17H_2O$; 4—$5CaO \cdot 6SiO_2 \cdot 3H_2O$; 5—$5CaO \cdot 6SiO_2 \cdot 5.5H_2O$; 6—$2CaO \cdot 3SiO_2 \cdot 2.5H_2O$; 7—$CaO \cdot SiO_2 \cdot H_2O$; 8—$6CaO \cdot 6SiO_2 \cdot H_2O$.

As shown in Figure 2, the thermodynamic calculation results show that temperature has an important influence on the growth process and crystal transformation of calcium silicate. At 275~350 K (37~77 °C), the Gibbs free energies for the synthesis of calcium silicate minerals with different Si/Ca ratios were all negative, which indicates that Equations (6)–(13) may all occur. Low temperature is conducive to the formation of calcium silicate hydrate containing more crystalline water, reaction Equations (9), (10), and (13) are more likely to occur than the others, that is, calcium silicate minerals with a silicon calcium ratio of 5:6 and 6:6 are more likely to be generated, which is consistent with the results of the XRD. With the gradual increase in the synthesis temperature, other C–S–H with a Si/Ca ratio gradually formed and transformed into the C–S–H crystal with a Si/Ca ratio of 5:6. With the continuous progress of the reaction, the xonotlite-type calcium silicate crystal with a Si/Ca ratio of 6:6 was finally formed. Calcium silicate minerals with different ratios, the degree of ease, or priority of formation will also vary with the reaction temperature, for example, $2CaO\cdot3SiO_2\cdot2.5H_2O$ is easier to generate than $2CaO\cdot SiO_2\cdot1.17H_2O$ when synthesized at room temperature. When the synthesis temperature is lower than 350 K, Equation (11) is more likely to occur than (8). In contrast, when the synthesis temperature was higher than 350 K, $2CaO\cdot SiO_2\cdot1.17H_2O$ is more likely to be generated. The relationship between the Gibbs free energy of various calcium silicates and temperature is shown in Table 5.

Table 5. The thermodynamic equations of various calcium silicates.

Number	Chemical Formula	The Equation for T(K) and ΔG(kJ/mol)
1	$3CaO\cdot2SiO_2\cdot3H_2O$	$\Delta G = -0.3455T + 41.1508$
2	$CaO\cdot2SiO_2\cdot2H_2O$	$\Delta G = 0.0257T - 38.66$
3	$2CaO\cdot SiO_2\cdot1.17H_2O$	$\Delta G = -0.2754T + 10.6294$
4	$5CaO\cdot6SiO_2\cdot3H_2O$	$\Delta G = -0.4426T - 52.4165$
5	$5CaO\cdot6SiO_2\cdot5.5H_2O$	$\Delta G = -0.3622T - 89.9723$
6	$2CaO\cdot3SiO_2\cdot2.5H_2O$	$\Delta G = -0.0891T - 52.8534$
7	$CaO\cdot SiO_2\cdot H_2O$	$\Delta G = -0.0953T - 6.6816$
8	$6CaO\cdot6SiO_2\cdot H_2O$	$\Delta G = -0.7291T + 6.1358$

3.2. Phase Analysis

In order to determine the effect of different reaction temperatures on the synthesis of microporous calcium silicate, as shown in Figure 3, XRD analysis of the microporous calcium silicate samples synthesized at different reaction temperatures was carried out. XRD results showed that when the reaction temperature was higher than 295 K (22 °C) in the range of 310 K (37 °C)~365 K (92 °C), C–S–H was the main phase. When the reaction temperature was 295 K (22 °C), there was a diffraction peak of calcite at 29.40°, and there was no diffraction peak of calcium silicate minerals. When 2θ was 18.06°, 34.11°, 48.61°, and 54.38°, the characteristic diffraction peaks of $Ca(OH)_2$ appeared. This shows that the sample still contained more residual $Ca(OH)_2$, which was not involved in the synthesis reaction. With the increase in reaction temperature, the diffraction peak gradually sharpened and reached a higher 2θ, which means that the average base spacing of the microporous calcium silicate decreased.

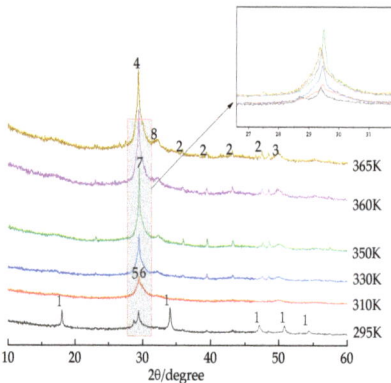

Figure 3. XRD patterns of the products at different temperatures. 1—$Ca(OH)_2$; 2—$CaCO_3$; 3—$CaO\cdot SiO_2\cdot H_2O$; 4—$2CaO\cdot SiO_2\cdot1.7H_2O$; 5—$5CaO\cdot6SiO_2\cdot5.5H_2O$; 6—$5CaO\cdot6SiO_2\cdot3H2O$; 7—$2CaO\cdot3SiO_2\cdot2.5H_2O$; 8—$CaO\cdot2SiO_2\cdot2H_2O$.

As shown in Figure 3, when the reaction temperature was 295 K (22 °C), a large amount of $Ca(OH)_2$ in the sample did not participate in the synthesis reaction, and there was no obvious diffraction peak of calcium silicate minerals in the XRD pattern. When the reaction temperature rose above 310 K (37 °C), most of the $Ca(OH)_2$ in the raw material participated in the synthesis reaction, the diffraction peak of $Ca(OH)_2$ gradually disappeared, and a small part of the calcium silicate with a silicon to calcium ratio of 5:6 was formed, but the diffraction fraction was diffuse and broad, and the crystallization condition was very poor, which is consistent with the thermodynamic theoretical calculation results, that when the reaction temperature reaches 350 K (77 °C) or above, the characteristic diffraction peaks of $CaCO_3$ appear at 36.02°, 39.46°, 43.22°, and 47.63°. The residual $Ca(OH)_2$ in the sample combined with CO_2 in the air during drying to form calcium carbonate, and the diffraction peak of $2CaO \cdot 3SiO_2 \cdot 2.5H_2O$ calcium silicate appeared at 29.49°, which confirms the previous thermodynamic theoretical calculation result. When the reaction temperature rose above 350 K (77 °C), the characteristic diffraction peaks of calcium silicate appeared at the 2θ positions of 29.42°, 32.25°, and 49.86°, respectively. The crystal characteristics of calcium silicate gradually became obvious, which was consistent with the TEM test results. When the reaction temperature was 360 K (87 °C), the diffraction peaks of $CaO \cdot 2SiO_2 \cdot 2H_2O$ and $CaO \cdot SiO_2 \cdot H_2O$ appeared at 32.32° and 49.76°. XRD analysis results combined with thermodynamic theoretical calculation showed that calcium silicate minerals with a silicon to calcium ratio of 5:6 were more readily produced than other types. With the progress of the synthesis reaction, calcium silicate products with different ratios of silicon to calcium were formed continuously, and all kinds of calcium silicate minerals formed in the early stage also transformed continuously, forming some single or multiple composite calcium silicate products. It is speculated that due to the relationship with reaction time, the formation of a xonotlite-type calcium silicate with a silicon calcium ratio of 6:6 could not be observed in the sample. The formation rate and crystallization degree of calcium silicate were positively correlated with temperature. With the continuous increase in reaction temperature, the diffraction peak of calcium silicate was significantly increased and enhanced, indicating that the amount of calcium silicate gradually increased, the peak value became gradually higher, and the peak type gradually steeper, which represents the gradual development of calcium silicate crystals. The diffraction peak of calcium silicate gradually shifted to a higher 2θ direction, indicating that the average substrate spacing of C–S–H decreased, and more calcium in the diffraction pattern was reflected in the XRD pattern in the form of $CaCO_3$, indicating that the crystallinity of calcium silicate was still poor.

3.3. Microstructure Analysis

When the reaction temperature was 310 K (37 °C), as shown in the SEM (Figure 4b), the morphology of the microporous calcium silicate initially appeared in the reactants, but it was not obvious. When the reaction temperature was 330 K (57 °C), the generated calcium silicate began to form in large quantities. In Figure 4c, obvious micropores appeared in some areas, and the calcium silicate particles showed an obvious agglomeration behavior. At this time, the pore size of the micropores was about 0.3 μm. Compared with the XRD analysis, it can be seen that the micropore calcium silicate crystals generated at this time were poor, and the corresponding TEM (Figure 5a) showed that no crystal characteristics were observed in the reaction products at this time, The synthesis temperature was 350 K (77 °C), as shown in Figure 4d, and the SEM diagram shows that the morphology of microporous calcium silicate appeared in more and more areas, and the honeycomb shape became more obvious. Combined with TEM (Figure 5b), the electron diffraction pattern showed a wide and diffuse diffraction ring, indicating that the microporous calcium silicate generated at this temperature still had a typical amorphous structure. With the increase in the reaction temperature, the amount of calcium silicate gradually increased, the crystal shape gradually developed, and the microporous morphology was more and more obvious. When the reaction temperature reached 360 K, the pore size increased to about 0.6–1 μm, as shown in Figure 4e, where its surface was porous and well-developed, and obvious diffraction spots appeared around the electron diffraction pattern ring, indicating that the microporous calcium silicate generated at this temperature had crystalline characteristics. In combination with the TEM (Figure 5c), it can be seen that obvious lattice patterns appeared during the synthesis of calcium silicate, indicating that its crystal form transformed from an irregular amorphous state to a regular crystal state. When the reaction temperature rose to 365 K, the morphology of the micropores showed a more obvious open petal. The results of the elemental analysis showed that the synthesized C–S–H was mainly composed of Si, Ca, O, and Na elements. In the high-alkali system, the Na element was the main impurity component in the microporous calcium silicate, which was consistent with the composition analysis in Table 4.

Figure 4. Micromorphology of the products at different temperatures: (**a**) 295 K; (**b**) 310 K; (**c**) 330 K; (**d**) 350 K; (**e**) 360 K; (**f**) 365 K.

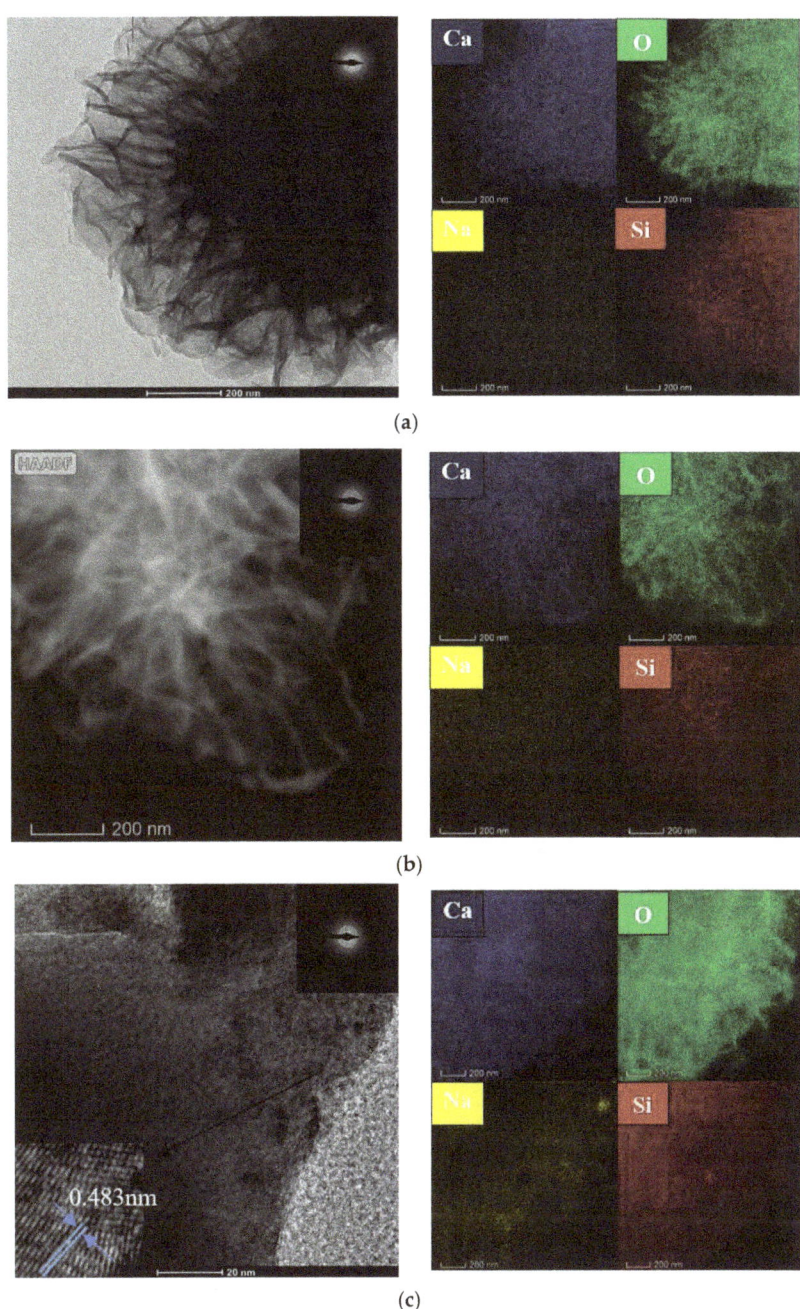

Figure 5. TEM of products at different temperatures: (**a**) 295 K; (**b**) 310 K; (**c**) 330 K.

3.4. FTIR and TG–DSC Analysis

The infrared spectrum is shown in Figure 6. In the high wave-number region, an absorption peak at 3643.6 cm^{-1} could be observed, which was the –OH stretching vibration peak, corresponding to the –OH bending vibration at 1632.8 cm^{-1}. It is speculated that the absorption peak is caused by the presence of a small amount of water in the sample and the presence of the O–H bond in the microporous calcium silicate sample. The bending vibration of the Si–O bond was at 664 cm^{-1} (Q2) and 870.9 cm^{-1} (Q1), and the stretching vibration of the Si–O bond was at 958.3 cm^{-1} (Q2) and 445.3 cm^{-1} (Q1), No Si–O (Q0), and Si–O (Q3) were found in the six samples. Therefore, it is speculated that the structure of the microporous calcium silicate synthesized is a Si–O tetrahedral single-chain, and each of the two adjacent Q2 [SiO$_4$$^{4-}$] tetrahedra is linked to the Si–O bond by the [O^{2-}] co-top link, and Ca^{2+} is embedded in the interlayer structure of calcium silicate, while 1415.3 cm^{-1} corresponds to the stretching vibration of C–O.

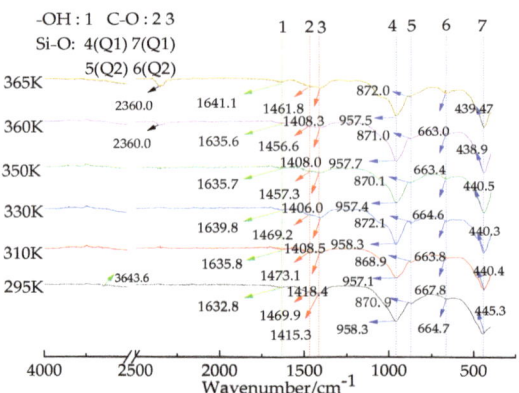

Figure 6. Infrared spectrum of the products at different temperatures.

The TG curves of the porous calcium silicate are shown in Figure 7, indicating that microporous calcium silicate had three stages of weight loss from room temperature to 1470 K. The first stage of weight loss occurred between 320 K and 570 K, and the weight loss rate was about 12–15%, which was mainly caused by a small amount of free water in the raw material and the removal of crystal water in the structure. In this temperature range, a large endothermic peak appeared on the differential scanning calorimetry curve (DSC curve) around 400 K corresponding to it. At this time, the crystal of the sample had not changed, and its main components were still amorphous. In the range of 570~1070 K, the weight loss of the second stage occurred, and the weight loss rate was about 5–7%, as a part of Ca(OH)$_2$ in the sample absorbs CO$_2$ in the air and turns into calcium carbonate, and the decomposition of CaCO$_3$ and the residual Ca(OH)$_2$ in the sample are endothermic decomposed into CaO and H$_2$O. In this temperature range, the bond fracture of Si–OH resulted in the transformation of dehydrated microporous calcium silicate into calcium silicate without –OH(Ca$_6$Si$_6$O$_{18}$). These three reasons together caused the TG curve in the range of 870~970 K to show obvious weight loss, while the dehydrated calcium silicate showed an endothermic reaction during dehydroxylation, leading to a relatively weak endothermic peak in the differential scanning calorimetry curve (DSC curve) at about 970 K. The thermogravimetric curve (TG) was relatively flat after 1070 K, indicating that there was no obvious weightlessness. The DSC curve showed a moderate exothermic peak at about 1370 K, which was due to the transformation of wollastonite in the sample into the high-temperature variant Ca$_3$Si$_3$O$_9$.The DSC curve showed a moderate exothermic peak at about 1370 K, which was due to the transformation of the sample from wollastonite to the high temperature variant Ca$_3$Si$_3$O$_9$, so no obvious weight loss occurred, and the final weight residue rate of the sample was between 76% and 81%.

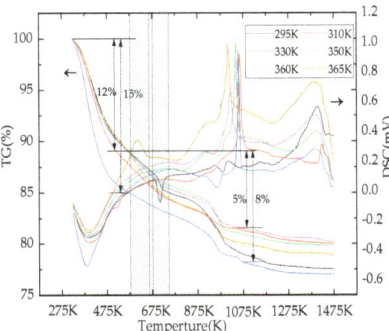

Figure 7. TG–DSC curves of the products at different temperatures.

3.5. Chemical Composition and Particle Size Analysis

The chemical composition of microporous calcium silicate synthesized at different synthesis temperatures is shown in Table 3. When the reaction temperature was 295 K (22 °C), the reaction was slow due to the low reaction temperature; as a result, the content of SiO_2 in the XRF detection of the synthetic species was only 37.31%. As the reaction temperature gradually increased, when the reaction temperature reached 310 K (37 °C), the chemical composition of the synthesized calcium silicate tended to be stable, with CaO accounting for about 40~42% and SiO_2 accounting for about 50%. The proportion of CaO was smaller than that of SiO_2, which may be caused by the substitution of Al^{3+} for Si^{4+} into the silico tetrahedron to form the Al–O tetrahedron in the high-alkali system.

The composition analysis of the synthesized microporous calcium silicate at different temperatures showed that Na and Al are the main components of the synthesized microporous calcium silicate impurities. The composition of ESS shows that in the process of fly ash desilication, a small part of Al^{3+} enter into ESS. It is speculated that this part of Al^{3+} replaces silicon atoms in the silicon oxygen tetrahedron in the synthesis of calcium silicate products and participates in the synthesis reaction. The surface of microporous calcium silicate is negatively charged. According to the principle of electric neutrality, Ca^{2+} acts as a balance charge and will exist in the interlayer structure of the microporous calcium silicate, Na^+, which is abundant in ESS, and can also play a similar role to Ca^{2+} in balancing charge, but its electrostatic effect is weaker than Ca^{2+}, thus replacing a small part of Ca^{2+} into calcium silicate synthesis. There is an obvious competitive relationship between Ca^{2+} and Na^+ in the microporous calcium silicate structure. As shown in Table 3, among the six microporous calcium silicate samples prepared, due to the low reaction temperature of sample no. 1, the composition of microporous calcium silicate in the sample was less, so the reference value was low. The proportion of Ca in the other five samples was in order from less to more with the synthesis temperature from low to high, while the proportion of Na showed an obvious opposite trend, which well proved that there was a competitive relationship in the structure of calcium silicate due to balancing the negative charge on the surface of the microporous calcium silicate.

The particle size of microporous calcium silicate has a very important influence on its application in papermaking, filtration, and so on. The particle size distribution of microporous calcium silicate generated by hydration at different temperatures is shown in Figure 8. The particle size of the synthesized microporous calcium silicate was between 5 μm and 20 μm, and the particle size frequency curve of the samples was an approximately normal distribution. The particle size of the synthesized microporous calcium silicate particles increased gradually with the increase in the synthesis temperature: the particle size was about 8 μm at 295 K (22 °C) and corresponded to 14 μm at 330 K (57 °C). As the synthesis temperature continued to rise, when the temperature reached more than 350 K (77 °C), the average particle size reached about 17 μm, indicating that the temperature had a significant impact on the particle size of the calcium silicate. As the average particle size increased with the increase in the synthesis temperature, a reasonable adjustment of the synthesis temperature can be used to control the particle size of the generated microporous calcium silicate.

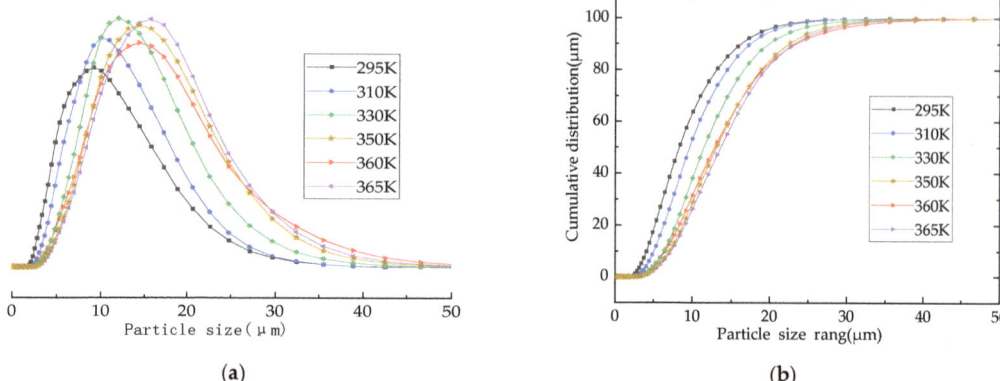

(a)   **(b)**

Figure 8. Particle size distribution at different temperatures: (**a**) particle size range; (**b**) cumulative distribution range.

3.6. Synthesis Process Analysis

According to the above experimental results and discussion, in the process of extracting the silicon element from fly ash, not only does the amorphous SiO_2 from fly ash react with NaOH to generate Na_2SiO_3, but also a small amount of minerals containing Al_2O_3 react with NaOH to generate $NaAlO_2$. The main components of the reaction solution are $Na2SiO_3$ and unreacted NaOH as well as a small amount of $NaAlO_2$ and other trace substances. As shown in Figure 9, in the dynamic synthesis of microporous calcium silicate with ESS and lime milk, first, the Si–O bond in SiO_2 is broken, which leads to the increase in freedom of the SiO_2 tetrahedron to form the $H_2SiO_4^{2-}$ group. The Si–O tetrahedron contained in it was linked to the Si–O tetrahedron contained in another $H_2SiO_4^{2-}$ by $[O^{2-}]$, formed by the Q1 form of the Si–O tetrahedron separated at both ends. Every two adjacent Q2 tetrahedrons by the $[O^{2-}]$ co-top link the Si–O tetrahedron single-chain polymerization structure, Ca^{2+} enters the interchain structure and is linked to the Si–O bond, thus in space forms a chain to chain interlayer structure.

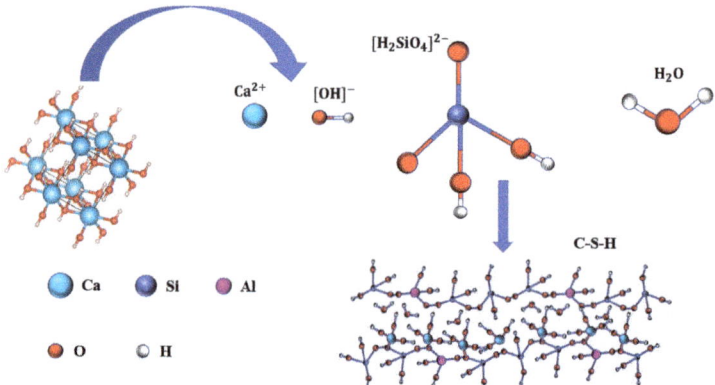

Figure 9. Molecular structure models of the microporous calcium silicate hydrate.

4. Conclusions

In this work, the influence of ESS as a silicon source on the synthesis of microporous calcium silicate in a high-alkali system was studied by combining thermodynamic theoretical calculation with experimental results. The major conclusions of this paper are summarized as follows:

(1) The XRD test results combined with thermodynamic calculation show that when ESS is used to synthesize microporous calcium silicate, calcium silicate products with different Si/Ca ratios are formed continuously with the progress of synthesis reaction. The crystalline forms of calcium silicate minerals synthesized in the early stage are also transformed, and $5CaO \cdot 6SiO_2 \cdot 5.5H_2O$ and $5CaO \cdot 6SiO_2 \cdot 3H_2O$ are more easily generated. When the synthesis temperature is 350 K (77 °C), $2CaO \cdot 3SiO_2 \cdot 2.5H_2O$ will be generated. As the synthesis temperature continues to rise, $2CaO \cdot SiO_2 \cdot 1.17H_2O$ will be generated when it reaches 330 K (57 °C), and $CaO \cdot SiO_2 \cdot H_2O$ will be generated, but will eventually transform into the xonotlite-type calcium silicate crystal with a silicon to calcium ratio of 6:6, which is consistent with the thermodynamic calculation results.

(2) The formation rate, crystallization degree, and particle size of the calcium silicate hydrate are positively correlated with temperature. The particle size of the calcium silicate hydrate was about 8 μm when the temperature was 295 K (22 °C). When the synthesis temperature was 330 K (57 °C), the particle size increased to about 14 μm. When the temperature reached above 350 K (77 °C), the average particle size reached about 17 μm, and the synthesis temperature was 360 K (87 °C). Therefore, the synthesized calcium silicate showed obvious crystal characteristics, which marks the transition from irregular amorphous to ordered crystalline.

(3) When using ESS to synthesize calcium silicate minerals, Na and Al are the main components of the impurities. The XRF test results showed that the proportion of Ca and Na in the sample showed an obvious inverse relationship. Na^+ will replace a small part of Ca^{2+} into the calcium silicate compound, forming a competitive relationship in the microporous calcium silicate structure. It is speculated that the reason is that both Na^+ and Ca^{2+} play a role in balancing charge, while the electrostatic effect of Na^+ is weaker than Ca^{2+}.

Author Contributions: Conceptualization, Z.Y.; Methodology, Z.Y.; Software, Z.Y. and D.K.; Validation, C.F., Y.J., and D.K.; Formal analysis, K.W. and D.K.; Investigation, Z.Y.; Resources, Z.Y.; Data curation, Y.J.; Writing—original draft preparation, D.K.; Writing—review and editing, Z.Y.; Visualization, supervision, investigation, D.Z.; Project administration, Z.Y.; Funding acquisition, Z.Y. All authors have read and agreed to the published version of the manuscript.

Funding: This research was funded by the Inner Mongolia Autonomous Region Science and Technology Plan Foundation (grant numbers 2020GG0287, 2020GG0257, and 2022YFHH0050), Inner Mongolia University of Technology Foundation (grant number ZZ201911), Inner Mongolia University of Technology Educational Reform Foundation (grant number 2022109), Inner Mongolia Autonomous Region Higher Education Basial Research Foundation (grant number JY20220380), and the Inner Mongolia Autonomous Region Graduate Student Innovation Research Foundation (grant number S20210189Z).

Institutional Review Board Statement: Not applicable.

Informed Consent Statement: Not applicable.

Data Availability Statement: The data presented in this study are available on request from the corresponding author. At the time the project was carried out, there was no obligation to make the data publicly available.

Acknowledgments: The authors gratefully acknowledge the Inner Mongolia Datang International Tuoketuo Power Generation Co., Ltd. for providing the raw materials.

Conflicts of Interest: The authors declare that they have no known competing financial interests or personal relationships that could have appeared to influence the work reported in this paper.

References

1. Ohenoja, K.; Rissanen, J.; Kinnunen, P. Direct carbonation of peat-wood fly ash for carbon capture and utilization in construction application. *J. CO2 Util.* **2020**, *40*, 101203. [CrossRef]
2. Xiaofeng, H.; Tova, J.; Andrey, K. Utilization of fly ash and waste lime from pulp and paper mills in the Argon Oxygen Decarburization process. *J. Clean. Prod.* **2020**, *261*, 121182.
3. Guojing, W.; Xiao, F.; Min, G. Resource utilization of municipal solid waste incineration fly ash in iron ore sintering process: A novel thermal treatment. *J. Clean. Prod.* **2020**, *263*, 121400.
4. Zhijie, Y.; De, Z.; Chenyang, F. Hydration Mechanisms of Alkali-Activated Cementitious Materials with Ternary Solid Waste Composition. *Materials* **2022**, *15*, 3616.
5. Kaushal, K.; Rishabh, A.; Sarah, K. Characterization of fly ash for potential utilization in green concrete. *Mater. Today: Proc.* **2022**, *56*, 1886–1890.
6. Rui, X.; Tao, L.; Lijing, W. Utilization of coal fly ash waste for effective recapture of phosphorus from waters. *Chemosphere* **2022**, *287*, 132431.
7. Jinliang, L.; Yao, W. Predicting the chloride diffusion in concrete incorporating fly ash by a multi-scale model. *J. Clean. Prod.* **2022**, *330*, 129767.
8. Zihao, L.; Koji, T.; Hidehiro, K. A study on engineering properties and environmental impact of sustainable concrete with fly ash or GGBS. *Constr. Build. Mater.* **2022**, *316*, 125776.
9. Cheng, F.; Baomin, W.; Hongmei, A. A comparative study on characteristics and leaching toxicity of fluidized bed and grate furnace MSWI fly ash. *J. Environ. Manag.* **2022**, *305*, 114345.
10. Teixeira, E.R.; Camões, A.; Branco, F.G.; Camões, A.; Branco, F.G. Synergetic effect of biomass fly ash on improvement of high-volume coal fly ash concrete properties. *Constr. Build. Mater.* **2022**, *314*, 125680. [CrossRef]
11. Cui, H.; Zhu, G.; Qiu, L.; Ye, X. Facile synthesis of Mg-doped calcium silicate porous nanoparticles for targeted drug delivery and ovarian cancer treatment. *Ceram. Int.* **2021**, *47*, 24942–24948. [CrossRef]
12. Yayoi, K.; Shiyang, C.; Takehisa, H. Adsorption of a poorly water-soluble drug onto porous calcium silicate by the sealed heating method. *Int. J. Pharm.* **2020**, *587*, 119637.
13. Yangyu, L.; Hongwei, J.; Guangxin, Z. Synthesis and humidity control performances of natural opoka based porous calcium silicate hydrate. *Adv. Powder Technol.* **2019**, *30*, 2733–2741.
14. Sonjia, H.; Mikkel, S.B.; Søren, R.H. Impact of amorphous micro silica on the C-S-H phase formation in porous calcium silicates. *J. Non-Cryst. Solids* **2018**, *481*, 556–561.
15. Kaihui, Z.; Lei, M.; Wei, M. Preparation of high strength lamellar porous calcium silicate ceramics by directional freeze casting. *Ceram. Int.* **2021**, *47*, 31187–31193.
16. Lin, L.; Tao, J.; Bojian, C. Recycling of Ti-extraction blast furnace slag: Preparation of calcium silicate board with high slag content by steam pressure curing. *Process Saf. Environ. Prot.* **2022**, *158*, 432–444.
17. Congqi, L.; Yong, Z.; Yongyi, L. Effects of nano-SiO_2, nano-$CaCO_3$ and nano-TiO_2 on properties and microstructure of the high content calcium silicate phase cement (HCSC). *Constr. Build. Mater.* **2022**, *314*, 125377.
18. Liu, Y.; Xie, M.; Xu, E.; Gao, X.; Yang, Y.; Deng, H. Development of calcium silicate-coated expanded clay based form-stable phase change materials for enhancing thermal and mechanical properties of cement-based composite. *Sol. Energy* **2018**, *174*, 24–34. [CrossRef]
19. Boissonnet, G.; Chalk, C.; Nicholls, J. Thermal insulation of CMAS (Calcium-Magnesium-Alumino-Silicates)-attacked plasma-sprayed thermal barrier coatings. *J. Eur. Ceram. Soc.* **2020**, *40*, 2042–2049. [CrossRef]
20. Leite, F.H.G.; Almeida, T.F.; FariaJr, R.T. Synthesis and characterization of calcium silicate insulating material using avian eggshell waste. *Ceram. Int.* **2017**, *43*, 4674–4679. [CrossRef]
21. Fang, Q.; Junya, C.; Ganyu, Z. Crystallization behavior of calcium silicate hydrate in highly alkaline system: Structure and kinetics. *J. Cryst. Growth* **2022**, *584*, 126578.
22. Zhijie, Y.; Dong, K.; De, Z. Crystal transformation of calcium silicate minerals synthesized by calcium silicate slag and silica fume with increase of C/S molar ratio. *J. Mater. Res. Technol.* **2021**, *15*, 4185–4192.
23. Sylvain, G.; Francies, C.; Yannick, L. X-ray diffraction: A powerful tool to probe and understand the structure of nanocrystalline calcium silicate hydrates. *Struct. Sci. Cryst. Eng. Mater.* **2013**, *69*, 465–473.
24. Shaw, S.; Clark, S.M.; Henderson, C.M.B. Hydrothermal formation of the calcium silicate hydrates, tobermorite ($Ca_5Si_6O_{16}(OH)_2 \cdot 4H_2O$) and xonotlite ($Ca_6Si_6O_{17}(OH)_2$): An in situ synchrotron study. *Chem. Geol.* **2000**, *167*, 129–140. [CrossRef]

materials

Review

Microplastics Derived from Food Packaging Waste—Their Origin and Health Risks

Kornelia Kadac-Czapska [1], **Eliza Knez** [1], **Magdalena Gierszewska** [2], **Ewa Olewnik-Kruszkowska** [2,*] and **Małgorzata Grembecka** [1,*]

[1] Department of Bromatology, Faculty of Pharmacy, Medical University of Gdańsk, 80-416 Gdańsk, Poland
[2] Department of Physical Chemistry and Physicochemistry of Polymers, Faculty of Chemistry, Nicolaus Copernicus University in Toruń, 87-100 Toruń, Poland
*   Correspondence: olewnik@umk.pl (E.O.-K.); malgorzata.grembecka@gumed.edu.pl (M.G.); Tel.: +48-56-611-22-10 (E.O.-K.); +48-58-349-10-93 (M.G.)

Abstract: Plastics are commonly used for packaging in the food industry. The most popular thermoplastic materials that have found such applications are polyethylene (PE), polypropylene (PP), poly(ethylene terephthalate) (PET), and polystyrene (PS). Unfortunately, most plastic packaging is disposable. As a consequence, significant amounts of waste are generated, entering the environment, and undergoing degradation processes. They can occur under the influence of mechanical forces, temperature, light, chemical, and biological factors. These factors can present synergistic or antagonistic effects. As a result of their action, microplastics are formed, which can undergo further fragmentation and decomposition into small-molecule compounds. During the degradation process, various additives used at the plastics' processing stage can also be released. Both microplastics and additives can negatively affect human and animal health. Determination of the negative consequences of microplastics on the environment and health is not possible without knowing the course of degradation processes of packaging waste and their products. In this article, we present the sources of microplastics, the causes and places of their formation, the transport of such particles, the degradation of plastics most often used in the production of packaging for food storage, the factors affecting the said process, and its effects.

Keywords: polymer; plastic; waste; degradation; microplastic; nanoplastic; environment pollution; food safety; human health

Citation: Kadac-Czapska, K.; Knez, E.; Gierszewska, M.; Olewnik-Kruszkowska, E.; Grembecka, M. Microplastics Derived from Food Packaging Waste—Their Origin and Health Risks. *Materials* **2023**, *16*, 674. https://doi.org/10.3390/ma16020674

Academic Editor: Daniela Fico

Received: 9 December 2022
Revised: 4 January 2023
Accepted: 6 January 2023
Published: 10 January 2023

1. Introduction

Plastics constitute a group of versatile synthetic materials with numerous applications. Their ubiquity has negative consequences in the form of extensive environmental pollution [1]. It is currently stated that 60–80% of garbage is plastic [2]. Due to improper environmental policies and little public awareness or ignorance, a significant amount of waste enters the environment and causes serious problems of uncontrolled pollution [3,4]. In the European Union (EU), 80–85% of marine waste is plastic, of which 50% are single-use products. These articles and their waste can slowly decompose and generate numerous smaller pieces of debris [5]. Plastic particles between 0.1 and 5000 μm in size are referred to as microplastics (MP) [6]. Particles smaller than MP, with sizes between 1 and 100 nm, are nanoplastics (NP) [7].

The sources of MPs, their properties, and potential harm are of widespread concern [8]. Studies have shown that such particles are present in both aquatic and terrestrial environments, posing a threat to the functioning of ecosystems [9]. Microplastics are found in soil [10], freshwater [8,11], seas and oceans [12], snow [13], wastewater [14,15], air [16], plants [17], and animal organisms [18]. The formation of MPs is a global threat [19], as they can travel as far as 6000 km [13] and enter the trophic chain [2]. Such particles can contaminate food and beverages [5,20] (Figure 1).

Figure 1. Scheme of food contamination due to MPs.

Microplastics have been found in fish [21], shellfish [22–24], poultry meat [25], eggs [26], salt [27–29], sugar [30], fruits [31], vegetables [32], water [33], milk [34,35], honey [35–37], beer [35], wine [38], tea [39–41], energy drinks, soft drinks [41], and infant formula [42] (Figure 2). Consumption of MPs negatively affects the digestive, respiratory, and circulatory systems [43]. They can accumulate in the body, causing inflammation. Contact with MPs is associated with the risk of oxidative stress, changes in cell division and viability, DNA damage, immune reactions, metabolic disruption, intestinal dysbiosis, and increased risk of cancer, respiratory, and neurodegenerative diseases [6,43].

Depending on their origin, MPs are divided into primary and secondary. Primary MPs are particles designed to be microscopic in size. They are used in the form in which they were produced. The degradation of plastics in the environment is considered one of the main processes contributing to the formation of secondary MPs [44]. Prolonged exposure of packaging waste to factors such as sunlight, water, temperature, and microbial action leads to its fragmentation into smaller pieces. These particles have the character of anthropogenic pollutants. The degradation of plastics is also very important with regard to forensic issues, as plastic packaging is one way to dispose of items or residues associated with criminal activity [45]. Primary and secondary MPs further degrade to NPs [46].

Determination of the MPs' negative consequences on the environment, as well as animal and human health, is not possible without knowledge about the degradation of packaging waste made of plastics and their products. In this work, we paid attention to the analysis of the MPs' sources and the reasons and circumstances for their formation. Moreover, the transport of such particles, the course of degradation processes of plastics most often used in the production of packaging for food storage, the factors affecting the said process, and its effects.

Figure 2. Occurrence of MPs in foods and beverages.

2. Methods

During the preparation of this article, we reviewed the literature and extracted the most relevant information regarding the locations and sources of MPs. We focused on scientific papers from ScienceDirect and Scopus databases. "Microplastic," "food," and "degradation" were used as the search terms in the title, keywords, and abstract. The full texts of the chosen articles were analyzed, and then the fundamental information was summarized. We devoted special attention to the papers concerning plastic degradation in the territorial and aquatic environment. The analysis of scientific articles focused on peer-reviewed papers written in English that were published as of the year 2020. Papers before 2020 were included due to their relevance to MP research. We reviewed more than 280 papers on microplastics and polymer degradation.

3. Food Packaging

Products made of plastics have gained popularity due to their low production costs, light weight, ease of use, and durability. It is estimated that about 39.6% of such materials are used for packaging [43]. The purpose of food packaging is protection, encapsulation, convenience, and communication with consumers. Packaging protects food from mechanical damage and microbiological and chemical contamination [47] and facilitates food storage, handling, and transportation [48].

Currently, there are several regulations on plastic products intended for food contact. Unfortunately, the topic of MPs is not explicitly addressed in them. They refer to polymers, plastics, and additives used at the processing stage. European Commission Regulation No. 10/2011 on plastic materials and articles intended to come into contact with food states that substances with a molecular weight of more than 1000 Da cannot be absorbed in the body, and possible health risks may be caused by unreacted monomers or said additives, which are transferred to food through migration from the material [49]. According to the legislation, the released substances must not adversely affect the organoleptic characteristics of food and exceed the permissible limits of global and specific migration. Global migration is understood as the mass of residues of all substances released from the product into food simulants. The global migration limit is equal to 10 mg per 1 dm^2. Specific migration, on the other hand, refers only to the specific substance released from the article into the model fluids under the test conditions. The specific migration limit from plastic products was set for selected elements, i.e., Ba (up to 1 mg/kg), Co (up to 0.05 mg/kg), Cu (up to 5 mg/kg), Fe (up to 48 mg/kg), Li (up to 0.6 mg/kg), Mn (up to 0.6 mg/kg), and Zn (up to 25 mg/kg).

In addition, European Commission Regulation No. 10/2011 states that the risk assessment for substances released from packaging should include the substance itself and the degradation products arising from the intended use [49]. This statement, therefore, eliminates secondary MPs arising from the degradation of plastic packaging waste in the environment from the area of concern.

The most popular packaging for food protection and storage are containers, bottles, films, pouches, and cups [43]. They are usually made of high-density polyethylene (HDPE), low-density polyethylene (LDPE), polypropylene (PP), polyesters (such as poly(ethylene terephthalate) (PET)), and polystyrene (PS) [48,50].

Each type of polymer is characterized by different properties, and thus, they find various applications. Polyethylene (PE) is mainly dedicated to film and bags. Water bottles are made of PET, and caps are usually made of PP [5]. The release of MPs from plastic bottles and cartons was investigated. Most of the particles in water from returnable bottles were identified as PET (84%) and PP (7%), while in water from beverage cartons, other MPs, such as PE. This can be explained by the fact that the cartons are coated with PE film on the inside. In both situations, the particles are smaller than 20 μm [5]. The last of the polymers described—PS, is most often applied in the foamed form. Until recently, PS was used for disposable food packaging with heat-insulating properties. However, according to Directive 2019/904 of the European Parliament and the Council, the marketing of food and beverage containers made of expanded PS has been restricted since 2021 [51].

4. Plastics Degradation

During the storage of packaging waste made of plastics, aging occurs, that is, the gradual loss of physical and mechanical properties of the material. Degradation can happen under the influence of mechanical forces (mechanical degradation), temperature (thermal degradation), light (photodegradation), various chemicals (chemical degradation), and biological factors (biological degradation) [52]. As a result of their action, materials can fragment into macroplastics, then MPs and smaller particles (Table 1), which undergo further decomposition into small-molecule compounds, CO_2 or CH_4 (Table 2) [43,53]. It should be noticed that sources of MPs can be food containers made of plastic when heated in a microwave oven [54].

Table 1. Microplastics degradation effects.

Polymer Type	Degradation Method	Effect	References
PE	Photodegradation	Oxygen functional groups on the surface; the increase of specific surface area	[55]
	Chemical degradation (prothioconazole)	Cracks	[56]
PP	Photodegradation	Reduction in microplastic particle volume	[57]
	Biological degradation (*Serratia marcescens* and *Enterobacter* spp.)	Surface changes (microcracks and corrugations)	[58]
	Biological degradation (*Rhodococcus* sp. And *Bacillus* sp.)	The reduction of the polymer mass; structural and morphological changes in PP	[59]
PE, PP	Biological degradation (*Spirulina* sp.)	Changes of functional groups; a decrease in carbon in PE and PP	[60]
PS	Biological degradation (*Bacillus cereus* CH6)	The surface morphology changes	[61]
PET	Chemical and thermal degradation	Changes in surface morphology, crystallinity, and carbonyl index	[62]
PS, PET	Biological degradation (*Bacillus* sp.)	Structural and surface changes; weight loss; a decrease in the carbon content	[63]
PE, PP, PS, PET	Chemical degradation (Fenton's reagent)	Wrinkles, voids, and holes on the surface; oxygen functional groups on the surface; increased hydrophilicity and acidity of the surface; reduced MP size	[64]

Table 2. Decomposition of plastics into small-molecule compounds.

Polymer Type	Degradation Method	Degradation Products	References
PE	Thermal degradation	H_2, CH_4, C_2H_4, and C_3H_6	[65]
	Photodegradation	CO_2, H_2O	[66]
	Photodegradation	C_2H_6, CO_2, H_2O, and formaldehyde	[67]
	Chemical degradation (O_3 and H_2O_2)	3-pentanol, 3-pentanone	[68]
PP	Photodegradation	Formaldehyde, acetaldehyde, 2-propynyl, hydroxypropyl, acetone, 2-propenyl, butanal, 4-pentyn-1-olate, 4-pentyn-1-olate, (2-ethoxyethyl)oxonium, and acetylacetonate	[57]
PS	Biological degradation (the microbially driven Fenton reaction)	2-isopropyl-5-methyl-1-heptanol, nonahexacontanoic acid	[69]
	Photodegradation	Benzene, toluene, phenol, styrene, and 2-propenylbenzene	[70]
	Photodegradation	Acrolein, benzene, propanal, methyl vinyl ketone, and methyl propenyl ketone	[71]
PET	Biological degradation (*Rhococcus* sp. *SSM1*)	Monomer—terephthalic acid (TPA)	[72]
	Biological degradation (petase)	Mono-(2-hydroxyethyl) terephthalate, bis-(2-hydroxyethyl) terephthalate and ethylene glycol	[73]
PE, PP, PET, PS	Chemical degradation (Fenton oxidation)	CO_2	[74]

Materials **2023**, *16*, 674

Unfortunately, most studies devoted to plastic degradation only address its early stages, probably because of the very long time required to achieve the goal (e.g., estimated half-lives ranging for HDPE bottles in the marine environment is 58 years) [75,76]. In the environment, however, the degradation of conventional plastics is a long-term process subjected to environmental conditions. The influence of a specific factor on the rate of degradation depends on the type of plastic, including the polymer structure, degree of cross-linking, molar mass, degree of crystallinity, and the presence of additives in the material that affect the processing and performance properties of the material (e.g., photostabilizers, heat stabilizers, plasticizers, flame retardants, nanoparticles, pigments). During aging, introduced additives can be released into food and exhibit harmful effects on organisms in contact with them (Table 3) [77]. Bisphenol A (BPA) is an example of a plasticizer [78], which release from MPs has been confirmed. A study of fishes (*Dicentrarchus labrax, Trachurus trachurus, Scomber colias*) from the Northeast Atlantic showed the presence of BPA, bisphenol B (BPB), and bisphenol E (BPE) in muscles, and BPA additionally in the liver [79]. Individuals in which MPs were detected had significantly higher concentrations of bisphenols than those whose bodies were not contaminated with plastic particles [80]. BPA (4.02 mg/L) was found to stimulate the production of reactive oxygen species, resulting in reduced biomass viability and even apoptosis [81]. Currently, BPA cannot be used in the production of baby bottles in the US. In the European Union, according to European Commission Regulation 10/2011, a specific migration limit of 0.6 mg/kg has been established for this compound [49].

Table 3. Health effects of additives used in plastics processing.

Additive Type	Example of the Chemical Compound	Health Effects	References
Plasticizers	Phthalates [di(2-ethylhexyl)phthalate (DEHP), diethyl phthalate (DEP), and dibutyl phthalate (DBP)]	Increased oxidative stress and inflammation: - inhibition of human salivary aldehyde dehydrogenase (hsALDH) - impact on peroxisome proliferator-activated receptor-α. Disruption in the endocrine system: - phthalates can affect the adrenal cortex (H295R cells) and cause significant disturbance of steroid hormones synthesis - decreased in testosterone and increased in 17β-estradiol. Impairment of the reproductive system: - reduction in ovulatory follicles - oocytes with poor maturation - DEHP decreased testicular function in rats - the perinatal DBP and DEP exposure may show significant growth retardations and also affect brain development and emotions like attention problems, anxiety, and depression.	[81–86]
	Acetyl tributyl citrate (ATBC)	Risk of interaction with drugs: - ATBC-induced cytochrome P4503A4. Reproductive system disorders: - ATBC decreased the number of primordial, primary, and secondary follicles present in the mice's ovary.	[87,88]

Table 3. *Cont.*

Additive Type	Example of the Chemical Compound	Health Effects	References
	Poly(ethylene glycol) (PEG)	Allergy: - PEG is a risk factor for IgE-mediated anaphylaxis.	[89,90]
	Bisphenol A	Endocrine effects: - increased α-chymotrypsin activity - increased oxidative stress (BPA produces reactive oxygen species (ROS)). Reproductive system disorders: - BPA decreased viability of bovine theca cells in vitro - higher BPA exposure was associated with lower semen quality in Chinese men and endometriosis in women. Neurodevelopmental disorder: - creatinine-adjusted BPA levels were associated with a 3.3–3.6% increase in attention-deficit hyperactivity symptoms (ADHD) rating scale IV. Renal function disorders: - BPA exposure may negatively impact on kidney function and structure.	[91–98]
Antioxidants	Arylamines	Pro-cancer activity: - exposure to arylamines is associated with a higher risk of bladder cancer (mainly *p*-phenylenediamine). Autoimmune diseases: - arylamines are reported to cause lupus-inducibility.	[99,100]
Light stabilizers and ultraviolet (UV) absorbents	Hindered amines light stabilizers (e.g., bis(2,2,6,6-tetramethyl-4-piperidyl) sebacate)	Cytotoxic effect: - decreased viability and activity in epithelial cells.	[101]
	Benzotriazole UV stabilizers [e.g., UV-328 (2-(2H-benzotriazol-2-yl)-4,6-di-tert-pentylphenol)]	Inflammation: - their metabolites in human blood increased oxidative stress. Gene expression profiling: - UV-320 was a strong Peroxisome Proliferator-Activated Receptor α (PPARα) agonist in mice.	[102,103]
Heat stabilizers	Vinyl chloride	Liver diseases: - exposure to vinyl chloride was associated with cirrhosis and hepatocellular carcinoma.	[104–107]
Flame retardants	Short, medium, and long chlorinated paraffins (SCCP/MCCP/LCCP)	Cytotoxic effect: - 10–15% lower relative cells viability - SCCPS caused cell membrane damage.	[108,109]
	Tris(1,3-dichloro-2-propyl) phosphate (TDCPP)	Cytotoxic effect: - TDCPP inhibited cell growth, decreased cell viability, and increased cell toxicity in vitro.	[110,111]

Table 3. *Cont.*

Additive Type	Example of the Chemical Compound	Health Effects	References
	Sb_2O_3	Pro-cancer activity: - increased the cancer risks for inhalation exposure - induced DNA damage. Inflammation: - increased oxidative stress.	[112,113]
	Polybrominated diphenyl (PBB) and polybrominated diphenyl ethers (PBDES)	Pro-cancer activity: - increased risk of thyroid cancer. Metabolic disorders: - diabetes and metabolic syndrome.	[114,115]
Pigments	TiO_2	Inflammation: - increased ROS production. Microbiota dysfunctions: - variations in microbiota abundance, gut dysfunctions, and reduction in short-chain fatty acids (SCFAS) levels.	[116,117]

4.1. Mechanical Degradation

Mechanical degradation refers to the breakdown of plastics due to external forces, collision, and abrasion of materials [118–120]. Microplastics can be introduced into food during its preparation. It has been estimated that 100–300 MPs/mm are formed on the cutting board when cuts are made during food preparation. On the other hand, in an aqueous environment, the freezing and thawing of plastics can also cause the mechanical degradation of polymers [121].

The effect of external forces depends on the mechanical properties of the materials [122]. Plastics with a low elongation value at break are more prone to fragmentation under external tensile forces. This leads to the tearing of polymer chains [123]. As a result of the mechanical degradation of primary and secondary MPs, smaller plastic particles (e.g., NPs) can be obtained [124].

4.2. Thermal Degradation

In addition to mechanical grinding, the temperature can also affect the course and efficiency of plastic degradation [125]. When enough heat is absorbed, long polymer chains can be broken, generating radicals [126]. These can react with oxygen and produce peroxides, which decompose to form free hydroxyl radicals and alkoxy radicals. The reaction can proceed spontaneously until the energy supply ceases or inert products are formed by the collision of two radicals. The temperature required for thermal degradation is related to the thermal properties of the plastics and the availability of oxygen [122]. Singh et al. concluded that the decomposition of PE occurs in one stage, between 230 and 510 °C [127]. Polypropylene has an onset degradation temperature of 286 °C [128], while PS has an onset degradation temperature of 370 °C [129]. However, the pyrolysis process of PET starts sharply and occurs at around 427–477 °C (around 90% of the process) [130].

4.3. Photodegradation

Photodegradation of plastics involves reactions initiated by solar radiation. As a result of the changes, plastics are gradually destroyed, with fragmentation into smaller particles and the formation of MPs [131]. As a result of UV radiation, new functional groups are formed, and the crystallinity, thermal and mechanical properties, and surface morphology of MPs change [132]. In the environment, during solar radiation, plastic waste is also affected by atmospheric oxygen, so the process is often referred to as oxidative photodegradation. Thermal oxidation of plastics occurs in conjunction with photodegradation, especially on beaches or sidewalks that are exposed directly to sunlight [133,134]. Polymers containing aromatic rings in their structure (PS and PET) were found to be more susceptible to oxidation compared to polymers formed by aliphatic chains (PE and PP) [64].

Currently, plastics in which photodegradation is an intended feature are gaining popularity. These are photodegradable materials containing sensitizers that degrade when exposed to UV light in the presence of oxygen. Polyolefins, intended for the manufacture of disposable packaging, are the largest contributor.

4.3.1. Course of Plastic Degradation

Photodegradation of polymers generally involves a free radical mechanism. There are three main steps: photoinitiation, propagation, and termination. Norrish type I and II reactions produce radicals, and ketone groups, which cause cleavage of the main chain. Free radicals can react with oxygen to form superoxide radicals, which are converted into superoxide molecules. The peroxide moiety dissociates into macroalkoxy and hydroxyl radicals, which catalyze the further reaction. During the reaction, aldehydes, ketones, carboxylic acids, esters, and alcohols can be formed; moreover, chain scission and crosslinking of polymers can be observed [135].

Some of the photochemical reactions can be ionic or ion radical in nature. This is especially observed in the case of polymers characterized by an ionic structure.

4.3.2. Effect of Various Factors on the Photodegradation of MP

The process of photodegradation of packaging waste depends on many factors. The wavelength of solar radiation, atmospheric oxygen concentration, O_3 formation, the presence of SO_2, NO_2, and metallic compounds, as well as mechanical factors, are of particular importance.

Effect of Radiation

Photodegradation of plastics in the environment is caused by solar radiation reaching the Earth. The efficiency of the photodegradation process depends on the wavelength of light. The shorter the wavelength, the higher the radiation energy. Short-wavelength radiation, compared to long-wavelength radiation, generally induces faster and more efficient changes in the chemical structure of macromolecules and in the physical properties of the polymer. UV radiation energy of 254 nm is already sufficient to break C-C and C-H chemical bonds. While the energy of visible radiation is much lower, so it breaks only the weakest chemical bonds [136]. The impact of UV radiation on plastic packaging waste results in the embrittlement of the material and the formation of cracks and fractures at their surfaces. Song et al. found that these factors did not directly affect the fragmentation of PE and PP [137]. The formation of plastic particles required subsequent mechanical abrasion. This implies that beyond the action of UV radiation, additional physical force is required for the formation of MPs.

Effect of Oxygen

Plastics can undergo decomposition reactions in the presence of oxygen. This agent is involved in the oxidation cycle of the irradiated polymer, reacting with macroradicals of various types.

Initially, photooxidation occurs in the thin surface layer of the sample, up to 100 μm [138]. The concentration of photoproducts is the greatest near the surface of the degraded plastic. For example, the effect of UV radiation on PET degradation was investigated. It was found that the process occurs at a depth of up to 20 μm [139].

As a result of the gradual diffusion of oxygen deep into the material, reactions can occur throughout the polymer. The efficiency of this process depends on the oxygen concentration and the properties of the polymer. Oxygen diffusion is possible in amorphous polymers. In the case of crystalline polymers, it is a limited process. Oxidation is favored by elevated temperatures and the presence of catalysts such as metals and metal ions.

This results in the formation of numerous oxidation products, such as peroxides, alcohols, ketones, aldehydes, acids, peroxyacids, peresters, or γ-lactones [140].

Effect of Ozone

It should be mentioned that O_3 can be formed from O_2 as a result of UV radiation and atmospheric discharge, which occurs naturally in low concentrations in the atmosphere. Ozone, even at low concentrations, can react with the polymer directly and attack unsaturated C=C double bonds. This reaction causes the destruction of polymer chains and the formation of carboxyl and ester groups. Ozone can also react with saturated polymers, but at a much slower rate [141].

Effect of Oxides

Compounds such as SO_2 and NO_2 are commonly found in the atmosphere. They can attack plastics directly or catalyze radical formation, which also leads to degradation. Sulfur(IV) oxide can be excited by UV radiation, forming a reactive singlet or triplet state that reacts with unsaturated C=C double bonds directly or produces O_3 through a photochemical reaction with O_2. Nitrogen(IV) oxide is very reactive due to the presence of odd electrons in the molecule, which can easily react with unsaturated C=C double bonds in the polymer. As with SO_2, the photochemical reaction of NO_2 with O_2 also produces ozone [142].

Effect of Metal Compounds

Pollution of the atmosphere, soil, and water can be a source of metal cations, such as Fe, Pb, Cu, Zn, Mn, and Hg. The changes caused by their presence depend primarily on the type of inorganic compound (from which they originate) and the structure of macromolecules. They can affect the degradation process of the plastic and absorb on the surface of MPs, which exhibit high specific surface area and hydrophobicity [143–147]. It was found that aged MPs showed higher sorption of heavy metals than plastic particles with an undegraded surface, indicating a higher environmental and health risk of degraded particles [148].

Effect of Mechanical Factors

It was found that in the presence of UV radiation, plastics are more susceptible to mechanical abrasion [137]. The action of mechanical forces on plastic packaging waste causes the fragmentation and formation of surface defects on the surface of MPs, the presence of which is associated with the breakage of polymer chains. As a result of this process, free radicals are generated, which are initiators of photodegradation. At the same time, the aforementioned microdefects increase the surface area of MPs and facilitate the diffusion of atmospheric oxygen into their depths.

4.3.3. Changes in the Properties of MPs

It was found that photodegradation in air causes a decrease in molecular weight, and the mechanical and physicochemical properties of plastics. Moreover, it changes the appearance and texture of the studied material [149]. As a result of UV radiation, MP particles become brittle, their surface properties switch, and roughness and porosity increase. Furthermore, the hydrophilicity and, thus, the adhesion and wettability properties change. Particles, which are colorless by nature, turn yellow under UV radiation due to the formation of sequences of conjugated double bonds of different lengths. In addition, excipients that modify the properties of the polymer can decompose when exposed to light and initiate degradation of the macromolecules.

4.3.4. Photodegradation of Selected Polymers

The probability of initiated photodegradation, C-H oxidation, and chain scission depends on the structure of the polymer [150,151]. Susceptibility to photodegradation is related to the presence of UV-absorbing chromophore groups in macromolecules. Polymers containing aromatic rings or carbonyl groups in their structure are sensitive to photochemical degradation. Macromolecules that do not contain chromophore groups in their structure also undergo photodegradation. However, it is caused by the presence of structural defects or trace amounts of impurities, including catalyst residues [151,152]. Macromolecules without tertiary hydrogen groups were found to be very stable.

A comparison of plastics with and without tertiary C-H bonds reveals that reactivity (i.e., bond dissociation energies) decreases as follows: PS > PP > PE [153]. Similar relationships were obtained by performing a study of the effect of simulated sunlight. Fragmentation initiation proceeded in the order PS (<1 year) > PP (<2 years) > LDPE (>3 years) [132]. The lowest degradability of polyolefins can be explained by the high level of hydrophobicity [154].

Polyethylene and Polypropylene

The photodegradation of PE and PP is similar. However, PP is less stable than LDPE and HDPE due to the presence of a tertiary carbon in the main chain, which is more susceptible to oxygen attack [155]. Taking into account the photochemical stability of the listed polyalkanes, they can be ranked in the following order: PP, LDPE, and HDPE.

The formation of radicals under UV radiation in PE is made possible by the presence of various types of inclusions (RH). These contaminants are residues from unreacted reagents used during polymerization or material processing (e.g., initiators, catalysts, solvents, pigments). As a result of UV exposure, the RH decomposes into R$^{\cdot}$ and H$^{\cdot}$ radicals, which react with the polymer and cause the formation of radicals in the macro-chains. The following reactions involving radicals lead to random chain disruption and the formation of lower molecular weight degradation products [156,157]. The oxygen diffusion coefficient in polyalkenes is about twice as low as in their low-molecular-weight homolog. During the photooxidation of PE (with RH participation), carbonyl and hydroxyl groups are formed (Figure 3), as well as H_2O, CH_4, methanol, propanone, CO, and CO_2. At the same time, conjugated double bonds can be formed. The oxidative degradation process of PE was carried out. It was found that C-H bonds oxidize and carbonyl groups are formed, which facilitates the formation of biofilms [158].

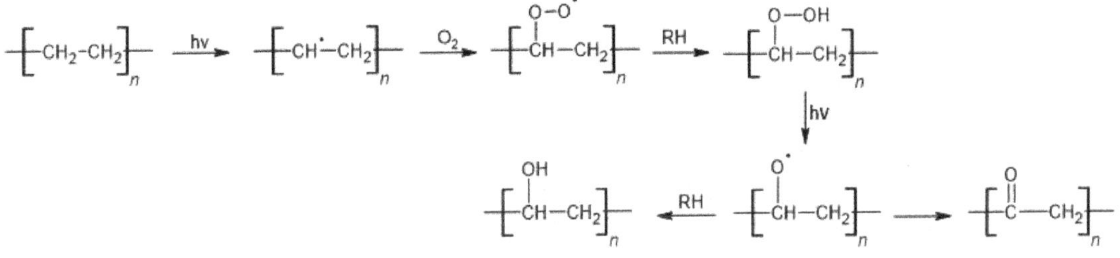

Figure 3. Photodegradation of PE (hν—to add energy via photons, RH—inclusions).

It was found that as a result of photo- and oxydegradation of PP, the morphology of the particles and their hydrophobic properties change [159], and a reduction of at least 65% volume is observed [57]. Polypropylene mainly undergoes chain breakage, depolymerization, and photo-oxidation reactions. It should be stressed that in the case of this polymer, isolated double bonds are formed.

A model study of isotactic PP films, commonly used in packaging, was conducted to simulate the process of MP formation using UV. Shredding of the tested material into sub-millimeter particles was observed in less than 48 h. This allowed an estimation of the lifetime of this type of product between 9 months and 3.2 years, depending on the place and climate in which the waste is located [19]. In another study, in a simulated beach environment, 12 months of UV exposure and 2 months of mechanical abrasion to PP and PE resulted in the formation of approximately 6084 and 20 particles, respectively [137].

Poly(Ethylene Terephthalate)

Poly(ethylene terephthalate) is a polymer that is resistant to environmental and biological factors [160]. However, it can be photodegraded through radical reactions. As a result of this process, PET packaging waste loses its mechanical properties; moreover, the formation of surface microcracks and color changes are observed. Poly(ethylene terephthalate) degradation is initiated by radiation with a wavelength of $\lambda < 315$ nm. Then, the alkyl and phenyl radicals created undergo reactions with oxygen, forming hydroxyl, aldehyde, and carboxyl groups at the ends of the chains. As a result of photooxidation, hydroxyl groups are also produced in aromatic rings. Hydroxyls can react with aromatic rings in the polymer backbone to make hydroxyterephthalate groups (Figure 4) [161,162]. Photodegradation of PET leads to the cleavage of the ester bond. As a consequence of this process, CO, CO_2, terephthalic acid, anhydrides, carboxylic acids, and esters can be created [152].

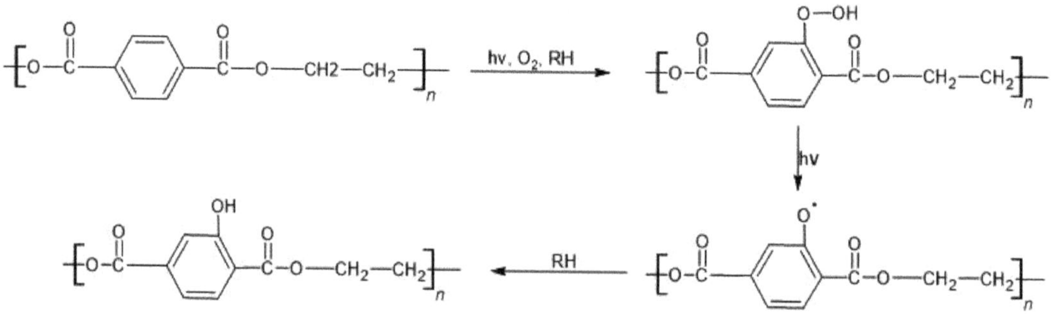

Figure 4. Photodegradation of PET (hν—to add energy via photons, RH—inclusions).

Polystyrene

Polystyrene is susceptible to photodegradation due to the presence of phenyl rings, which under UV radiation (200–300 nm), become excited and form a triplet state. As a result of UV absorption, the following changes are observed: main chain breakage, hydrogen atom stripping, and phenyl ring stripping. During the degradation process, macroradicals are formed, which in the next step, undergo oxidation in the presence of atmospheric oxygen, with the formation of superoxide radicals. Subsequent reactions lead to the formation of hydroperoxide, hydroxyl, and carboxyl groups. Eventually, chain scission occurs, forming carbonyl compounds, benzene, styrene, and olefin (Figure 5) [163,164]. In summary, polystyrene MPs can be formed by photodegradation. These particles will be further decomposed. It was shown that 12-month UV exposure and 2-month mechanical treatment of expanded PS (EPS) allowed the observation of 12,152 MPs [137]. The quantities of small-molecule degradation products released during irradiation can show both an upward trend (e.g., benzene and toluene) and a downward trend (e.g., styrene and 2-propenylbenzene) [70].

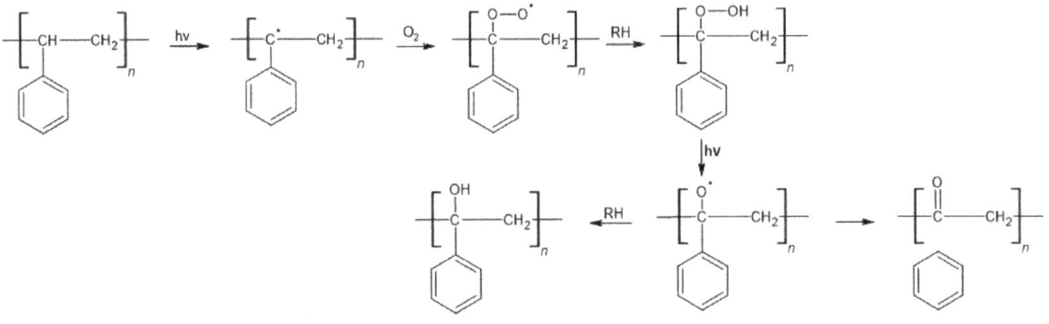

Figure 5. Photodegradation of PS (hv—to add energy via photons, RH—inclusions).

4.4. Chemical Degradation

The most important chemical factors affecting the degradation of plastics in an aqueous environment are the pH value and salinity of the water. High concentrations of H^+ or OH^- in the aqueous environment can catalyze the degradation of plastics that are susceptible to hydrolysis, such as polyamide (PA) [165]. These two factors can also affect the surface of MPs, their properties in aqueous environments, and their affinity for other contaminants. Polyethylene and PS in the form of MPs were studied by Liu et al. [166]. In the mentioned work, it was found that the presence of NaCl and $CaCl_2$ increases the sorption of both diethyl phthalate (DEP) and dibutyl phthalate (DBP) [166].

4.5. Biological Degradation

The biological degradation of plastics is determined by organisms (e.g., bacteria, fungi, and insects) that can destroy materials physically through biting, chewing [167], or biochemical processes [122,168]. The ingested plastics can be retained in the stomach, where fragmentation will occur, subsequently releasing particles [169]. This process can be accelerated by abiotic degradation, resulting in the formation of low molecular weight degradation products and the formation of cracks and pores on the surface of the plastic [170]. The biological degradation of PP with *Bacillus* sp. strain 27 and *Rhodococcus* sp. strain 36 made it possible to conclude that this is a process dependent on the type of microorganisms. In the first case, the weight loss was found to be 4.0%, and in the second, 6.4% [59]. In terms of degradation potential, the type of polymer is also important. It was found that bacteria degraded PP more easily than PE, while fungi degraded PE more easily than PP [58]. However, among synthetic polymers, aliphatic polyesters are the most susceptible to microbial degradation. It is widely believed that the ability of microorganisms to degrade synthetic polyesters is due to their chemical similarity to natural polyhydroxybutyrate (PHB), which is the backup material of many bacterial strains. Depending on the absence or presence of ester and amide groups, plastics can be attacked by various extracellular hydrolases. It is assumed that polyesters are degraded by enzymes, such as proteases, esterases, lipases, and cutinases.

The degradation of polymers that do not contain ester and amide groups by extracellular enzymes is a very complicated process. These polymers can be oxidized by O_2 with hydrolase catalysis, resulting in low-molecular-weight degradation products. Laccase enzyme has played a major role in PE degradation by *Rhodococcus ruber*. The activity of laccase is improved by the presence of copper. Hydroquinone peroxidase, on the other hand, was found to be responsible for PS degradation by *Azotobacter beijerinckii* HM121 [171]. The biological degradation of plastics can also be caused by algal enzymes [172].

Degradation with the participation of extracellular enzymes breaks polymer chains with the yield of shorter-chain polymers as well as oligomers, dimers, and single molecules [173]. Ultimately, plastics can be mineralized to CO_2 and H_2O under aerobic conditions and to CH_4, CO_2, organic acids, H_2O, and NH_4. Degradation of plastics under anaerobic conditions is energetically disadvantageous compared to degradation under aerobic conditions, and complete mineralization can take much longer [174].

5. Packaging Waste Dump

Packaging waste is ubiquitous. It can be transported from land to water and from water to land [175–178]. Microplastics formed in land and water can also move between different ecosystems [179,180]. These environments are commonly thought of as independent, but in fact, they are closely interconnected [181].

5.1. Terrestrial Environment

Microplastic in the terrestrial environment is formed by the fragmentation of larger plastics into smaller pieces due to exposure to UV radiation, wind action, agricultural activities, oxidation processes, and chemical and biological interactions [9,180,182]. The combined effects of the aforementioned factors can accelerate the aging of MPs, manifested by changes in color, crystallinity, chemical composition, and surface properties [183]. Microplastic in the terrestrial environment affects soil quality and biota [146,184]. For example, it has been found that the presence of MPs can significantly reduce the volume of phosphates available in the soil [185].

5.1.1. Sources and Transport of Microplastics in the Terrestrial Environment

Significant amounts of MPs are generated in landfills, peri-road areas, and agricultural areas [186,187]. Soil contamination can come from many sources, including compost [147], mulch film [188], greenhouse materials, irrigation tools [189], plant protection products, fertilizers [190], municipal solid waste, sewage treatment plants [191], used tires [119,164], and precipitation [192]. The presence of plastic particles in soils from China [193–195], Iran [196], Brazil [197], and Spain [198] was confirmed. It was found that the distribution of MPs in soils showed differences not only regionally but also in-depth [10]. The movement of MPs with groundwater can cause pollution of freshwater ecosystems, also contributing to marine pollution [199].

Plastic particles that reach the soil surface are transported to deeper layers of the soil through cultivation, infiltration, and animal activity [9,189,200]. Polyethylene beads can be transported from the soil surface down the soil profile by *Lumbricus terrestris* [200].

Polyethylene is the most commonly used polymer to study the degradation of plastics in soil. The degradation of this material was found to be increased by elevated pH and humidity. Polyethylene bags buried in soil for 2 years showed an increase in surface roughness. A nearly 5% decrease in weight was found for commercial carrier bags made of PE stored in mangrove soil over 8 weeks. This was due to the action of heterotrophic bacteria capable of producing hydrolytic enzymes [201]. In an experiment on the degradation of plastics buried in soil for 32 years, significant bleaching of LDPE film was observed, but no evidence of PS degradation was observed [202]. Thus, further studies are needed to determine the effects of individual polymers on soil properties and functions. These analyses should consider a wide range of particle sizes and shapes, as well as different types of substrates.

gastrointestinal tract has been proven by studying human feces [259–261]. Currently, it is considered that the sole measure of human exposure assessment on MPs is the discovery and quantification of plastic particles in these samples. The optical method is used to estimate MPs in human feces through involuntary ingestion [262]. Based on the conducted analysis, it has been determined that the concentration of PET in the feces of infants is ten times higher than in samples taken from adults [261]. Microplastic has also been detected in meconium samples. This fact, while disconcerting, is no surprise at all within the context of research that has confirmed the presence of MPs such as PE, PP, polyurethane (PU), PS, PVC, and poly(butylene succinate) (PBS) in the human placenta [263,264].

Table 4. The origin of MPs from various individual sources.

Sources of MPs	Quantity of MPs	Polymer Types	References
Effluents	50–86 MPs/dm^3	PE, PP, and PS	[243]
	840–3116 µg/dm^3	PE, PP, PET, PVC, and PMMA	[244]
	1.2–23.1 µg/ dm^3	PET	[245]
Bees	–	Polyester, PE, PVC, PU, epoxy resin, PAN, POM, PP, PS, PSU, PTFE, and PA	[246]
Boat ropes	11–822 MPs/m	PP, polysteel (a blend of PE and PP)	[247]
Take-out food containers	3–29 MPs/container	PS, polyester, rayon, acrylic, nylon, PE, PP, and PET (depending on the container type)	[239]
Disposable cups (pe-coated paper cups)	675–5984 MPs/dm^3	PE	[240]
Disposable cups (PP cups)	781–4951 MPs/dm^3	PP	
Disposable cups (PS cups)	838–5215 MPs/dm^3	PS	
Plastic bottles for children	16.2 million MPs/dm^3	PP (bottle material)	[241]
Cutting board	100–300 MPs/mm per cut	–	[242]

poly(vinyl chloride)—PVC; poly(methyl methacrylate)—PMMA; polyurethane—PU; polyacrylonitrile—PAN; polyoxymethylene—POM; polysulfone—PSU; polytetrafluoroethylene—PTFE; polyamide—PA.

While the vast literature shows that MP accumulates in living organisms [6,255,257,265], information regarding the harmfulness of this type of particle for people is limited (Figure 6). Since mice constitute a common mammal model, one should very precisely follow research regarding the influence of this pollution on the functioning of their organisms. It allows for an extrapolation of the results into humans with a perspective to assess health risks. Research on mice proved that such particles might cross the brain-blood barrier [257]. This information is particularly alarming because MPs presence was confirmed in human blood [255].

The impact of MPs on human health depends on many factors [266]. The size, shape, and chemical composition of plastic particles are the most significant ones [262]. However, personal characteristics are also important, including age, organism size, demographic features, and lifestyle [267].

Microplastics can induce oxidative stress in the cells. The immune system recognizes MP as an enemy, first reacting violently, the increase of antioxidation defense is observed, then the organism is weakened [268]. Thus, performing long-term research is important to discover the true effects of MPs on human health and the environment [269].

 **HaCaT keratinocyte cells**
Dermal fibroblasts

Pro-inflammatory effects
The reduction of viability

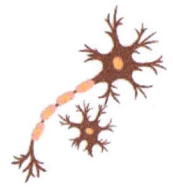 **Microglial HMC-3 cells**

Apoptosis

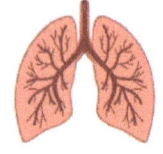 **Lung epithelial A549 cells**

Reduction of cell viability

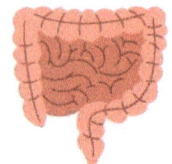 **Intestinal CCD-18Co cells**
Intestinal epithelial Caco-2

Metabolic changes
Mitochondrial depolarization

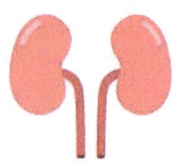 **Kidney Proximal Tubular Epithelial Cells HK-2**

Inflammation

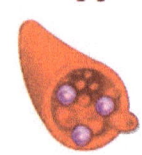 **Peripheral blood mononuclear cells**
Peripheral blood lymphocytes

Cytotoxic effects
Genotoxic effects

Figure 6. Toxicological studies and the implication of MPs contamination on human health.

8. Identification Methods of MPs

MPs are analyzed through several stages, such as separation, identification, visualization, and quantification. Techniques used to characterize MPs are mainly microscopic (optical microscopy, fluorescence microscopy, Scanning Electron Microscopy—SEM, Transmission Electron Microscopy—TEM, and Atomic Force Microscopy—AFM) and spectroscopic (Fourier Transform Infrared Spectroscopy—FT-IR, Raman Spectroscopy—RS, Nuclear Magnetic Resonance—NMR) methods [270–285] (Figure 7). They are mostly used to identify the polymeric composition of MPs, analysis of the shape, color, and size of the particles, as well as their quantity in test samples.

The combination of FTIR spectroscopy and optical microscopy (μ-FTIR) as well as Raman spectroscopy and microscopy (μ-Raman), are the two popular techniques used to identify MPs due to their sensitivity to small particles and accuracy in characterization. μ-FTIR methods are time-consuming, but sample preparation is relatively simple, making it a useful tool for the identification of particles up to 10 μm [274,276–280]. However, μ-Raman spectroscopy constitutes a reliable approach for analyzing particles as small as 1 μm [273]. With the application of these two methods, it is possible to characterize MPs accurately and reliably, making them invaluable tools for the analysis of plastic particles. These techniques are used to analyze environmental samples [272–282] (Table 5).

(a)

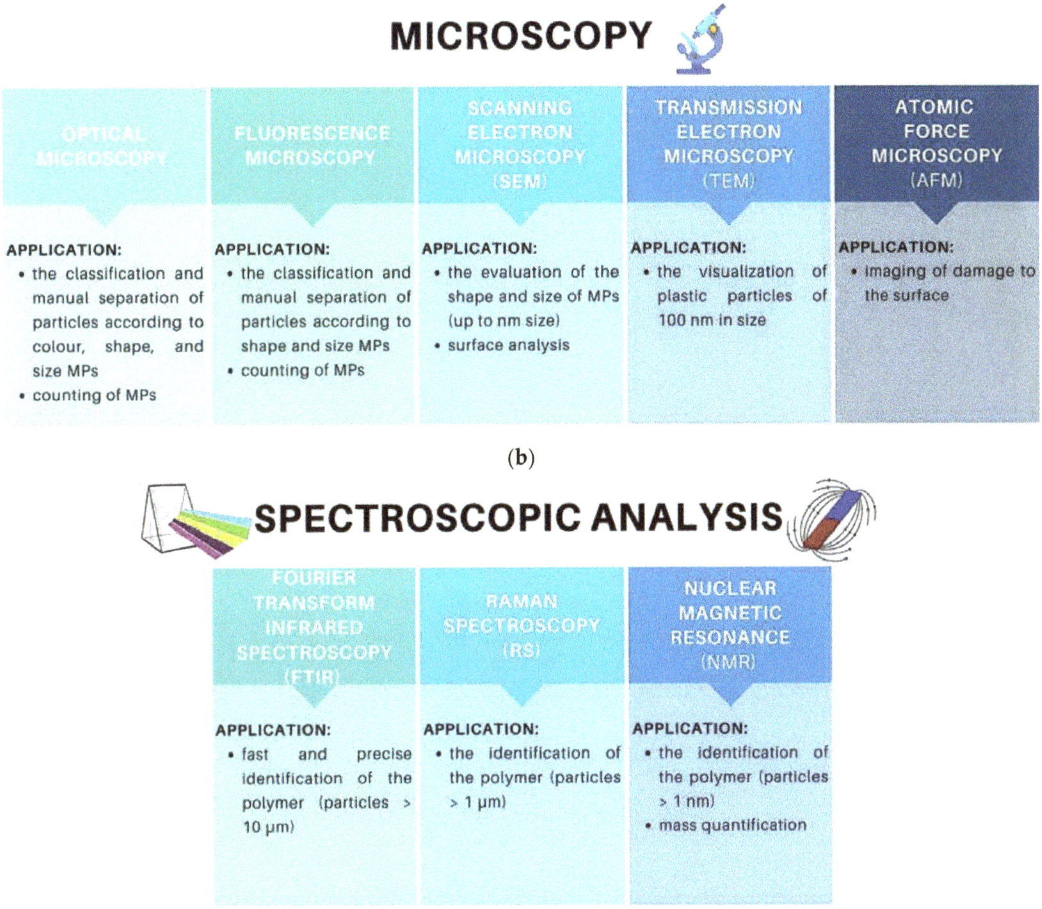

MICROSCOPY

OPTICAL MICROSCOPY	FLUORESCENCE MICROSCOPY	SCANNING ELECTRON MICROSCOPY (SEM)	TRANSMISSION ELECTRON MICROSCOPY (TEM)	ATOMIC FORCE MICROSCOPY (AFM)
APPLICATION: • the classification and manual separation of particles according to colour, shape, and size MPs • counting of MPs	**APPLICATION:** • the classification and manual separation of particles according to shape and size MPs • counting of MPs	**APPLICATION:** • the evaluation of the shape and size of MPs (up to nm size) • surface analysis	**APPLICATION:** • the visualization of plastic particles of 100 nm in size	**APPLICATION:** • imaging of damage to the surface

(b)

SPECTROSCOPIC ANALYSIS

FOURIER TRANSFORM INFRARED SPECTROSCOPY (FTIR)	RAMAN SPECTROSCOPY (RS)	NUCLEAR MAGNETIC RESONANCE (NMR)
APPLICATION: • fast and precise identification of the polymer (particles > 10 μm)	**APPLICATION:** • the identification of the polymer (particles > 1 μm)	**APPLICATION:** • the identification of the polymer (particles > 1 nm) • mass quantification

Figure 7. Identification methods of MPs (**a**) microscopy (**b**) spectroscopic analysis.

Table 5. Occurrence, analysis, and abundance of MPs in the environment.

Occurrence of MPs	Methods of MPs Analysis	Abundance of MPs	References
The Chukchi Sea, western Arctic Ocean	FTIR	0–18,815 MPs/km^2; 0–445 g/km^2	[272]
The open Baltic Sea	Optical microscope, FTIR	79 ± 18 MPs/m^3	[273]
Surface waters of the Kattegat/Skagerrak, Denmark	μ-FTIR	11–87 MPs/m^3	[274]
Estuarine surface water in Mauritius	Optical microscope, FTIR	249–412 MPs/dm^3	[275]
The water columns of catchments in Kamniška Bistrica, Slovenia	Optical microscope, FTIR	59 ± 16 MPs/m^3	[276]
The water columns of catchments in Ljubljanica, Slovenia	Optical microscope, FTIR, μ-FTIR	31 ± 14 MPs/m^3	[276]
Groundwater in the Haean Basin of Korea	μ-FTIR	0.02–3.48 MPs/dm^3	[277]
Estuarine sediments in Mauritius	Optical microscope, FTIR	74–235 MPs/kg	[275]
The sediments of catchments in Kamniška Bistrica, Slovenia	Optical microscope, FTIR	22 ± 20 MPs/kg	[276]
The sediments of catchments in Ljubljanica, Slovenia	Optical microscope, FTIR, μ-FTIR	23 ± 25 MPs/kg	[276]
The sediments of the Weser River catchment, Germany	μ-FTIR	99 ± 85 MPs/m^2	[278]
The atmosphere of the Northwestern Pacific Ocean	Optical microscope, μ-FTIR	0.0046–0.064 MPs/m^3	[279]
Air in the Gdańsk harbour	Optical microscope, μ-Raman	161 ± 75 MPs/m^3	[273]
Air of Baltic Sea	Optical microscope, μ-Raman	24 ± 9 MPs/m^3	[273]
Air of the Gotland Island	Optical microscope, μ-Raman	45 ± 20 MPs/m^3	[273]
Air in the Weser River catchment, Germany	RS	91 ± 47 MPs/m^3	[278]
Soil of the green park in Coimbra, Portugal	Optical microscope, μ-FTIR	158,000 MPs/kg	[280]
Soil of the landfill in Coimbra, Portugal	Optical microscope, μ-FTIR	150,000 MPs/kg	[280]
Soil of industrial area in Coimbra, Portugal	Optical microscope, μ-FTIR	127,000 MPs/kg	[280]
Soil of dump in Coimbra, Portugal	Optical microscope, μ-FTIR	126,000 MPs/kg	[280]
Soil of the forest in Coimbra, Portugal	Optical microscope, μ-FTIR	55,000 MPs/kg	[280]
Soil of Bhopal, India	Optical microscope, FTIR	2.5 ± 0.71–180 ± 13.44 MPs/kg	[281]

9. Conclusions

Plastic packaging waste is subjected to abiotic and biotic degradation processes. Factors affecting the said process can have synergistic or antagonistic effects. However, an analysis of the cases described in the scientific literature allows us to conclude that they are mostly convergent processes. They cause oxidation and disruption of polymer chains and lead to fragmentation, with the formation of MPs. The degradation of plastic packaging waste is a long and complex process that depends on the material's composition, physicochemical and mechanical properties, and its interaction with the environment. It is believed that the most important variables associated with plastic degradation are visible light and the presence of NO_2 and O_3. The changes in properties observed due to photodegradation and thermal and chemical degradation affect mechanical properties, particularly their elongation at break and tensile modulus. Degradation of the plastic in the environment is able to reduce the values of elongation at break, which decreases the value of external forces required for fragmentation. Moreover, it facilitates the fragmentation and formation of MPs.

Microplastic is ubiquitous. It is found in both terrestrial and aquatic environments. These particles accumulate in natural ecosystems and adversely impact some animals and plants, as well as affect soil functions. Microplastics can take up to 292 years to degrade in the deep sea. They can have adverse effects on the aquatic environment and especially on fish digestion, reproduction, and development. The particles of plastics cause changes in the size and distribution of soil aggregates, increase the evaporation rate of water and lead to groundwater contamination. Moist soils can cause them to release plasticizers. In addition, MPs affect plant growth and reduce seed germination, shoot length, and root biomass. They penetrate the intestinal walls of soil nematodes, resulting in oxidative stress and influencing gene expression.

Research results published to date show the prevalence of MP in food and beverage products. However, the evaluation of food contamination by plastic particles is still at a very early stage. In order to conduct it effectively, gaps in analytical methodologies and toxicity studies of such particles must be identified and eliminated. However, a correct analysis of the harmfulness of MP will not be possible without understanding the causes and products of plastic degradation.

Over the past 10 years, the number of scientific and popular science publications on MPs has increased dramatically. The topic of MPs is gaining increasing attention from scientists, the public, policymakers, and regulators. It is a problem already recognized internationally. Its solution may lie in conducting appropriate environmental education, introducing more efficient packaging waste management, and searching for, developing, and implementing effective and economically viable technologies for removing MPs from our environment.

Author Contributions: Conceptualization, K.K.-C. and M.G. (Małgorzata Grembecka); methodology, K.K.-C. and E.K.; formal analysis, E.O.-K. and M.G. (Magdalena Gierszewska); investigation, K.K.-C.; data curation, E.O.-K. and M.G. (Magdalena Gierszewska); writing—original draft preparation, K.K.-C. and E.K.; writing—review and editing, K.K.-C. and M.G. (Małgorzata Grembecka); visualization, K.K.-C.; supervision, K.K.-C. and M.G. (Małgorzata Grembecka). All authors have read and agreed to the published version of the manuscript.

Funding: This research received no external funding.

Institutional Review Board Statement: Not applicable.

Informed Consent Statement: Not applicable.

Data Availability Statement: Not applicable.

Conflicts of Interest: The authors declare no conflict of interest.

References

1. Rillig, M.C. Microplastic in Terrestrial Ecosystems and the Soil? *Environ. Sci. Technol.* **2012**, *46*, 6453–6454. [CrossRef] [PubMed]
2. García Rellán, A.; Vázquez Ares, D.; Vázquez Brea, C.; Francisco López, A.; Bello Bugallo, P.M. Sources, Sinks and Transformations of Plastics in Our Oceans: Review, Management Strategies and Modelling. *Sci. Total Environ.* **2023**, *854*, 158745. [CrossRef] [PubMed]
3. Jambeck, J.R.; Geyer, R.; Wilcox, C.; Siegler, T.R.; Perryman, M.; Andrady, A.; Narayan, R.; Law, K.L. Plastic Waste Inputs from Land into the Ocean. *Science* **2015**, *347*, 768–771. [CrossRef] [PubMed]
4. Geyer, R.; Jambeck, J.R.; Law, K.L. Production, Use, and Fate of All Plastics Ever Made. *Sci. Adv.* **2017**, *3*, e1700782. [CrossRef] [PubMed]
5. Schymanski, D.; Goldbeck, C.; Humpf, H.U.; Fürst, P. Analysis of Microplastics in Water by Micro-Raman Spectroscopy: Release of Plastic Particles from Different Packaging into Mineral Water. *Water Res.* **2018**, *129*, 154–162. [CrossRef]
6. Kadac-Czapska, K.; Knez, E.; Grembecka, M. Food and Human Safety: The Impact of Microplastics. *Crit. Rev. Food Sci. Nutr.* **2022**. [CrossRef]

7. Mattsson, K.; Jocic, S.; Doverbratt, I.; Hansson, L.-A. Nanoplastics in the Aquatic Environment. In *Microplastic Contamination in Aquatic Environments: An Emerging Matter of Environmental Urgency*; Zeng, E.Y., Ed.; Elsevier: Amsterdam, The Netherlands, 2018; Volume 13, pp. 379–399. [CrossRef]

8. Li, X.; Liang, R.; Li, Y.; Zhang, Y.; Wang, Y.; Li, K. Microplastics in Inland Freshwater Environments with Different Regional Functions: A Case Study on the Chengdu Plain. *Sci. Total Environ.* **2021**, *789*, 147938. [CrossRef]

9. Guo, J.J.; Huang, X.P.; Xiang, L.; Wang, Y.Z.; Li, Y.W.; Li, H.; Cai, Q.Y.; Mo, C.H.; Wong, M.H. Source, Migration and Toxicology of Microplastics in Soil. *Environ. Int.* **2020**, *137*, 105263. [CrossRef]

10. Zhao, S.; Zhang, Z.; Chen, L.; Cui, Q.; Cui, Y.; Song, D.; Fang, L. Review on Migration, Transformation and Ecological Impacts of Microplastics in Soil. *Appl. Soil Ecol.* **2022**, *176*, 104486. [CrossRef]

11. Wang, Y.; Zhou, B.; Chen, H.; Yuan, R.; Wang, F. Distribution, Biological Effects and Biofilms of Microplastics in Freshwater Systems—A Review. *Chemosphere* **2022**, *299*, 134370. [CrossRef]

12. Gao, L.; Wang, Z.; Peng, X.; Su, Y.; Fu, P.; Ge, C.; Zhao, J.; Yang, L.; Yu, H.; Peng, L. Occurrence and Spatial Distribution of Microplastics, and Their Correlation with Petroleum in Coastal Waters of Hainan Island, China. *Environ. Pollut.* **2022**, *294*, 118636. [CrossRef]

13. Aves, A.R.; Revell, L.E.; Gaw, S.; Ruffell, H.; Schuddeboom, A.; Wotherspoon, N.E.; LaRue, M.; McDonald, A.J. First Evidence of Microplastics in Antarctic Snow. *Cryosphere* **2022**, *16*, 2127–2145. [CrossRef]

14. Bao, R.; Wang, Z.; Qi, H.; Mehmood, T.; Cai, M.; Zhang, Y.; Yang, R.; Peng, L.; Liu, F. Occurrence and Distribution of Microplastics in Wastewater Treatment Plant in a Tropical Region of China. *J. Clean. Prod.* **2022**, *349*, 131454. [CrossRef]

15. Šaravanja, A.; Pušić, T.; Dekanić, T. Microplastics in Wastewater by Washing Polyester Fabrics. *Materials* **2022**, *15*, 2683. [CrossRef]

16. Yao, Y.; Glamoclija, M.; Murphy, A.; Gao, Y. Characterization of Microplastics in Indoor and Ambient Air in Northern New Jersey. *Environ. Res.* **2022**, *207*, 112142. [CrossRef]

17. Yu, Z.-f.; Song, S.; Xu, X.-l.; Ma, Q.; Lu, Y. Sources, Migration, Accumulation and Influence of Microplastics in Terrestrial Plant Communities. *Environ. Exp. Bot.* **2021**, *192*, 104635. [CrossRef]

18. Ding, J.; Li, J.; Sun, C.; Jiang, F.; Ju, P.; Qu, L.; Zheng, Y.; He, C. Detection of Microplastics in Local Marine Organisms Using a Multi-Technology System. *Anal. Methods* **2019**, *11*, 78–87. [CrossRef]

19. Huber, M.; Archodoulaki, V.M.; Pomakhina, E.; Pukánszky, B.; Zinöcker, E.; Gahleitner, M. Environmental Degradation and Formation of Secondary Microplastics from Packaging Material: A Polypropylene Film Case Study. *Polym. Degrad. Stab.* **2022**, *195*, 109794. [CrossRef]

20. Peixoto, D.; Pinheiro, C.; Amorim, J.; Oliva-Teles, L.; Guilhermino, L.; Vieira, M.N. Microplastic Pollution in Commercial Salt for Human Consumption: A Review. *Estuar. Coast. Shelf Sci.* **2019**, *219*, 161–168. [CrossRef]

21. Wang, Q.; Zhu, X.; Hou, C.; Wu, Y.; Teng, J.; Zhang, C.; Tan, H.; Shan, E.; Zhang, W.; Zhao, J. Microplastic Uptake in Commercial Fishes from the Bohai Sea, China. *Chemosphere* **2021**, *263*, 127962. [CrossRef]

22. Daniel, D.B.; Ashraf, P.M.; Thomas, S.N.; Thomson, K.T. Microplastics in the Edible Tissues of Shellfishes Sold for Human Consumption. *Chemosphere* **2021**, *264*, 128554. [CrossRef] [PubMed]

23. Ding, J.; Li, J.; Sun, C.; Jiang, F.; He, C.; Zhang, M.; Ju, P.; Ding, N.X. An Examination of the Occurrence and Potential Risks of Microplastics across Various Shellfish. *Sci. Total Environ.* **2020**, *739*, 139887. [CrossRef]

24. Chinfak, N.; Sompongchaiyakul, P.; Charoenpong, C.; Shi, H.; Yeemin, T.; Zhang, J. Abundance, Composition, and Fate of Microplastics in Water, Sediment, and Shellfish in the Tapi-Phumduang River System and Bandon Bay, Thailand. *Sci. Total Environ.* **2021**, *781*, 146700. [CrossRef] [PubMed]

25. Kedzierski, M.; Lechat, B.; Sire, O.; le Maguer, G.; le Tilly, V.; Bruzaud, S. Microplastic Contamination of Packaged Meat: Occurrence and Associated Risks. *Food Packag. Shelf Life* **2020**, *24*, 100489. [CrossRef]

26. Liu, Q.; Chen, Z.; Chen, Y.; Yang, F.; Yao, W.; Xie, Y. Microplastics Contamination in Eggs: Detection, Occurrence and Status. *Food Chem.* **2022**, *397*, 133771. [CrossRef]

27. Kapukotuwa, R.W.M.G.K.; Jayasena, N.; Weerakoon, K.C.; Abayasekara, C.L.; Rajakaruna, R.S. High Levels of Microplastics in Commercial Salt and Industrial Salterns in Sri Lanka. *Mar. Pollut. Bull.* **2022**, *174*, 113239. [CrossRef]

28. Lee, H.J.; Song, N.S.; Kim, J.S.; Kim, S.K. Variation and Uncertainty of Microplastics in Commercial Table Salts: Critical Review and Validation. *J. Hazard. Mater.* **2021**, *402*, 123743. [CrossRef]

29. Manimozhi, N.; Rani, V.; Sudhan, C.; Manimekalai, D.; Shalini, R.; Abarna, K.M. Spatiotemporal Occurrence, Distribution, and Characterization of Microplastics in Salt Pans of the Coastal Region of the Gulf of Mannar, Southeast Coast of India. *Reg. Stud. Mar. Sci.* **2022**, *53*, 102350. [CrossRef]

30. Afrin, S.; Rahman, M.M.; Hossain, M.N.; Uddin, M.K.; Malafaia, G. Are There Plastic Particles in My Sugar? A Pioneering Study on the Characterization of Microplastics in Commercial Sugars and Risk Assessment. *Sci. Total Environ.* **2022**, *837*, 155849. [CrossRef]

31. Oliveri Conti, G.; Ferrante, M.; Banni, M.; Favara, C.; Nicolosi, I.; Cristaldi, A.; Fiore, M.; Zuccarello, P. Micro- and Nano-Plastics in Edible Fruit and Vegetables. The First Diet Risks Assessment for the General Population. *Environ. Res.* **2020**, *187*, 109677. [CrossRef]

32. Tympa, L.-E.; Katsara, K.; Moschou, P.N.; Kenanakis, G.; Papadakis, V.M. Do Microplastics Enter Our Food Chain Via Root Vegetables? A Raman Based Spectroscopic Study on Raphanus Sativus. *Materials* **2021**, *14*, 2329. [CrossRef]

33. Koelmans, A.A.; Mohamed Nor, N.H.; Hermsen, E.; Kooi, M.; Mintenig, S.M.; de France, J. Microplastics in Freshwaters and Drinking Water: Critical Review and Assessment of Data Quality. *Water Res.* **2019**, *155*, 410–422. [CrossRef]

34. Kutralam-Muniasamy, G.; Pérez-Guevara, F.; Elizalde-Martínez, I.; Shruti, V.C. Branded Milks—Are They Immune from Microplastics Contamination? *Sci. Total Environ.* **2020**, *714*, 136823. [CrossRef]

35. Diaz-Basantes, M.F.; Conesa, J.A.; Fullana, A. Microplastics in Honey, Beer, Milk and Refreshments in Ecuador as Emerging Contaminants. *Sustainability* **2020**, *12*, 5514. [CrossRef]

36. al Naggar, Y.; Brinkmann, M.; Sayes, C.M.; Al-Kahtani, S.N.; Dar, S.A.; El-Seedi, H.R.; Grünewald, B.; Giesy, J.P. Are Honey Bees at Risk from Microplastics? *Toxics* **2021**, *9*, 109. [CrossRef]

37. Balzani, P.; Galeotti, G.; Scheggi, S.; Masoni, A.; Santini, G.; Baracchi, D. Acute and Chronic Ingestion of Polyethylene (PE) Microplastics Has Mild Effects on Honey Bee Health and Cognition. *Environ. Pollut.* **2022**, *305*, 119318. [CrossRef]

38. Prata, J.C.; Paço, A.; Reis, V.; da Costa, J.P.; Fernandes, A.J.S.; da Costa, F.M.; Duarte, A.C.; Rocha-Santos, T. Identification of Microplastics in White Wines Capped with Polyethylene Stoppers Using Micro-Raman Spectroscopy. *Food Chem.* **2020**, *331*, 127323. [CrossRef]

39. Hernandez, L.M.; Xu, E.G.; Larsson, H.C.E.; Tahara, R.; Maisuria, V.B.; Tufenkji, N. Plastic Teabags Release Billions of Microparticles and Nanoparticles into Tea. *Environ. Sci. Technol.* **2019**, *53*, 12300–12310. [CrossRef]

40. Afrin, S.; Rahman, M.; Akbor, A.; Siddique, A.B.; Uddin, K.; Malafaia, G. Is There Tea Complemented with the Appealing Flavor of Microplastics? A Pioneering Study on Plastic Pollution in Commercially Available Tea Bags in Bangladesh. *Sci. Total Environ.* **2022**, *837*, 155833. [CrossRef]

41. Shruti, V.C.; Pérez-Guevara, F.; Elizalde-Martínez, I.; Kutralam-Muniasamy, G. First Study of Its Kind on the Microplastic Contamination of Soft Drinks, Cold Tea and Energy Drinks—Future Research and Environmental Considerations. *Sci. Total Environ.* **2020**, *726*, 138580. [CrossRef]

42. Liu, S.; Guo, J.; Liu, X.; Yang, R.; Wang, H.; Sun, Y.; Chen, B.; Dong, R. Detection of Various Microplastics in Placentas, Meconium, Infant Feces, Breastmilk and Infant Formula: A Pilot Prospective Study. *Sci. Total Environ.* **2023**, *854*, 158699. [CrossRef] [PubMed]

43. Jadhav, E.B.; Sankhla, M.S.; Bhat, R.A.; Bhagat, D.S. Microplastics from Food Packaging: An Overview of Human Consumption, Health Threats, and Alternative Solutions. *Environ. Nanotechnol. Monit. Manag.* **2021**, *16*, 100608. [CrossRef]

44. Ivleva, N.P.; Wiesheu, A.C.; Niessner, R. Microplastic in Aquatic Ecosystems. *Angew. Chem. Int.* **2017**, *56*, 1720–1739. [CrossRef] [PubMed]

45. Sullivan, C.; Thomas, P.; Stuart, B. An Atomic Force Microscopy Investigation of Plastic Wrapping Materials of Forensic Relevance Buried in Soil Environments. *Aust. J. Forensic Sci.* **2019**, *51*, 596–605. [CrossRef]

46. Zhao, K.; Wei, Y.; Dong, J.; Zhao, P.; Wang, Y.; Pan, X.; Wang, J. Separation and Characterization of Microplastic and Nanoplastic Particles in Marine Environment. *Environ. Pollut.* **2022**, *297*, 118773. [CrossRef] [PubMed]

47. Rydz, J.; Musiol, M.; Zawidlak-Wegrzyńska, B.; Sikorska, W. Present and Future of Biodegradable Polymers for Food Packaging Applications. In *Biopolymers for Food Design*; Grumezescu, A.M., Holban, A.M., Eds.; Academic Press: Cambridge, MA, USA, 2018; Volume 14, pp. 431–467. [CrossRef]

48. Geueke, B.; Groh, K.; Muncke, J. Food Packaging in the Circular Economy: Overview of Chemical Safety Aspects for Commonly Used Materials. *J. Clean. Prod.* **2018**, *193*, 491–505. [CrossRef]

49. European Commission. Commission Regulation (EU) No 10/2011 of 14 January 2011 on Plastic Materials and Articles Intended to Come into Contact with Food. *Off. J. Eur. Union* **2011**, *12*, 1–89.

50. Hahladakis, J.N.; Iacovidou, E. Closing the Loop on Plastic Packaging Materials: What Is Quality and How Does It Affect Their Circularity? *Sci. Total Environ.* **2018**, *630*, 1394–1400. [CrossRef]

51. European Union. Directive (EU) 2019/904 of the European Parliament and of the Council of 5 June 2019 on the Reduction of the Impact of Certain Plastic Products on the Environment. *Off. J. Eur. Union* **2019**, *155*, 1–19.

52. Zettler, E.R.; Mincer, T.J.; Amaral-Zettler, L.A. Life in the "Plastisphere": Microbial Communities on Plastic Marine Debris. *Environ. Sci. Technol.* **2013**, *47*, 7137–7146. [CrossRef]

53. Fueser, H.; Mueller, M.T.; Weiss, L.; Höss, S.; Traunspurger, W. Ingestion of Microplastics by Nematodes Depends on Feeding Strategy and Buccal Cavity Size. *Environ. Pollut.* **2019**, *255*, 113227. [CrossRef]

54. He, Y.J.; Qin, Y.; Zhang, T.L.; Zhu, Y.Y.; Wang, Z.J.; Zhou, Z.S.; Xie, T.Z.; Luo, X.D. Migration of (Non-) Intentionally Added Substances and Microplastics from Microwavable Plastic Food Containers. *J. Hazard. Mater.* **2021**, *417*, 126074. [CrossRef]

55. Guan, Y.; Gong, J.; Song, B.; Li, J.; Fang, S.; Tang, S.; Cao, W.; Li, Y.; Chen, Z.; Ye, J.; et al. The Effect of UV Exposure on Conventional and Degradable Microplastics Adsorption for Pb (II) in Sediment. *Chemosphere* **2022**, *286*, 131777. [CrossRef]

56. Li, R.; Liu, Y.; Sheng, Y.; Xiang, Q.; Zhou, Y.; Cizdziel, J.V. Effect of Prothioconazole on the Degradation of Microplastics Derived from Mulching Plastic Film: Apparent Change and Interaction with Heavy Metals in Soil. *Environ. Pollut.* **2020**, *260*, 113988. [CrossRef]

57. Uheida, A.; Mejía, H.G.; Abdel-Rehim, M.; Hamd, W.; Dutta, J. Visible Light Photocatalytic Degradation of Polypropylene Microplastics in a Continuous Water Flow System. *J. Hazard. Mater.* **2021**, *406*, 124299. [CrossRef]

58. Wróbel, M.; Szymańska, S.; Kowalkowski, T.; Hrynkiewicz, K. Selection of Microorganisms Capable of Polyethylene (PE) and Polypropylene (PP) Degradation. *Microbiol. Res.* **2023**, *267*, 127251. [CrossRef]

59. Auta, H.S.; Emenike, C.U.; Jayanthi, B.; Fauziah, S.H. Growth Kinetics and Biodeterioration of Polypropylene Microplastics by *Bacillus* sp. and *Rhodococcus* sp. Isolated from Mangrove Sediment. *Mar. Pollut. Bull.* **2018**, *127*, 15–21. [CrossRef]

60. Hadiyanto, H.; Khoironi, A.; Dianratri, I.; Suherman, S.; Muhammad, F.; Vaidyanathan, S. Interactions between Polyethylene and Polypropylene Microplastics and *Spirulina* sp. Microalgae in Aquatic Systems. *Heliyon* **2021**, *7*, e07676. [CrossRef]
61. Yuan, J.; Cao, J.; Yu, F.; Ma, J. Microbial Degradation of Polystyrene Microplastics by a Novel Isolated Bacterium in Aquatic Ecosystem. *Sustain. Chem. Pharm.* **2022**, *30*, 100873. [CrossRef]
62. Dilara Hatinoglu, M.; Dilek Sanin, F. Fate and Effects of Polyethylene Terephthalate (PET) Microplastics during Anaerobic Digestion of Alkaline-Thermal Pretreated Sludge. *Waste Manag.* **2022**, *153*, 376–385. [CrossRef]
63. Auta, H.S.; Abioye, O.P.; Aransiola, S.A.; Bala, J.D.; Chukwuemeka, V.I.; Hassan, A.; Aziz, A.; Fauziah, S.H. Enhanced Microbial Degradation of PET and PS Microplastics under Natural Conditions in Mangrove Environment. *J. Environ. Manag.* **2022**, *304*, 114273. [CrossRef] [PubMed]
64. Ortiz, D.; Munoz, M.; Nieto-Sandoval, J.; Romera-Castillo, C.; de Pedro, Z.M.; Casas, J.A. Insights into the Degradation of Microplastics by Fenton Oxidation: From Surface Modification to Mineralization. *Chemosphere* **2022**, *309*, 136809. [CrossRef] [PubMed]
65. Jing, X.; Dong, J.; Huang, H.; Deng, Y.; Wen, H.; Xu, Z.; Ceylan, S. Interaction between Feedstocks, Absorbers and Catalysts in the Microwave Pyrolysis Process of Waste Plastics. *J. Clean. Prod.* **2021**, *291*, 125857. [CrossRef]
66. Kamalian, P.; Khorasani, S.N.; Abdolmaleki, A.; Karevan, M.; Khalili, S.; Shirani, M.; Neisiany, R.E. Toward the Development of Polyethylene Photocatalytic Degradation. *J. Polym. Eng.* **2020**, *40*, 181–191. [CrossRef]
67. Tofa, T.S.; Kunjali, K.L.; Paul, S.; Dutta, J. Visible Light Photocatalytic Degradation of Microplastic Residues with Zinc Oxide Nanorods. *Environ. Chem. Lett.* **2019**, *17*, 1341–1346. [CrossRef]
68. Amelia, D.; Fathul Karamah, E.; Mahardika, M.; Syafri, E.; Mavinkere Rangappa, S.; Siengchin, S.; Asrofi, M. Effect of Advanced Oxidation Process for Chemical Structure Changes of Polyethylene Microplastics. *Mater. Today Proc.* **2022**, *52*, 2501–2504. [CrossRef]
69. Yang, Y.; Chen, J.; Chen, Z.; Yu, Z.; Xue, J.; Luan, T.; Chen, S.; Zhou, S. Mechanisms of Polystyrene Microplastic Degradation by the Microbially Driven Fenton Reaction. *Water Res.* **2022**, *223*, 118979. [CrossRef]
70. Wu, X.; Chen, X.; Jiang, R.; You, J.; Ouyang, G. New Insights into the Photo-Degraded Polystyrene Microplastic: Effect on the Release of Volatile Organic Compounds. *J. Hazard. Mater.* **2022**, *431*, 128523. [CrossRef]
71. Lomonaco, T.; Manco, E.; Corti, A.; la Nasa, J.; Ghimenti, S.; Biagini, D.; di Francesco, F.; Modugno, F.; Ceccarini, A.; Fuoco, R.; et al. Release of Harmful Volatile Organic Compounds (VOCs) from Photo-Degraded Plastic Debris: A Neglected Source of environmental pollution. *J. Hazard. Mater.* **2020**, *394*, 122596. [CrossRef]
72. Kumar, V.; Maitra, S.S.; Singh, R.; Burnwal, D.K. Acclimatization of a Newly Isolated Bacteria in Monomer Tere-Phthalic Acid (TPA) May Enable It to Attack the Polymer Poly-Ethylene Tere-Phthalate(PET). *J. Environ. Chem. Eng.* **2020**, *8*, 103977. [CrossRef]
73. Khairul Anuar, N.F.S.; Huyop, F.; Ur-Rehman, G.; Abdullah, F.; Normi, Y.M.; Sabullah, M.K.; Abdul Wahab, R. An Overview into Polyethylene Terephthalate (PET) Hydrolases and Efforts in Tailoring Enzymes for Improved Plastic Degradation. *Int. J. Mol. Sci.* **2022**, *23*, 12644. [CrossRef]
74. Lin, Z.; Jin, T.; Zou, T.; Xu, L.; Xi, B.; Xu, D.; He, J.; Xiong, L.; Tang, C.; Peng, J.; et al. Current Progress on Plastic/Microplastic Degradation: Fact Influences and Mechanism. *Environ. Pollut.* **2022**, *304*, 119159. [CrossRef]
75. Garnai Hirsch, S.; Barel, B.; Segal, E. Characterization of Surface Phenomena: Probing Early Stage Degradation of Low-Density Polyethylene Films. *Polym. Eng. Sci.* **2019**, *59*, E129–E137. [CrossRef]
76. Chamas, A.; Moon, H.; Zheng, J.; Qiu, Y.; Tabassum, T.; Jang, J.H.; Abu-Omar, M.; Scott, S.L.; Suh, S. Degradation Rates of Plastics in the Environment. *ACS Sustain. Chem. Eng.* **2020**, *8*, 3494–3511. [CrossRef]
77. Boyle, D.; Catarino, A.I.; Clark, N.J.; Henry, T.B. Polyvinyl Chloride (PVC) Plastic Fragments Release Pb Additives That Are Bioavailable in Zebrafish. *Environ. Pollut.* **2020**, *263*, 114422. [CrossRef]
78. Rios-Fuster, B.; Alomar, C.; Paniagua González, G.; Garcinuño Martínez, R.M.; Soliz Rojas, D.L.; Fernández Hernando, P.; Deudero, S. Assessing Microplastic Ingestion and Occurrence of Bisphenols and Phthalates in Bivalves, Fish and Holothurians from a Mediterranean Marine Protected Area. *Environ. Res.* **2022**, *214*, 114034. [CrossRef]
79. Chen, H.; Zou, Z.; Tang, M.; Yang, X.; Tsang, Y.F. Polycarbonate Microplastics Induce Oxidative Stress in Anaerobic Digestion of Waste Activated Sludge by Leaching Bisphenol A. *J. Hazard. Mater.* **2023**, *443*, 130158. [CrossRef]
80. Barboza, L.G.A.; Cunha, S.C.; Monteiro, C.; Fernandes, J.O.; Guilhermino, L. Bisphenol A and Its Analogs in Muscle and Liver of Fish from the North East Atlantic Ocean in Relation to Microplastic Contamination. Exposure and Risk to Human Consumers. *J. Hazard. Mater.* **2020**, *393*, 122419. [CrossRef]
81. Ahmad, S.; Arsalan, A.; Hashmi, A.; Khan, M.A.; Siddiqui, W.A.; Younus, H. A Comparative Study Based on Activity, Conformation and Computational Analysis on the Inhibition of Human Salivary Aldehyde Dehydrogenase by Phthalate Plasticizers: Implications in Assessing the Safety of Packaged Food Items. *Toxicology* **2021**, *462*, 152947. [CrossRef]
82. Sheikh, I.A. Stereoselectivity and the Potential Endocrine Disrupting Activity of Di-(2-Ethylhexyl)Phthalate (DEHP) against Human Progesterone Receptor: A Computational Perspective. *J. Appl. Toxicol.* **2016**, *36*, 741–747. [CrossRef]
83. Duan, C.; Fang, Y.; Sun, J.; Li, Z.; Wang, Q.; Bai, J.; Peng, H.; Liang, J.; Gao, Z. Effects of Fast Food Packaging Plasticizers and Their Metabolites on Steroid Hormone Synthesis in H295R Cells. *Sci. Total Environ.* **2020**, *726*, 138500. [CrossRef] [PubMed]
84. Sree, C.G.; Buddolla, V.; Lakshmi, B.A.; Kim, Y.-J. Phthalate Toxicity Mechanisms: An Update. *Comp. Biochem. Physiol. Part-C Toxicol. Pharmacol.* **2023**, *263*, 109498. [CrossRef] [PubMed]

85. Zhang, J.; Yao, Y.; Pan, J.; Guo, X.; Han, X.; Zhou, J.; Meng, X. Maternal Exposure to Di-(2-Ethylhexyl) Phthalate (DEHP) Activates the PI3K/Akt/MTOR Signaling Pathway in F1 and F2 Generation Adult Mouse Testis. *Exp. Cell Res.* **2020**, *394*, 112151. [CrossRef] [PubMed]
86. Lucaccioni, L.; Trevisani, V.; Passini, E.; Righi, B.; Plessi, C.; Predieri, B.; Iughetti, L. Perinatal Exposure to Phthalates: From Endocrine to Neurodevelopment Effects. *Int. J. Mol. Sci.* **2021**, *22*, 4063. [CrossRef] [PubMed]
87. Sheikh, I.A.; Beg, M.A. Structural Characterization of Potential Endocrine Disrupting Activity of Alternate Plasticizers Di-(2-Ethylhexyl) Adipate (DEHA), Acetyl Tributyl Citrate (ATBC) and 2,2,4-Trimethyl 1,3-Pentanediol Diisobutyrate (TPIB) with Human Sex Hormone-Binding Globulin. *Reprod. Toxicol.* **2019**, *83*, 46–53. [CrossRef]
88. Rasmussen, L.M.; Sen, N.; Liu, X.; Craig, Z.R. Effects of Oral Exposure to the Phthalate Substitute Acetyl Tributyl Citrate on Female Reproduction in Mice. *J. Appl. Toxicol.* **2017**, *37*, 668–675. [CrossRef]
89. Giavina-Bianchi, P.; Kalil, J. Polyethylene Glycol Is a Cause of IgE-Mediated Anaphylaxis. *J. Allergy Clin. Immunol. Pract.* **2019**, *7*, 1874–1875. [CrossRef]
90. Garvey, L.H.; Nasser, S. Anaphylaxis to the First COVID-19 Vaccine: Is Polyethylene Glycol (PEG) the Culprit? *Br. J. Anaesth.* **2021**, *126*, e106–e108. [CrossRef]
91. Kim, J.I.; Lee, Y.A.; Shin, C.H.; Hong, Y.C.; Kim, B.N.; Lim, Y.H. Association of Bisphenol A, Bisphenol F, and Bisphenol S with ADHD Symptoms in Children. *Environ. Int.* **2022**, *161*, 107093. [CrossRef]
92. Sirohi, D.; al Ramadhani, R.; Knibbs, L.D. Environmental Exposures to Endocrine Disrupting Chemicals (EDCs) and Their Role in Endometriosis: A Systematic Literature Review. *Rev. Environ. Health* **2021**, *36*, 101–115. [CrossRef]
93. Yoo, M.H.; Lee, S.J.; Kim, W.; Kim, Y.; Kim, Y.B.; Moon, K.S.; Lee, B.S. Bisphenol A Impairs Renal Function by Reducing Na+/K+-ATPase and F-Actin Expression, Kidney Tubule Formation In Vitro and In Vivo. *Ecotoxicol. Environ. Saf.* **2022**, *246*, 114141. [CrossRef]
94. Ďurovcová, I.; Kyzek, S.; Fabová, J.; Makuková, J.; Gálová, E.; Ševčovičová, A. Genotoxic Potential of Bisphenol A: A Review. *Environ. Pollut.* **2022**, *306*, 119346. [CrossRef]
95. Tyner, M.D.W.; Maloney, M.O.; Kelley, B.J.B.; Combelles, C.M.H. Comparing the Effects of Bisphenol A, C, and F on Bovine Theca Cells in Vitro. *Reprod. Toxicol.* **2022**, *111*, 27–33. [CrossRef]
96. Chen, P.-P.; Liu, C.; Zhang, M.; Miao, Y.; Cui, F.-P.; Deng, Y.-L.; Luo, Q.; Zeng, J.-Y.; Shi, T.; Lu, T.-T.; et al. Associations between Urinary Bisphenol A and Its Analogues and Semen Quality: A Cross-Sectional Study among Chinese Men from an Infertility Clinic. *Environ. Int.* **2022**, *161*, 107132. [CrossRef]
97. Zhang, L.; Zhang, J.; Fan, S.; Zhong, Y.; Li, J.; Zhao, Y.; Ni, S.; Liu, J.; Wu, Y. A Case-Control Study of Urinary Concentrations of Bisphenol A, Bisphenol F, and Bisphenol S and the Risk of Papillary Thyroid Cancer. *Chemosphere* **2023**, *312*, 137162. [CrossRef]
98. Guo, S.; Zhao, Q.; Li, Y.; Chu, S.; He, F.; Li, X.; Sun, N.; Zong, W.; Liu, R. Potential Toxicity of Bisphenol A to α-Chymotrypsin and the Corresponding Mechanisms of Their Binding. *Spectrochim. Acta A Mol. Biomol. Spectrosc.* **2023**, *285*, 121910. [CrossRef]
99. Kolli, R.T.; Xu, Z.; Panduri, V.; Taylor, J.A. Differential Gene Expression in Bladder Tumors from Workers Occupationally Exposed to Arylamines. *Biomed. Res. Int.* **2021**, *2021*, 2624433. [CrossRef]
100. Chung, K.-T. Carcinogenicity Allergenicity and Lupus-Inducibility of Arylamines. *Front. Biosci.* **2016**, *8*, 748. [CrossRef]
101. Tipton, D.A.; Lewis, J.W. Effects of a Hindered Amine Light Stabilizer and a UV Light Absorber Used in Maxillofacial Elastomers on Human Gingival Epithelial Cells and Fibroblasts. *J. Prosthet. Dent.* **2008**, *100*, 220–231. [CrossRef]
102. Hirata-Koizumi, M.; Ise, R.; Kato, H.; Matsuyama, T.; Nishimaki-Mogami, T.; Takahashi, M.; Ono, A.; Ema, M.; Hirose, A. Transcriptome Analyses Demonstrate That Peroxisome Proliferator-Activated Receptor α (PPARα) Activity of an Ultraviolet Absorber, 2-(2'-Hydroxy-3',5'-Di-Tert-Butylphenyl)Benzotriazole, as Possible Mechanism of Their Toxicity and the Gender Differences. *J. Toxicol. Sci.* **2016**, *41*, 693–700. [CrossRef]
103. Denghel, H.; Göen, T. Determination of the UV Absorber 2-(2H-Benzotriazol-2-Yl)-4,6-Di-Tert-Pentylphenol (UV 328) and Its Oxidative Metabolites in Human Urine by Dispersive Liquid-Liquid Microextraction and GC–MS/MS. *J. Chromatogr. B* **2020**, *1144*, 122071. [CrossRef] [PubMed]
104. Bhutta, A.; Tao, J.; Li, J.; Arteel, G.E.; Monga, S.S.; Beier, J.I. Tu1286: Environmental Vinyl Chloride Exposure Aggravates Tumorigenesis in a Murine Model of Hepatocellular Cancer. *Gastroenterology* **2022**, *162*, S-1261. [CrossRef]
105. Lotti, M. Do Occupational Exposures to Vinyl Chloride Cause Hepatocellular Carcinoma and Cirrhosis? *Liver Int.* **2017**, *37*, 630–633. [CrossRef] [PubMed]
106. Frullanti, E.; la Vecchia, C.; Boffetta, P.; Zocchetti, C. Vinyl Chloride Exposure and Cirrhosis: A Systematic Review and Meta-Analysis. *Dig. Liver Dis.* **2012**, *44*, 775–779. [CrossRef] [PubMed]
107. Fedeli, U.; Girardi, P.; Mastrangelo, G. Occupational Exposure to Vinyl Chloride and Liver Diseases. *World J. Gastroenterol.* **2019**, *25*, 4885–4891. [CrossRef]
108. Liu, W.; Zhou, H.; Qiu, Z.; Liu, T.; Yuan, Y.; Guan, R.; Li, N.; Wang, W.; Li, X.; Zhao, C. Effect of Short-Chain Chlorinated Paraffins (SCCPs) on Lipid Membranes: Combination of Molecular Dynamics and Membrane Damage Experiments. *Sci. Total Environ.* **2021**, *775*, 144906. [CrossRef] [PubMed]
109. Ren, X.; Geng, N.; Zhang, H.; Wang, F.; Gong, Y.; Song, X.; Luo, Y.; Zhang, B.; Chen, J. Comparing the Disrupting Effects of Short-, Medium- and Long-Chain Chlorinated Paraffins on Cell Viability and Metabolism. *Sci. Total Environ.* **2019**, *685*, 297–307. [CrossRef]

110. Li, J.; Giesy, J.P.; Yu, L.; Li, G.; Liu, C. Effects of Tris(1,3-Dichloro-2-Propyl) Phosphate (TDCPP) in Tetrahymena Thermophila: Targeting the Ribosome. *Sci. Rep.* **2015**, *5*, 10562. [CrossRef]
111. Killilea, D.W.; Chow, D.; Xiao, S.Q.; Li, C.; Stoller, M.L. Flame Retardant Tris(1,3-Dichloro-2-Propyl)Phosphate (TDCPP) Toxicity Is Attenuated by N -Acetylcysteine in Human Kidney Cells. *Toxicol. Rep.* **2017**, *4*, 260–264. [CrossRef]
112. el Shanawany, S.; Foda, N.; Hashad, D.I.; Salama, N.; Sobh, Z. The Potential DNA Toxic Changes among Workers Exposed to Antimony Trioxide. *Environ. Sci. Pollut. Res.* **2017**, *24*, 12455–12461. [CrossRef]
113. Schildroth, S.; Osborne, G.; Smith, A.R.; Yip, C.; Collins, C.; Smith, M.T.; Sandy, M.S.; Zhang, L. Occupational Exposure to Antimony Trioxide: A Risk Assessment. *Occup. Environ. Med.* **2021**, *78*, 413–418. [CrossRef]
114. Lim, J.-S.; Lee, D.-H.; Jacobs, D.R. Association of Brominated Flame Retardants With Diabetes and Metabolic Syndrome in the U.S. Population, 2003–2004. *Diabetes Care* **2008**, *31*, 1802–1807. [CrossRef]
115. Deziel, N.C.; Alfonso-Garrido, J.; Warren, J.L.; Huang, H.; Sjodin, A.; Zhang, Y. Exposure to Polybrominated Diphenyl Ethers and a Polybrominated Biphenyl and Risk of Thyroid Cancer in Women: Single and Multi-Pollutant Approaches. *Cancer Epidemiol. Biomarkers Prev.* **2019**, *28*, 1755–1764. [CrossRef]
116. Rinninella, E.; Cintoni, M.; Raoul, P.; Mora, V.; Gasbarrini, A.; Mele, M.C. Impact of Food Additive Titanium Dioxide on Gut Microbiota Composition, Microbiota-Associated Functions, and Gut Barrier: A Systematic Review of In Vivo Animal Studies. *Int. J. Environ. Res. Public Health* **2021**, *18*, 2008. [CrossRef]
117. Hwang, J.-S.; Yu, J.; Kim, H.-M.; Oh, J.-M.; Choi, S.-J. Food Additive Titanium Dioxide and Its Fate in Commercial Foods. *Nanomaterials* **2019**, *9*, 1175. [CrossRef]
118. Cesa, F.S.; Turra, A.; Checon, H.H.; Leonardi, B.; Baruque-Ramos, J. Laundering and Textile Parameters Influence Fibers Release in Household Washings. *Environ. Pollut.* **2020**, *257*, 113553. [CrossRef]
119. Wagner, S.; Hüffer, T.; Klöckner, P.; Wehrhahn, M.; Hofmann, T.; Reemtsma, T. Tire Wear Particles in the Aquatic Environment— A Review on Generation, Analysis, Occurrence, Fate and Effects. *Water Res.* **2018**, *139*, 83–100. [CrossRef]
120. Sommer, F.; Dietze, V.; Baum, A.; Sauer, J.; Gilge, S.; Maschowski, C.; Gieré, R. Tire Abrasion as a Major Source of Microplastics in the Environment. *Aerosol Air Qual. Res.* **2018**, *18*, 2014–2028. [CrossRef]
121. Pal, P.; Pandey, J.P.; Sen, G. Synthesis and Application as Programmable Water Soluble Adhesive of Polyacrylamide Grafted Gum Tragacanth (GT-g-PAM). In *Biopolymer Grafting: Applications*; Thakur, V.K., Ed.; Elsevier: Amsterdam, The Netherlands, 2018; Volume 4, pp. 153–203. [CrossRef]
122. Crawford, C.B.; Quinn, B. Physiochemical Properties and Degradation. In *Microplastic Pollutants*; Crawford, C.B., Quinn, B., Eds.; Elsevier: Amsterdam, The Netherlands, 2017; Volume 4, pp. 57–100. [CrossRef]
123. Sohma, J. Mechanochemical Degradation. In *Comprehensive Polymer Science and Supplements*; Allen, G., Bevington, J.C., Eds.; Pergamon: Oxford, UK, 1989; Volume 23, pp. 621–644. [CrossRef]
124. el Hadri, H.; Gigault, J.; Maxit, B.; Grassl, B.; Reynaud, S. Nanoplastic from Mechanically Degraded Primary and Secondary Microplastics for Environmental Assessments. *NanoImpact* **2020**, *17*, 100206. [CrossRef]
125. Ammala, A.; Bateman, S.; Dean, K.; Petinakis, E.; Sangwan, P.; Wong, S.; Yuan, Q.; Yu, L.; Patrick, C.; Leong, K.H. An Overview of Degradable and Biodegradable Polyolefins. *Prog. Polym. Sci.* **2011**, *36*, 1015–1049. [CrossRef]
126. Pirsaheb, M.; Hossini, H.; Makhdoumi, P. Review of Microplastic Occurrence and Toxicological Effects in Marine Environment: Experimental Evidence of Inflammation. *Process Saf. Environ. Prot.* **2020**, *142*, 1–14. [CrossRef]
127. Singh, S.; Patil, T.; Tekade, S.P.; Gawande, M.B.; Sawarkar, A.N. Studies on Individual Pyrolysis and Co-Pyrolysis of Corn Cob and Polyethylene: Thermal Degradation Behavior, Possible Synergism, Kinetics, and Thermodynamic Analysis. *Sci. Total Environ.* **2021**, *783*, 147004. [CrossRef] [PubMed]
128. Maubane, L.; Lekalakala, R.; Orasugh, J.T.; Letwaba, J. Effect of Short-Chain Architecture on the Resulting Thermal Properties of Polypropylene. *Polymer* **2023**, *264*, 125533. [CrossRef]
129. Ahmed, L.; Zhang, B.; Hawkins, S.; Mannan, M.S.; Cheng, Z. Study of Thermal and Mechanical Behaviors of Flame Retardant Polystyrene-Based Nanocomposites Prepared Via In-Situ Polymerization Method. *J. Loss Prev. Process Ind.* **2017**, *49*, 228–239. [CrossRef]
130. Martín-Gullón, I.; Esperanza, M.; Font, R. Kinetic Model for the Pyrolysis and Combustion of Poly-(Ethylene Terephthalate) (PET). *J. Anal. Appl. Pyrolysis* **2001**, *58–59*, 635–650. [CrossRef]
131. Song, Y.K.; Hong, S.H.; Eo, S.; Shim, W.J. The Fragmentation of Nano- and Microplastic Particles from Thermoplastics Accelerated by Simulated-Sunlight-Mediated Photooxidation. *Environ. Pollut.* **2022**, *311*, 119847. [CrossRef] [PubMed]
132. Ainali, N.M.; Bikiaris, D.N.; Lambropoulou, D.A. Aging Effects on Low- and High-Density Polyethylene, Polypropylene and Polystyrene under UV Irradiation: An Insight into Decomposition Mechanism by Py-GC/MS for Microplastic Analysis. *J. Anal. Appl. Pyrolysis* **2021**, *158*, 105207. [CrossRef]
133. Kamweru, P.K.; Ndiritu, F.G.; Kinyanjui, T.K.; Muthui, Z.W.; Ngumbu, R.G.; Odhiambo, P.M. Study of Temperature and UV Wavelength Range Effects on Degradation of Photo-Irradiated Polyethylene Films Using DMA. *J. Macromol. Sci. Phys.* **2011**, *50*, 1338–1349. [CrossRef]
134. Andrady, A.L.; Hamid, H.S.; Torikai, A. Effects of Climate Change and UV-B on Materials. *Photochem. Photobiol. Sci.* **2003**, *2*, 68–72. [CrossRef]
135. Torikai, A.; Takeuchi, A.; Nagaya, S.; Fueki, K. Photodegradation of Polyethylene: Effect of Crosslinking on the Oxygenated Products and Mechanical Properties. *Polym. Photochem.* **1986**, *7*, 199–211. [CrossRef]

136. Liu, K.; Wang, Z.; Zhang, Y.; Xu, D.; Gao, J.; Ma, Z.; Wang, Y. Vapour-Liquid Equilibrium Measurements and Extractive Distillation Process Design for Separation of Azeotropic Mixture (Dimethyl Carbonate + ethanol). *J. Chem. Thermodyn.* **2019**, *133*, 10–18. [CrossRef]

137. Song, Y.K.; Hong, S.H.; Jang, M.; Han, G.M.; Jung, S.W.; Shim, W.J. Combined Effects of UV Exposure Duration and Mechanical Abrasion on Microplastic Fragmentation by Polymer Type. *Environ. Sci. Technol.* **2017**, *51*, 4368–4376. [CrossRef]

138. ter Halle, A.; Ladirat, L.; Gendre, X.; Goudouneche, D.; Pusineri, C.; Routaboul, C.; Tenailleau, C.; Duployer, B.; Perez, E. Understanding the Fragmentation Pattern of Marine Plastic Debris. *Environ. Sci. Technol.* **2016**, *50*, 5668–5675. [CrossRef]

139. Lin, C.-C.; Krommenhoek, P.J.; Watson, S.S.; Gu, X. Depth Profiling of Degradation of Multilayer Photovoltaic Backsheets after Accelerated Laboratory Weathering: Cross-Sectional Raman Imaging. *Sol. Energy Mater. Sol. Cells* **2016**, *144*, 289–299. [CrossRef]

140. Zweifel, H. Principles of Oxidative Degradation. In *Stabilization of Polymeric Materials. Macromolecular Systems—Materials Approach*; Abe, A., Monnerie, L., Shibaev, V., Suter, U.W., Tirrell, D., Ward, I.M., Eds.; Springer: Berlin/Heidelberg, Germany, 1998; Volume 1, pp. 1–40. [CrossRef]

141. Cheremisinoff, N.P.O. *Condensed Encyclopedia of Polymer Engineering Terms*, 1st ed.; Butterworth-Heinemann: Oxford, UK, 2001; pp. 193–199. [CrossRef]

142. McKeen, L.W. Introduction to the Weathering of Plastics. In *The Effect of UV Light and Weather on Plastics and Elastomers*, 3rd ed.; McKeen, L.W., Ed.; William Andrew: Norwich, CT, USA, 2019; Volume 2, pp. 17–41. [CrossRef]

143. Bank, M.S.; Hansson, S.V. The Plastic Cycle: A Novel and Holistic Paradigm for the Anthropocene. *Environ. Sci. Technol.* **2019**, *53*, 7177–7179. [CrossRef]

144. Carr, S.A.; Liu, J.; Tesoro, A.G. Transport and Fate of Microplastic Particles in Wastewater Treatment Plants. *Water Res.* **2016**, *91*, 174–182. [CrossRef]

145. Nizzetto, L.; Futter, M.; Langaas, S. Are Agricultural Soils Dumps for Microplastics of Urban Origin? *Environ. Sci. Technol.* **2016**, *50*, 10777–10779. [CrossRef]

146. Dissanayake, P.D.; Kim, S.; Sarkar, B.; Oleszczuk, P.; Sang, M.K.; Haque, M.N.; Ahn, J.H.; Bank, M.S.; Ok, Y.S. Effects of Microplastics on the Terrestrial Environment: A Critical Review. *Environ. Res.* **2022**, *209*, 112734. [CrossRef]

147. Bradney, L.; Wijesekara, H.; Palansooriya, K.N.; Obadamudalige, N.; Bolan, N.S.; Ok, Y.S.; Rinklebe, J.; Kim, K.H.; Kirkham, M.B. Particulate Plastics as a Vector for Toxic Trace-Element Uptake by Aquatic and Terrestrial Organisms and Human Health Risk. *Environ. Int.* **2019**, *131*, 104937. [CrossRef]

148. Peng, L.; Fu, D.; Qi, H.; Lan, C.Q.; Yu, H.; Ge, C. Micro- and Nano-Plastics in Marine Environment: Source, Distribution and Threats—A Review. *Sci. Total Environ.* **2020**, *698*, 134254. [CrossRef]

149. Ojeda, T.; Freitas, A.; Birck, K.; Dalmolin, E.; Jacques, R.; Bento, F.; Camargo, F. Degradability of Linear Polyolefins under Natural Weathering. *Polym. Degrad. Stab.* **2011**, *96*, 703–707. [CrossRef]

150. Hahladakis, J.N.; Velis, C.A.; Weber, R.; Iacovidou, E.; Purnell, P. An Overview of Chemical Additives Present in Plastics: Migration, Release, Fate and Environmental Impact during Their Use, Disposal and Recycling. *J. Hazard. Mater.* **2018**, *344*, 179–199. [CrossRef] [PubMed]

151. Rånby, B. Basic Reactions in the Photodegradation of Some Important Polymers. *J. Macromol. Sci.-Pure Appl. Chem.* **1993**, *30*, 583–594. [CrossRef]

152. Fairbrother, A.; Hsueh, H.C.; Kim, J.H.; Jacobs, D.; Perry, L.; Goodwin, D.; White, C.; Watson, S.; Sung, L.P. Temperature and Light Intensity Effects on Photodegradation of High-Density Polyethylene. *Polym. Degrad. Stab.* **2019**, *165*, 153–160. [CrossRef]

153. Gewert, B.; Plassmann, M.M.; MacLeod, M. Pathways for Degradation of Plastic Polymers Floating in the Marine Environment. *Environ. Sci. Process. Impacts* **2015**, *17*, 1513–1521. [CrossRef]

154. Takada, H.; Karapanagioti, H.K. (Eds.) *Hazardous Chemicals Associated with Plastics in the Marine Environment*, 1st ed.; Springer: Cham, Switzerland, 2019; Volume 78. [CrossRef]

155. Weber, R.; Watson, A.; Forter, M.; Oliaei, F. Review Article: Persistent Organic Pollutants and Landfills—A Review of Past Experiences and Future Challenges. *Waste Manag. Res.* **2011**, *29*, 107–121. [CrossRef]

156. He, P.; Chen, L.; Shao, L.; Zhang, H.; Lü, F. Municipal Solid Waste (MSW) Landfill: A Source of Microplastics?—Evidence of Microplastics in Landfill Leachate. *Water Res.* **2019**, *159*, 38–45. [CrossRef]

157. Su, Y.; Zhang, Z.; Wu, D.; Zhan, L.; Shi, H.; Xie, B. Occurrence of Microplastics in Landfill Systems and Their Fate with Landfill Age. *Water Res.* **2019**, *164*, 114968. [CrossRef]

158. Karlsson, T.M.; Hassellöv, M.; Jakubowicz, I. Influence of Thermooxidative Degradation on the in Situ Fate of Polyethylene in Temperate Coastal Waters. *Mar. Pollut. Bull.* **2018**, *135*, 187–194. [CrossRef]

159. Khoironi, A.; Hadiyanto, H.; Anggoro, S.; Sudarno, S. Evaluation of Polypropylene Plastic Degradation and Microplastic Identification in Sediments at Tambak Lorok Coastal Area, Semarang, Indonesia. *Mar. Pollut. Bull.* **2020**, *151*, 110868. [CrossRef]

160. Eubeler, J.P.; Bernhard, M.; Knepper, T.P. Environmental Biodegradation of Synthetic Polymers II. Biodegradation of Different Polymer Groups. *Trends Anal. Chem.* **2010**, *29*, 84–100. [CrossRef]

161. Wong, J.K.H.; Lee, K.K.; Tang, K.H.D.; Yap, P.S. Microplastics in the Freshwater and Terrestrial Environments: Prevalence, Fates, Impacts and Sustainable Solutions. *Sci. Total Environ.* **2020**, *719*, 137512. [CrossRef]

162. Pico, Y.; Alfarhan, A.; Barcelo, D. Nano- and Microplastic Analysis: Focus on Their Occurrence in Freshwater Ecosystems and Remediation Technologies. *Trends Anal. Chem.* **2019**, *113*, 409–425. [CrossRef]

163. Dris, R.; Gasperi, J.; Mirande, C.; Mandin, C.; Guerrouache, M.; Langlois, V.; Tassin, B. A First Overview of Textile Fibers, Including Microplastics, in Indoor and Outdoor Environments. *Environ. Pollut.* **2017**, *221*, 453–458. [CrossRef]
164. Kumar, M.; Xiong, X.; He, M.; Tsang, D.C.W.; Gupta, J.; Khan, E.; Harrad, S.; Hou, D.; Ok, Y.S.; Bolan, N.S. Microplastics as Pollutants in Agricultural Soils. *Environ. Pollut.* **2020**, *265*, 114980. [CrossRef]
165. Hocker, S.; Rhudy, A.K.; Ginsburg, G.; Kranbuehl, D.E. Polyamide Hydrolysis Accelerated by Small Weak Organic Acids. *Polymer* **2014**, *55*, 5057–5064. [CrossRef]
166. Liu, F.-f.; Liu, G.-z.; Zhu, Z.-l.; Wang, S.-c.; Zhao, F.-f. Interactions between Microplastics and Phthalate Esters as Affected by Microplastics Characteristics and Solution Chemistry. *Chemosphere* **2019**, *214*, 688–694. [CrossRef]
167. Mateos-Cárdenas, A.; O'Halloran, J.; van Pelt, F.N.A.M.; Jansen, M.A.K. Rapid Fragmentation of Microplastics by the Freshwater Amphipod Gammarus Duebeni (Lillj.). *Sci. Rep.* **2020**, *10*, 12799. [CrossRef]
168. Danso, D.; Chow, J.; Streit, W.R. Plastics: Environmental and Biotechnological Perspectives on Microbial Degradation. *Appl. Environ. Microbiol.* **2019**, *85*, e01095-19. [CrossRef]
169. Cau, A.; Avio, C.G.; Dessì, C.; Moccia, D.; Pusceddu, A.; Regoli, F.; Cannas, R.; Follesa, M.C. Benthic Crustacean Digestion Can Modulate the Environmental Fate of Microplastics in the Deep Sea. *Environ. Sci. Technol.* **2020**, *54*, 4886–4892. [CrossRef]
170. Wu, X.; Pan, J.; Li, M.; Li, Y.; Bartlam, M.; Wang, Y. Selective Enrichment of Bacterial Pathogens by Microplastic Biofilm. *Water Res.* **2019**, *165*, 114979. [CrossRef] [PubMed]
171. Nakamiya, K.; Sakasita, G.; Ooi, T.; Kinoshita, S. Enzymatic Degradation of Polystyrene by Hydroquinone Peroxidase of Azotobacter Beijerinckii HM121. *J. Ferment. Bioeng.* **1997**, *84*, 480–482. [CrossRef]
172. Priya, A.K.; Jalil, A.A.; Dutta, K.; Rajendran, S.; Vasseghian, Y.; Karimi-Maleh, H.; Soto-Moscoso, M. Algal Degradation of Microplastic from the Environment: Mechanism, Challenges, and Future Prospects. *Algal Res.* **2022**, *67*, 102848. [CrossRef]
173. Chen, X.; Xiong, X.; Jiang, X.; Shi, H.; Wu, C. Sinking of Floating Plastic Debris Caused by Biofilm Development in a Freshwater Lake. *Chemosphere* **2019**, *222*, 856–864. [CrossRef] [PubMed]
174. Gu, J.D. Microbiological Deterioration and Degradation of Synthetic Polymeric Materials: Recent Research Advances. *Int. Biodeterior. Biodegrad.* **2003**, *52*, 69–91. [CrossRef]
175. Moreira, F.T.; Prantoni, A.L.; Martini, B.; de Abreu, M.A.; Stoiev, S.B.; Turra, A. Small-Scale Temporal and Spatial Variability in the Abundance of Plastic Pellets on Sandy Beaches: Methodological Considerations for Estimating the Input of Microplastics. *Mar. Pollut. Bull.* **2016**, *102*, 114–121. [CrossRef]
176. Turrell, W.R. A Simple Model of Wind-Blown Tidal Strandlines: How Marine Litter Is Deposited on a Mid-Latitude, Macro-Tidal Shelf Sea Beach. *Mar. Pollut. Bull.* **2018**, *137*, 315–330. [CrossRef]
177. Zhang, K.; Chen, X.; Xiong, X.; Ruan, Y.; Zhou, H.; Wu, C.; Lam, P.K.S. The Hydro-Fluctuation Belt of the Three Gorges Reservoir: Source or Sink of Microplastics in the Water? *Environ. Pollut.* **2019**, *248*, 279–285. [CrossRef]
178. Zhang, K.; Su, J.; Xiong, X.; Wu, X.; Wu, C.; Liu, J. Microplastic Pollution of Lakeshore Sediments from Remote Lakes in Tibet Plateau, China. *Environ. Pollut.* **2016**, *219*, 450–455. [CrossRef]
179. Liu, P.; Shi, Y.; Wu, X.; Wang, H.; Huang, H.; Guo, X.; Gao, S. Review of the Artificially-Accelerated Aging Technology and Ecological Risk of Microplastics. *Sci. Total Environ.* **2021**, *768*, 144969. [CrossRef]
180. Karbalaei, S.; Hanachi, P.; Walker, T.R.; Cole, M. Occurrence, Sources, Human Health Impacts and Mitigation of Microplastic Pollution. *Environ. Sci. Pollut. Res.* **2018**, *25*, 36046–36063. [CrossRef]
181. Horton, A.A.; Dixon, S.J. Microplastics: An Introduction to Environmental Transport Processes. *WIREs Water* **2018**, *5*, e1268. [CrossRef]
182. Benítez, A.; Sánchez, J.J.; Arnal, M.L.; Müller, A.J.; Rodríguez, O.; Morales, G. Abiotic Degradation of LDPE and LLDPE Formulated with a Pro-Oxidant Additive. *Polym. Degrad. Stab.* **2013**, *98*, 490–501. [CrossRef]
183. Ren, Z.; Gui, X.; Xu, X.; Zhao, L.; Qiu, H.; Cao, X. Microplastics in the Soil-Groundwater Environment: Aging, Migration, and Co-Transport of Contaminants—A Critical Review. *J. Hazard. Mater.* **2021**, *419*, 126455. [CrossRef]
184. Zhang, K.; Hamidian, A.H.; Tubić, A.; Zhang, Y.; Fang, J.K.H.; Wu, C.; Lam, P.K.S. Understanding Plastic Degradation and Microplastic Formation in the Environment: A Review. *Environ. Pollut.* **2021**, *274*, 116554. [CrossRef]
185. Li, H.; Liu, L. Short-Term Effects of Polyethene and Polypropylene Microplastics on Soil Phosphorus and Nitrogen Availability. *Chemosphere* **2022**, *291*, 132984. [CrossRef]
186. Ng, E.L.; Huerta Lwanga, E.; Eldridge, S.M.; Johnston, P.; Hu, H.W.; Geissen, V.; Chen, D. An Overview of Microplastic and Nanoplastic Pollution in Agroecosystems. *Sci. Total Environ.* **2018**, *627*, 1377–1388. [CrossRef]
187. Wan, Y.; Wu, C.; Xue, Q.; Hui, X. Effects of Plastic Contamination on Water Evaporation and Desiccation Cracking in Soil. *Sci. Total Environ.* **2019**, *654*, 576–582. [CrossRef]
188. Boots, B.; Russell, C.W.; Green, D.S. Effects of Microplastics in Soil Ecosystems: Above and Below Ground. *Environ. Sci. Technol.* **2019**, *53*, 11496–11506. [CrossRef]
189. Bläsing, M.; Amelung, W. Plastics in Soil: Anal. Methods and Possible Sources. *Sci. Total Environ.* **2018**, *612*, 422–435. [CrossRef]
190. Katsumi, N.; Kusube, T.; Nagao, S.; Okochi, H. The Input–Output Balance of Microplastics Derived from Coated Fertilizeri Paddy Fields and the Timing of Their Discharge during the Irrigation Season. *Chemosphere* **2021**, *279*, 130574. [CrossRef] [PubMed]
191. Galafassi, S.; Nizzetto, L.; Volta, P. Plastic Sources: A Survey across Scientific and Grey Literature for Their Inventory and Relative Contribution to Microplastics Pollution in Natural Environments, with an Emphasis on Surface Water. *Sci. Total Environ.* **2019**, *693*, 133499. [CrossRef] [PubMed]

192. Dris, R.; Gasperi, J.; Saad, M.; Mirande, C.; Tassin, B. Synthetic Fibers in Atmospheric Fallout: A Source of Microplastics in the Environment? *Mar. Pollut. Bull.* **2016**, *104*, 290–293. [CrossRef] [PubMed]
193. Liu, Z.; Cai, L.; Dong, Q.; Zhao, X.; Han, J. Effects of Microplastics on Water Infiltration in Agricultural Soil on the Loess Plateau, China. *Agric. Water Manag.* **2022**, *271*, 107818. [CrossRef]
194. Hu, J.; He, D.; Zhang, X.; Li, X.; Chen, Y.; Wei, G.; Zhang, Y.; Ok, Y.S.; Luo, Y. National-Scale Distribution of Micro(Meso)Plastics in Farmland Soils across China: Implications for Environmental Impacts. *J. Hazard. Mater.* **2022**, *424*, 127283. [CrossRef]
195. Li, W.; Wang, S.; Wufuer, R.; Duo, J.; Pan, X. Distinct Soil Microplastic Distributions under Various Farmland-Use Types around Urumqi, China. *Sci. Total Environ.* **2023**, *857*, 159573. [CrossRef]
196. Nematollahi, M.J.; Keshavarzi, B.; Mohit, F.; Moore, F.; Busquets, R. Microplastic Occurrence in Urban and Industrial Soils of Ahvaz Metropolis: A City with a Sustained Record of Air Pollution. *Sci. Total Environ.* **2022**, *819*, 152051. [CrossRef]
197. da Silva Paes, E.; Gloaguen, T.V.; Silva, H. dos A. da C.; Duarte, T.S.; de Almeida, M. da C.; Costa, O.D.A.V.; Bomfim, M.R.; Santos, J.A.G. Widespread Microplastic Pollution in Mangrove Soils of Todos Os Santos Bay, Northern Brazil. *Environ. Res.* **2022**, *210*, 112952. [CrossRef]
198. Pérez-Reverón, R.; González-Sálamo, J.; Hernández-Sánchez, C.; González-Pleiter, M.; Hernández-Borges, J.; Díaz-Peña, F.J. Recycled Wastewater as a Potential Source of Microplastics in Irrigated Soils from an Arid-Insular Territory (Fuerteventura, Spain). *Sci. Total Environ.* **2022**, *817*, 152830. [CrossRef]
199. Rochman, C.M. Microplastics Research—From Sink to Source. *Science* **2018**, *360*, 28–29. [CrossRef]
200. Rillig, M.C.; Ziersch, L.; Hempel, S. Microplastic Transport in Soil by Earthworms. *Sci. Rep.* **2017**, *7*, 1362. [CrossRef]
201. Kumar, S.; Hatha, A.A.M.; Christi, K.S. Diversity and Effectiveness of Tropical Mangrove Soil Microflora on the Degradation of Polythene Carry Bags. *Rev. Biol. Trop.* **2007**, *55*, 777–786. [CrossRef]
202. Otake, Y.; Kobayashi, T.; Asabe, H.; Murakami, N.; Ono, K. Biodegradation of Low-Density Polyethylene, Polystyrene, Polyvinyl Chloride, and Urea Formaldehyde Resin Buried under Soil for over 32 Years. *J. Appl. Polym. Sci.* **1995**, *56*, 1789–1796. [CrossRef]
203. Ding, L.; Mao, R.F.; Guo, X.; Yang, X.; Zhang, Q.; Yang, C. Microplastics in Surface Waters and Sediments of the Wei River, in the Northwest of China. *Sci. Total Environ.* **2019**, *667*, 427–434. [CrossRef]
204. Liu, F.; Olesen, K.B.; Borregaard, A.R.; Vollertsen, J. Microplastics in Urban and Highway Stormwater Retention Ponds. *Sci. Total Environ.* **2019**, *671*, 992–1000. [CrossRef]
205. Huerta Lwanga, E.; Gertsen, H.; Gooren, H.; Peters, P.; Salánki, T.; van der Ploeg, M.; Besseling, E.; Koelmans, A.A.; Geissen, V. Incorporation of Microplastics from Litter into Burrows of Lumbricus Terrestris. *Environ. Pollut.* **2017**, *220*, 523–531. [CrossRef]
206. Qiu, Y.; Zhou, S.; Zhang, C.; Zhou, Y.; Qin, W. Soil Microplastic Characteristics and the Effects on Soil Properties and Biota: A Systematic Review and Meta-Analysis. *Environ. Pollut.* **2022**, *313*, 120183. [CrossRef]
207. Eriksen, M.; Mason, S.; Wilson, S.; Box, C.; Zellers, A.; Edwards, W.; Farley, H.; Amato, S. Microplastic Pollution in the Surface Waters of the Laurentian Great Lakes. *Mar. Pollut. Bull.* **2013**, *77*, 177–182. [CrossRef]
208. Mani, T.; Hauk, A.; Walter, U.; Burkhardt-Holm, P. Microplastics Profile along the Rhine River. *Sci. Rep.* **2016**, *5*, 17988. [CrossRef]
209. Rowley, K.H.; Cucknell, A.C.; Smith, B.D.; Clark, P.F.; Morritt, D. London's River of Plastic: High Levels of Microplastics in the Thames Water Column. *Sci. Total Environ.* **2020**, *740*, 140018. [CrossRef]
210. Ivar Do Sul, J.A.; Costa, M.F. The Present and Future of Microplastic Pollution in the Marine Environment. *Environ. Pollut.* **2014**, *185*, 352–364. [CrossRef] [PubMed]
211. Pham, C.K.; Ramirez-Llodra, E.; Alt, C.H.S.; Amaro, T.; Bergmann, M.; Canals, M.; Company, J.B.; Davies, J.; Duineveld, G.; Galgani, F.; et al. Marine Litter Distribution and Density in European Seas, from the Shelves to Deep Basins. *PLoS ONE* **2014**, *9*, e95839. [CrossRef] [PubMed]
212. Ho, N.H.E.; Not, C. Selective Accumulation of Plastic Debris at the Breaking Wave Area of Coastal Waters. *Environ. Pollut.* **2019**, *245*, 702–710. [CrossRef] [PubMed]
213. Pabortsava, K.; Lampitt, R.S. High Concentrations of Plastic Hidden beneath the Surface of the Atlantic Ocean. *Nat. Commun.* **2020**, *11*, 4073. [CrossRef] [PubMed]
214. Peeken, I.; Primpke, S.; Beyer, B.; Gütermann, J.; Katlein, C.; Krumpen, T.; Bergmann, M.; Hehemann, L.; Gerdts, G. Arctic Sea Ice Is an Important Temporal Sink and Means of Transport for Microplastic. *Nat. Commun.* **2018**, *9*, 1505. [CrossRef]
215. Beltrán-Sanahuja, A.; Casado-Coy, N.; Simó-Cabrera, L.; Sanz-Lázaro, C. Monitoring Polymer Degradation under Different Conditions in the Marine Environment. *Environ. Pollut.* **2020**, *259*, 113836. [CrossRef]
216. Andrady, A.L. Microplastics in the Marine Environment. *Mar. Pollut. Bull.* **2011**, *62*, 1596–1605. [CrossRef]
217. Zhao, Y.; Xiong, X.; Wu, C.; Xia, Y.; Li, J.; Wu, Y. Influence of Light and Temperature on the Development and Denitrification Potential of Periphytic Biofilms. *Sci. Total Environ.* **2018**, *613–614*, 1430–1437. [CrossRef]
218. Iñiguez, M.E.; Conesa, J.A.; Fullana, A. Recyclability of Four Types of Plastics Exposed to UV Irradiation in a Marine Environment. *Waste Manag.* **2018**, *79*, 339–345. [CrossRef]
219. Ioakeimidis, C.; Fotopoulou, K.N.; Karapanagioti, H.K.; Geraga, M.; Zeri, C.; Papathanassiou, E.; Galgani, F.; Papatheodorou, G. The Degradation Potential of PET Bottles in the Marine Environment: An ATR-FTIR Based Approach. *Sci. Rep.* **2016**, *6*, 23501. [CrossRef]
220. Williams, A.T.; Simmons, S.L. The Degradation of Plastic Litter in Rivers: Implications for Beaches. *J. Coast. Conserv.* **1996**, *2*, 63–72. [CrossRef]
221. Zhang, X.; Peng, X. How Long for Plastics to Decompose in the Deep Sea? *Geochem. Perspect. Lett.* **2022**, *22*, 20–25. [CrossRef]

222. Collard, F.; Gasperi, J.; Gabrielsen, G.W.; Tassin, B. Plastic Particle Ingestion by Wild Freshwater Fish: A Critical Review. *Environ. Sci. Technol.* **2019**, *53*, 12974–12988. [CrossRef]
223. Koongolla, J.B.; Lin, L.; Pan, Y.F.; Yang, C.P.; Sun, D.R.; Liu, S.; Xu, X.R.; Maharana, D.; Huang, J.S.; Li, H.X. Occurrence of Microplastics in Gastrointestinal Tracts and Gills of Fish from Beibu Gulf, South China Sea. *Environ. Pollut.* **2020**, *258*, 113734. [CrossRef]
224. Nanninga, G.B.; Scott, A.; Manica, A. Microplastic Ingestion Rates Are Phenotype-Dependent in Juvenile Anemonefish. *Environ. Pollut.* **2020**, *259*, 113855. [CrossRef]
225. Windsor, F.M.; Tilley, R.M.; Tyler, C.R.; Ormerod, S.J. Microplastic Ingestion by Riverine Macroinvertebrates. *Sci. Total Environ.* **2019**, *646*, 68–74. [CrossRef]
226. Li, B.; Su, L.; Zhang, H.; Deng, H.; Chen, Q.; Shi, H. Microplastics in Fishes and Their Living Environments Surrounding a Plastic Production Area. *Sci. Total Environ.* **2020**, *727*, 138662. [CrossRef]
227. Prata, J.C.; Venâncio, C.; Girão, A.V.; da Costa, J.P.; Lopes, I.; Duarte, A.C.; Rocha-Santos, T. Effects of Virgin and Weathered Polystyrene and Polypropylene Microplastics on Raphidocelis Subcapitata and Embryos of Danio Rerio under Environmental Concentrations. *Sci. Total Environ.* **2022**, *816*, 151642. [CrossRef]
228. Haider, T.P.; Völker, C.; Kramm, J.; Landfester, K.; Wurm, F.R. Plastics of the Future? The Impact of Biodegradable Polymers on the Environment and on Society. *Angew. Chem. Int.* **2019**, *58*, 50–62. [CrossRef]
229. Lee, S.-H.; Kim, M.-N. Isolation of Bacteria Degrading Poly(Butylene Succinate-Co-Butylene Adipate) and Their Lip A Gene. *Int. Biodeterior. Biodegrad.* **2010**, *64*, 184–190. [CrossRef]
230. al Hosni, A.S.; Pittman, J.K.; Robson, G.D. Microbial Degradation of Four Biodegradable Polymers in Soil and Compost Demonstrating Polycaprolactone as an Ideal Compostable Plastic. *Waste Manag.* **2019**, *97*, 105–114. [CrossRef]
231. Goto, T.; Kishita, M.; Sun, Y.; Sako, T.; Okajima, I. Degradation of Polylactic Acid Using Sub-Critical Water for Compost. *Polymers* **2020**, *12*, 2434. [CrossRef] [PubMed]
232. Itävaara, M.; Karjomaa, S.; Selin, J.-F. Biodegradation of Polylactide in Aerobic and Anaerobic Thermophilic Conditions. *Chemosphere* **2002**, *46*, 879–885. [CrossRef] [PubMed]
233. Bao, R.; Cheng, Z.; Hou, Y.; Xie, C.; Pu, J.; Peng, L.; Gao, L.; Chen, W.; Su, Y. Secondary Microplastics Formation and Colonized Microorganisms on the Surface of Conventional and Degradable Plastic Granules during Long-Term UV Aging in Various Environmental Media. *J. Hazard. Mater.* **2022**, *439*, 129686. [CrossRef] [PubMed]
234. Su, Y.; Cheng, Z.; Hou, Y.; Lin, S.; Gao, L.; Wang, Z.; Bao, R.; Peng, L. Biodegradable and Conventional Microplastics Posed Similar Toxicity to Marine Algae Chlorella Vulgaris. *Aquat. Toxicol.* **2022**, *244*, 106097. [CrossRef]
235. Fan, X.; Zou, Y.; Geng, N.; Liu, J.; Hou, J.; Li, D.; Yang, C.; Li, Y. Investigation on the Adsorption and Desorption Behaviors of Antibiotics by Degradable MPs with or without UV Ageing Process. *J. Hazard. Mater.* **2021**, *401*, 123363. [CrossRef]
236. Pironti, C.; Ricciardi, M.; Motta, O.; Miele, Y.; Proto, A.; Montano, L. Microplastics in the Environment: Intake through the Food Web, Human Exposure and Toxicological Effects. *Toxics* **2021**, *9*, 224. [CrossRef]
237. Senathirajah, K.; Attwood, S.; Bhagwat, G.; Carbery, M.; Wilson, S.; Palanisami, T. Estimation of the Mass of Microplastics Ingested—A Pivotal First Step Towards Human Health Risk Assessment. *J. Hazard. Mater.* **2021**, *404*, 124004. [CrossRef]
238. Cox, K.D.; Covernton, G.A.; Davies, H.J.; Dower, J.F.; Juanes, F.; Dudas, S.E. Human Consumption of Microplastics. *Environ. Sci. Technol.* **2019**, *53*, 7068–7074. [CrossRef]
239. Du, F.; Cai, H.; Zhang, Q.; Chen, Q.; Shi, H. Microplastics in Take-Out Food Containers. *J. Hazard. Mater.* **2020**, *399*, 122969. [CrossRef]
240. Chen, H.; Xu, L.; Yu, K.; Wei, F.; Zhang, M. Release of Microplastics from Disposable Cups in Daily Use. *Sci. Total Environ.* **2023**, *854*, 158606. [CrossRef]
241. Li, D.; Shi, Y.; Yang, L.; Xiao, L.; Kehoe, D.K.; Gun'ko, Y.K.; Boland, J.J.; Wang, J.J. Microplastic Release from the Degradation of Polypropylene Feeding Bottles During Infant Formula Preparation. *Nat. Food* **2020**, *1*, 746–754. [CrossRef]
242. Luo, Y.; Chuah, C.; Amin, A.; Khoshyan, A.; Gibson, C.T.; Tang, Y.; Naidu, R.; Fang, C. Assessment of Microplastics and Nanoplastics Released from a Chopping Board Using Raman Imaging in Combination with Three Algorithms. *J. Hazard. Mater.* **2022**, *431*, 128636. [CrossRef]
243. Hajji, S.; Ben-Haddad, M.; Abelouah, M.R.; De-la-Torre, G.E.; Alla, A.A. Occurrence, Characteristics, and Removal of Microplastics in Wastewater Treatment Plants Located on the Moroccan Atlantic: The Case of Agadir Metropolis. *Sci. Total Environ.* **2023**, *862*, 160815. [CrossRef]
244. Okoffo, E.D.; Rauert, C.; Thomas, K.V. Mass Quantification of Microplastic at Wastewater Treatment Plants by Pyrolysis-Gas Chromatography–Mass Spectrometry. *Sci. Total Environ.* **2023**, *856*, 159251. [CrossRef]
245. Tian, L.; Skoczynska, E.; van Putten, R.-J.; Leslie, H.A.; Gruter, G.-J.M. Quantification of Polyethylene Terephthalate Micro- and Nanoplastics in Domestic Wastewater Using a Simple Three-Step Method. *Sci. Total Environ.* **2023**, *857*, 159209. [CrossRef]
246. Carlos Edo, C.; Fernández-Alba, A.R.; Vejsnæs, F.; van der Steen, J.J.M.; Fernández-Piñas, F.; Rosal, R. Honeybees as Active Samplers for Microplastics. *Sci. Total Environ.* **2021**, *767*, 144481. [CrossRef]
247. Napper, I.E.; Wright, L.S.; Barrett, A.C.; Parker-Jurd, F.N.F.; Thompson, R.C. Potential Microplastic Release from the Maritime Industry: Abrasion of Rope. *Sci. Total Environ.* **2022**, *804*, 150155. [CrossRef]
248. Gautam, R.; Jo, J.-H.; Acharya, M.; Maharjan, A.; Lee, D.-E.; Bahadur, P.; Kim, C.-Y.; Kim, K.; Kim, H.-A.; Heo, Y. Evaluation of Potential Toxicity of Polyethylene Microplastics on Human Derived Cell Lines. *Sci. Total Environ.* **2022**, *838*, 156089. [CrossRef]

249. Zhang, Y.; Wang, S.; Olga, V.; Xue, Y.; Lv, S.; Diao, X.; Zhang, Y.; Han, Q.; Zhou, H. The Potential Effects of Microplastic Pollution on Human Digestive Tract Cells. *Chemosphere* **2022**, *291*, 132714. [CrossRef]
250. Bonanomi, M.; Salmistraro, N.; Porro, D.; Pinsino, A.; Colangelo, A.M.; Gaglio, D. Polystyrene Micro and Nano-Particles Induce Metabolic Rewiring in Normal Human Colon Cells: A Risk Factor for Human Health. *Chemosphere* **2022**, *303*, 134947. [CrossRef] [PubMed]
251. Wang, Y.-L.; Lee, Y.-H.; Hsu, Y.-H.; Chiu, I.-J.; Huang, C.C.-Y.; Huang, C.-C.; Chia, Z.-C.; Lee, C.-P.; Lin, Y.-F.; Chiu, H.-W. The Kidney-Related Effects of Polystyrene Microplastics on Human Kidney Proximal Tubular Epithelial Cells HK-2 and Male C57BL/6 Mice. *Environ. Health Perspect.* **2021**, *129*, 57003. [CrossRef] [PubMed]
252. Hwang, J.; Choi, D.; Han, S.; Choi, J.; Hong, J. An Assessment of the Toxicity of Polypropylene Microplastics in Human Derived Cells. *Sci. Total Environ.* **2019**, *684*, 657–669. [CrossRef] [PubMed]
253. Hwang, J.; Choi, D.; Han, S.; Jung, S.Y.; Choi, J.; Hong, J. Potential Toxicity of Polystyrene Microplastic Particles. *Sci. Rep.* **2020**, *10*, 7391. [CrossRef] [PubMed]
254. Lee, H.-S.; Amarakoon, D.; Wei, C.-I.; Choi, K.Y.; Smolensky, D.; Lee, S.-H. Adverse effect of polystyrene microplastics (PS-MPs) on tube formation and viability of human umbilical vein endothelial cells. *Food Chem. Toxicol.* **2021**, *154*, 112356. [CrossRef] [PubMed]
255. Leslie, H.A.; van Velzen, M.J.M.; Brandsma, S.H.; Vethaak, A.D.; Garcia-Vallejo, J.J.; Lamoree, M.H. Discovery and Quantification of Plastic Particle Pollution In Human Blood. *Environ. Int.* **2022**, *163*, 107199. [CrossRef]
256. Çobanoğlu, H.; Belivermiş, M.; Sıkdokur, E.; Kılıç, Ö.; Çayır, A. Genotoxic and Cytotoxic Effects of Polyethylene Microplastics on Human Peripheral Blood Lymphocytes. *Chemosphere* **2021**, *272*, 129805. [CrossRef]
257. Kwon, W.; Kim, D.; Kim, H.-Y.; Jeong, S.W.; Lee, S.-G.; Kim, H.-C.; Lee, Y.-J.; Kwon, M.K.; Hwang, J.-S.; Han, J.E.; et al. Microglial Phagocytosis of Polystyrene Microplastics Results in Immune Alteration and Apoptosis In Vitro and In Vivo. *Sci. Total Environ.* **2022**, *807*, 150817. [CrossRef]
258. Wu, B.; Wu, X.; Liu, S.; Wang, Z.; Chen, L. Size-Dependent Effects of Polystyrene Microplastics on Cytotoxicity and Efflux Pump Inhibition in Human Caco-2 Cells. *Chemosphere* **2019**, *221*, 333–341. [CrossRef]
259. Luqman, A.; Nugrahapraja, H.; Wahyuono, R.A.; Islami, I.; Haekal, M.H.; Fardiansyah, Y.; Putri, B.Q.; Amalludin, F.I.; Rofiqa, E.A.; Götz, F.; et al. Microplastic Contamination in Human Stools, Foods, and Drinking Water Associated with Indonesian Coastal Population. *Environments* **2021**, *8*, 138. [CrossRef]
260. Schwabl, P.; Köppel, S.; Königshofer, P.; Bucsics, T.; Trauner, M.; Reiberger, T.; Liebmann, B. Detection of Various Microplastics in Human Stool. *Ann. Intern. Med.* **2019**, *171*, 453–457. [CrossRef]
261. Zhang, J.; Wang, L.; Trasande, L.; Kannan, K. Occurrence of Polyethylene Terephthalate and Polycarbonate Microplastics in Infant and Adult Feces. *Environ. Sci. Technol. Lett.* **2021**, *8*, 989–994. [CrossRef]
262. Cho, Y.M.; Choi, K.H. The Current Status of Studies of Human Exposure Assessment of Microplastics and Their Health Effects: A Rapid Systematic Review. *Environ. Anal. Health Toxicol.* **2021**, *36*, e2021004. [CrossRef]
263. Braun, T.; Ehrlich, L.; Henrich, W.; Koeppel, S.; Lomako, I.; Schwabl, P.; Liebmann, B. Detection of Microplastic in Human Placenta and Meconium in a Clinical Setting. *Pharmaceutics* **2021**, *13*, 921. [CrossRef]
264. Zhu, L.; Zhu, J.; Zuo, R.; Xu, Q.; Qian, Y.; Lihui, A.N. Identification of Microplastics in Human Placenta Using Laser Direct Infrared Spectroscopy. *Sci. Total Environ.* **2023**, *856*, 159060. [CrossRef]
265. Li, W.; Chen, X.; Li, M.; Cai, Z.; Gong, H.; Yan, M. Microplastics as an aquatic pollutant affect gut microbiota within aquatic animals. *J. Hazard. Mater.* **2022**, *423*, 127094. [CrossRef]
266. Dusza, H.M.; Katrukha, E.A.; Nijmeijer, S.M.; Akhmanova, A.; Vethaak, A.D.; Walker, D.I.; Legler, J. Uptake, Transport, and Toxicity of Pristine and Weathered Micro- and Nanoplastics in Human Placenta Cells. *Environ. Health Perspect.* **2022**, *130*, 097006. [CrossRef]
267. Udovicki, B.; Andjelkovic, M.; Cirkovic-Velickovic, T.; Rajkovic, A. Microplastics in Food: Scoping Review on Health Effects, Occurrence, and Human Exposure. *Int. J. Food Contam.* **2022**, *9*, 7. [CrossRef]
268. Santana, M.F.M.; Moreira, F.T.; Turra, A. Trophic Transference of Microplastics under a Low Exposure Scenario: Insights on the Likelihood of Particle Cascading along Marine Food-Webs. *Mar. Pollut. Bull.* **2017**, *121*, 154–159. [CrossRef]
269. Boháčková, J.; Havlíčková, L.; Semerád, J.; Titov, I.; Trhlíková, O.; Beneš, H.; Cajthaml, T. In Vitro Toxicity Assessment of Polyethylene Terephthalate and Polyvinyl Chloride Microplastics Using Three Cell Lines From Rainbow Trout (*Oncorhynchus Mykiss*). *Chemosphere* **2023**, *312*, 136996. [CrossRef]
270. Bai, C.L.; Liu, L.Y.; Hu, Y.-B.; Zeng, E.Y.; Guo, Y. Microplastics: A Review of Analytical Methods, Occurrence and Characteristics in Food, and Potential Toxicities to Biota. *Sci. Total Environ.* **2022**, *806*, 150263. [CrossRef] [PubMed]
271. Lievens, S.; Slegers, T.; Mees, M.A.; Thielemans, W.; Poma, G.; Covaci, A.; Van Der Borght, M. A Simple, Rapid and Accurate Method for the Sample Preparation and Quantification of Meso- and Microplastics in Food and Food Waste Streams. *Environ. Pollut.* **2022**, *307*, 119511. [CrossRef] [PubMed]
272. Ikenoue, T.; Nakajima, R.; Fujiwara, A.; Onodera, J.; Itoh, M.; Toyoshima, J.; Watanabe, E.; Murata, A.; Nishino, S.; Kikuchi, T. Horizontal Distribution of Surface Microplastic Concentrations and Water-Column Microplastic Inventories in the Chukchi Sea, Western Arctic Ocean. *Sci. Total Environ.* **2023**, *855*, 159564. [CrossRef] [PubMed]

273. Ferrero, L.; Scibetta, L.; Markuszewski, P.; Mazurkiewicz, M.; Drozdowska, V.; Makuch, P.; Jutrzenka-Trzebiatowska, P.; Zaleska-Medynska, A.; Andò, S.; Saliu, F.; et al. Airborne and Marine Microplastics from an Oceanographic Survey at the Baltic Sea: An Emerging Role of Air-Sea Interaction? *Sci. Total Environ.* **2022**, *824*, 153709. [CrossRef] [PubMed]

274. Gunaalan, K.; Almeda, R.; Lorenz, C.; Vianello, A.; Iordachescu, L.; Papacharalampos, K.; Kiær, K.M.R.; Vollertsen, J.; Nielsen, T.G. Abundance and Distribution of Microplastics in Surface Waters of the Kattegat/Skagerrak (Denmark). *Environ. Pollut.* **2023**, *318*, 120853. [CrossRef]

275. Ragoobur, D.; Amode, N.S.; Somaroo, G.D.; Nazurally, N. Microplastics in Estuarine Water and Sediment in Mauritius. *Reg. Stud. Mar. Sci.* **2023**, *57*, 102766. [CrossRef]

276. Matjašič, T.; Mori, N.; Hostnik, I.; Bajt, O.; Viršek, M.K. Microplastic Pollution in Small Rivers Along Rural–Urban Gradients: Variations Across Catchments and Between Water Column and Sediments. *Sci. Total Environ.* **2023**, *858*, 160043. [CrossRef]

277. Cha, J.; Lee, J.-Y.; Chia, R.W. Microplastics Contamination and Characteristics of Agricultural Groundwater in Haean Basin of Korea. *Sci. Total Environ.* **2023**, *864*, 161027. [CrossRef]

278. Kernchen, S.; Löder, M.G.J.; Fischer, F.; Fischer, D.; Moses, S.R.; Georgi, C.; Nölscher, A.C.; Held, A.; Laforsch, C. Airborne Microplastic Concentrations and Deposition Across the Weser River Catchment. *Sci. Total Environ.* **2022**, *818*, 151812. [CrossRef]

279. Ding, J.; Sun, C.; He, C.; Zheng, L.; Dai, D.; Li, F. Atmospheric Microplastics in the Northwestern Pacific Ocean: Distribution, Source, and Deposition. *Sci. Total Environ.* **2022**, *829*, 154337. [CrossRef]

280. Leitão, I.A.; van Schaik, L.; Ferreira, A.J.D.; Alexandre, N.; Geissen, V. The Spatial Distribution of Microplastics in Topsoils of an Urban Environment—Coimbra City Case-Study. *Environ. Res.* **2023**, *218*, 114961. [CrossRef]

281. Singh, S.; Chakma, S.; Alawa, B.; Kalyanasundaram, M.; Diwan, V. Identification, Characterization, and Implications of Microplastics in Soil—A Case Study of Bhopal, Central India. *J. Hazard. Mater. Adv.* **2023**, *9*, 100225. [CrossRef]

282. Ricciardi, M.; Pironti, C.; Motta, O.; Miele, Y.; Proto, A.; Montano, L. Microplastics in the Aquatic Environment: Occurrence, Persistence, Analysis, and Human Exposure. *Water* **2021**, *13*, 973. [CrossRef]

283. Peez, N.; Rinesch, T.; Kolz, J.; Imhof, W. Applicable and cost-efficient microplastic analysis by quantitative ^{1}H-NMR spectroscopy using benchtop NMR and NoD methods. *Magn. Reson. Chem.* **2022**, *60*, 172–183. [CrossRef]

284. Mauel, A.; Pötzschner, B.; Meides, N.; Siegel, R.; Strohriegl, P.; Senker, J. Quantification of photooxidative defects in weathered microplastics using ^{13}C multiCP NMR spectroscopy. *RSC Adv.* **2022**, *12*, 10875–10885. [CrossRef]

285. Meides, N.; Mauel, A.; Menzel, T.; Altstädt, V.; Ruckdäschel, H.; Senker, J.; Strohriegl, P. Quantifying the fragmentation of polypropylene upon exposure to accelerated weathering. *Micropl. Nanopl.* **2022**, *2*, 23. [CrossRef]

Review

Alternative to Conventional Solutions in the Development of Membranes and Hydrogen Evolution Electrocatalysts for Application in Proton Exchange Membrane Water Electrolysis: A Review

Klara Perović *, Silvia Morović, Ante Jukić and Krešimir Košutić *

Faculty of Chemical Engineering and Technology, University of Zagreb, Marulićev trg 19, 10000 Zagreb, Croatia; smorovic@fkit.unizg.hr (S.M.); ajukic@fkit.unizg.hr (A.J.)
* Correspondence: kperovic@fkit.unizg.hr (K.P.); kkosutic@fkit.unizg.hr (K.K.)

Abstract: Proton exchange membrane water electrolysis (PEMWE) represents promising technology for the generation of high-purity hydrogen using electricity generated from renewable energy sources (solar and wind). Currently, benchmark catalysts for hydrogen evolution reactions in PEMWE are highly dispersed carbon-supported Pt-based materials. In order for this technology to be used on a large scale and be market competitive, it is highly desirable to better understand its performance and reduce the production costs associated with the use of expensive noble metal cathodes. The development of non-noble metal cathodes poses a major challenge for scientists, as their electrocatalytic activity still does not exceed the performance of the benchmark carbon-supported Pt. Therefore, many published works deal with the use of platinum group materials, but in reduced quantities (below 0.5 mg cm^{-2}). These Pd-, Ru-, and Rh-based electrodes are highly efficient in hydrogen production and have the potential for large-scale application. Nevertheless, great progress is needed in the field of water electrolysis to improve the activity and stability of the developed catalysts, especially in the context of industrial applications. Therefore, the aim of this review is to present all the process features related to the hydrogen evolution mechanism in water electrolysis, with a focus on PEMWE, and to provide an outlook on recently developed novel electrocatalysts that could be used as cathode materials in PEMWE in the future. Non-noble metal options consisting of transition metal sulfides, phosphides, and carbides, as well as alternatives with reduced noble metals content, will be presented in detail. In addition, the paper provides a brief overview of the application of PEMWE systems at the European level and related initiatives that promote green hydrogen production.

Keywords: proton exchange membrane water electrolysis (PEMWE); green hydrogen; renewable energy; electrocatalysts; noble metals; non-noble metals

Citation: Perović, K.; Morović, S.; Jukić, A.; Košutić, K. Alternative to Conventional Solutions in the Development of Membranes and Hydrogen Evolution Electrocatalysts for Application in Proton Exchange Membrane Water Electrolysis: A Review. *Materials* **2023**, *16*, 6319. https://doi.org/10.3390/ma16186319

Academic Editors: Enrico Negro and Daniela Fico

Received: 17 August 2023
Revised: 5 September 2023
Accepted: 19 September 2023
Published: 20 September 2023

1. Introduction

Excessive consumption of fossil fuels must be replaced with renewable ones, especially because the worldwide power demand will reach 24 or 26 TW by 2040, with emitted CO$_2$ emissions of 37–44 GT per year by 2040 [1]. Among the various fuel sources, hydrogen is often referred to as the "fuel of the future", with a high energy density of 140 MJ kg^{-1}, which is more than twice that of conventional solid fuels (50 MJ kg^{-1}) [2]. Other benefits of promoting the hydrogen economy include energy security by reducing oil imports, less pollution and better urban airy quality, sustainability by taking advantage of renewable energy sources, and economic viability by potentially shaping future global energy markets [3].

Currently, total global hydrogen production is about 500 billion cubic meters (bcm), with the produced hydrogen being extensively used in petroleum refining processes, in the petrochemical and chemical industries, in fuel cells, and in the fertilizer industry. Most of the hydrogen (about 96%) is produced from non-renewable fossil fuels, mostly by steam reforming of methane. Therefore, the hydrogen produced in this way is characterized by a lower purity and a high concentration of harmful greenhouse gases [2]. Only 4% of the hydrogen produced worldwide is produced in a renewable way by water electrolysis, which is a promising way of producing high-purity hydrogen without carbon emissions [1].

Synergies between hydrogen, electricity, and renewable energy sources are urgently needed [4]. Over the past decades, the increasing prices of electricity have postponed and hindered the production of electrolytic hydrogen, but this low percentage of its application is expected to increase with recent growth in energy capacity based on renewable sources like wind turbines and photovoltaics [5]. This is also supported by the revised Renewable Energy Directive 2018/2001/EU [6], which has established a new binding target for renewable energy in the EU for 2030 of at least 32%.

Proton exchange membrane electrolysis (PEMWE) is considered the most promising form of hydrogen production based on high efficiencies and suitable current densities even at moderate temperatures [4]. Combined with renewable energy sources, PEM electrolyzers can produce electrolytic hydrogen that can work as an energy vector/carrier and energy storage medium and thus overcome the intermittency of typical renewable energy sources [5]. After being stored, the hydrogen produced by electrolysis can be converted back into electricity when needed or used to refill fuel-cell-based cars [4].

Highly dispersed carbon-supported Pt- and Ir-based materials are currently being benchmark catalysts for the hydrogen evolution reaction (HER) and oxygen evolution reaction (OER) in PEMWE, but practical development to satisfy growing demand requires the use of cheaper catalysts [5,7]. The capital costs that currently make this technology less attractive could be lowered by reducing loading and/or substituting the expensive noble materials used for the fabrication of catalyst layers and hardware [5]. Furthermore, this article focuses on an overview of recent novel electrocatalysts that represent promising cathode materials for hydrogen production by PEM water electrolysis, considering non-noble metal alternatives together with reduced noble catalysts loading. The general mechanism of hydrogen evolution in water electrolysis, the specific performance of PEM water electrolyzers, and a summary of the novel electrocatalyst materials used with substituted and reduced noble metal contents are summarized.

2. Hydrogen Economy in Europe

Hydrogen represents the cornerstone of clean energy production; therefore, the European Commission proposed a Hydrogen Strategy for a Climate Neutral Europe in 2020 [8] in order to increase hydrogen supply and demand as well as to make the widespread production of hydrogen possible in 2050. Likewise, the strategy sets targets for the installation of at least 40 GW of renewable hydrogen electrolyzers by 2030. Furthermore, the European Commission has put together the "Fit for 55" packages, which relate to the EU's plan to reduce greenhouse gas emissions by at least 55% from 1990 levels by 2030, as required by European Climate Law [9,10]. The European Commission's proposed REPowerEU plan [11] aims to rapidly reduce the European Union's dependence on Russian fuel by setting a target of 10 million tons of renewable hydrogen produced domestically annually with an additional 10 million tons of hydrogen imported annually by 2030. To achieve these goals, the EU must rapidly increase the production of electrolyzers. According to industry estimates, to achieve the target of 10 million tons of renewable hydrogen produced in the EU, an electrolyzer capacity of 90–100 GW_{LHV} should be installed [12]. The electrolyzer manufacturing capacity should be scaled up significantly since the current capacity of electrolyzer manufacturers in Europe is estimated at 1.75 GW_{LHV} per year [12]. The development of innovative low-carbon technologies such as electrolyzers is supported by the EU Innovation Fund, which focuses on projects that can lead to significant emissions reductions. Hydrogen project promotors can also be supported by the European Investment Bank (EIB) [13] which, in recent years, has financed EUR 550 million in hydrogen projects.

Focusing exclusively on the application of PEM water electrolysis and the fulfillment of the European 2030 goals, the current operational state of PEM water electrolyzers and the goals that need to be achieved (Table 1) are given as part of the *Strategic Research and Innovation Agenda 2021–2027* published by the Clean Hydrogen Partnership in 2022 [14]. The set goals are primarily related to the reduction in capital and operating costs which will enable the achievement of market competitiveness of the mentioned technology for industrial application.

Table 1. Current operational state of PEM water electrolyzers and the predicted goals for 2024 and 2030. Modified from [14].

No.	Parameter	Unit	SoA	Targets	
			2020	2024	2030
1	Electricity consumption @ nominal capacity	kWh/kg	55	52	48
2	Capital cost	EUR/(kg/d) EUR/kW	2100 900	1550 700	1000 500
3	O&M cost	EUR/(kg/d)/y	41	30	21
4	Hot idle ramp time	s	2	1	1
5	Cold start ramp time	s	30	10	10
6	Degradation	%/1000 h	0.19	0.15	0.12
7	Current density	A/cm^2	2.2	2.4	3
8	Use of critical raw materials as catalysts	mg/W	2.5	1.25	0.25

Apart from the proposed plans [10,11], the European Commission has prepared a Hydrogen Public Funding Compass [15] to guide stakeholders to access information on the most important public funding programs and funds for renewable and low-carbon hydrogen. To accelerate Europe's 2030 hydrogen goals, potential hydrogen supply corridors are envisioned, including South Central Europe, the Iberian Peninsula, the North Sea, the Nordic-Baltic region, and the Eastern, and Southeastern H_2 supplying corridors (Figure 1) [16].

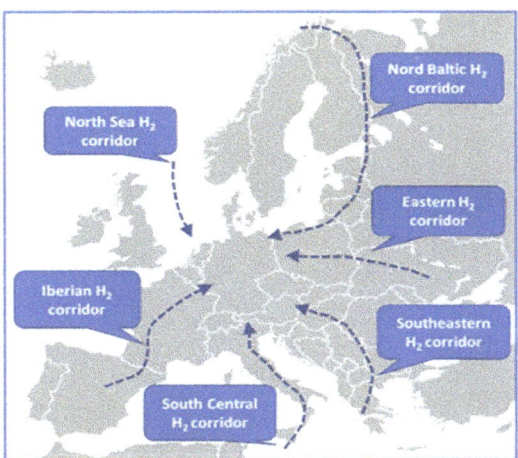

Figure 1. Hydrogen supply corridors in Europe. Modified from [16].

At the European Union level, there are currently several alliances and organizations promoting the development of the hydrogen economy. Examples include the European Clean Hydrogen Alliance [17], Hydrogen Europe [18], and Clean Hydrogen Joint Undertaking (Clean Hydrogen Partnership) [19]. In January 2023, the Clean Hydrogen Partnership, the successor of Fuel Cells and Hydrogen Joint Undertaking (FCH JU), selected nine hydrogen valley projects with a total funding of EUR 105.4 million [20]. Hydrogen valleys are regional ecosystems that demonstrate how hydrogen technologies work synergistically, but also complementarily, with other elements such as electricity grid, batteries, renewable energy production, gas infrastructure, etc.

As already mentioned, the rapid development of water electrolyzers and the intensification of scientific research in this field are closely related to the European Commission's intention to reduce energy dependence on Russia by expanding electrolyzer production capacity in the EU. In line with EU policy, electrolyzer manufacturers in Europe have agreed to increase their production capacities tenfold to 17.5 GW per year [21]. Table 2 provides a list of the main companies involved in the production of PEM electrolyzers in the EU.

Table 2. List of main companies involved in the manufacturing of PEM water electroyzer systems in Europe. Modified from [22].

Company	Manufacturing Site	Electrolyzer Type
AREVA H$_2$	France, Germany	PEM
CarboTech	Germany	PEM
Cummins—Hydrogenics	Belgium, Canada, Germany	PEM and ALKALINE
DeNora	Italy, Japan, USA	PEM and ALKALINE
iGas	Germany	PEM
ITM	UK	PEM
Nel Hydrogen	Denmark, Norway, USA	PEM and ALKALINE
Siemens Energy	Germany	PEM

3. Proton Exchange Membrane for Water Electrolysis (PEMWE)

The first water electrolyzer based on the concept of a solid polymer electrolyte was idealized by Thomas Grubb at General Electric in the 1960s, using a solid sulfonated polystyrene membrane as the electrolyte. This term also refers to proton exchange membrane or polymer electrolyte membrane (both with the acronym PEM) water electrolysis and, less commonly, to solid polymer electrolyte (SPE) water electrolysis [5]. The first PEM electrolysis journal was published in 1973 by Russell and co-workers [23] at General Electric, who used a PEM electrolysis cell to produce hydrogen from water splitting. It is interesting to note that at that time, the authors proposed the idea of using hydrogen as an energy storage device for off-peak periods in the power grid, as well as the possibility for its distribution and use for automotives. Nowadays, these are the main drivers for PEM electrolysis technology [24].

Water electrolysis technologies are divided into three categories depending on the electrolyte used: alkaline water electrolysis (AWE), PEMWE, and solid oxide water electrolysis (SOWE) [25]. PEM systems offer several advantages over the other two electrolysis technologies, such as higher hydrogen production rates, more compact design, and higher energy efficiency. Compared to alkaline electrolysis, which usually uses potassium hydroxide (KOH) solution as the electrolyte, the solid electrolyte membrane (thickness 50–250 μm) of the PEM electrolyzer significantly reduces gas crossover, allowing operation at high pressures [26].

In terms of commercial availability, AWE is emerging as the main competitor to PEM technology. A later section describes in more detail the commercial production and the use of PEM electrolyzers in the EU. Table 3 shows the basic differences between the mentioned technologies (AWE, PEMWE and SOWE) as well as the basic advantages and disadvantages of PEM technology in relation to the selected alkaline water electrolysis. It is also important to emphasize that, recently, more attention has been devoted to the electrolysis of seawater instead of scarce freshwater in the production of clean hydrogen. Although the mentioned technology is not the focus of this paper and will not be discussed in detail, it is important to mention that PEM electrolyzers and cells with a liquid electrolyte are considered the most suitable for the production of hydrogen from seawater electrolysis [27]. The most commonly used electrocatalysts for HER in seawater are various noble-metal-based materials, including noble-metal-based chalcogenides with cation vacancies, graphene-supported noble-metal-containing alloys, etc. [28].

Table 3. Comparison of water electrolysis technologies. Modified from [22,29].

Comparison between Technologies			
	AWE	PEMWE	SOWE
Operating temperature	70–90 °C	50–80 °C	700–850 °C
Operating pressure	1–30 bar	<70 bar	1 bar
Electrolyte	Potassium hydroxide (KOH) 5–7 mol L^{-1}	PFSA membranes	Yttria-stabilized zirconia (YSZ)
Separator	ZrO_2 stabilized with PPS mesh	Solid electrolyte (above)	Solid electrolyte (above)
Electrode/catalyst (oxygen side)	Nickel-coated perforated stainless steel	Iridium oxide	Perovskite-type (e.g., LSCF, LSM)
Electrode/catalyst (hydrogen side)	Nickel-coated perforated stainless steel	Platinum nanoparticles on carbon black	Ni/YSZ
Porous transport layer anode	Nickel mesh (not always present)	Platinum-coated sintered porous titanium	Coarse nickel mesh or foam
Porous transport layer cathode	Nickel mesh	Sintered porous titanium or carbon cloth	None
Bipolar plate anode	Nickel-coated stainless steel	Platinum-coated titanium	None
Bipolar plate cathode	Nickel-coated stainless steel	Gold-coated titanium	Cobalt-coated stainless steel
Frames and sealing	PSU, PTFE, EPDM	PTFE, PSU, ETFE	PTFE, silicon
PEMWE vs. AWE			
Advantages		**Disadvantages**	

Advantages	Disadvantages
Compact system design • fast heat-up and cool-off time, short response time; • low gas cross-permeation; • withstands higher operating pressures across the membrane; • higher purity of hydrogen and higher thermodynamic voltage; • easier hydrogen compression facilitates hydrogen storage. *Solid, thin electrolyte* • shorter proton transport route, lower ohmic loss; • operates under wide range of power input. *Operation at higher current density* • lower operational costs; • differential pressure across the electrolyte; • pressurized hydrogen side alone (avoidance of danger related to pressurized oxygen).	*Acidic electrolyte* • higher manufacturing cost due to expensive materials and components; • limited choices of stable earth abundant electrocatalysts for the oxygen evolution reaction (OER). *Solid, thin electrolyte* • it can be easily damaged by inappropriate operation (e.g., overheating) and cell design; • sensitive to imperfections, dust, and impurities.

The production of so-called green hydrogen by water electrolysis with the use of renewable electricity costs on average two to three times more than the production of so-called blue hydrogen obtained from natural gas by steam reforming. If the electricity input is added, the electrolyzer itself is the second largest cost component [22].

Compared to AWEs for hydrogen production, PEM electrolyzers are characterized by higher efficiency and current densities, as well as stable surrounding hydration conditions due to constant membrane exposure to the liquid phase of water, which makes the electrolytic membrane fully hydrated. There are several drawbacks that slow down the widespread application of this technology. First, a PEM water electrolyzer operating at pressures up to 70 bar can produce electrolytic-grade hydrogen and oxygen with high efficiency. As the pressure increases, the concentrations of hydrogen and oxygen produced can reach critical levels, increasing the risk of explosive gas mixture formation. To avoid this, the gas transfer of the generated gases must be reduced. The use of chemically and mechanically robust PFSA membranes can also be improved, considering their degradation, aging, and susceptibility to contamination.

Furthermore, commercialization of PEMWE technology requires testing of PEMWE cells under real (more aggressive) working conditions, since most of the current PEM water electrolyzers have been tested in the laboratory under stationary conditions (temperature, current density, and pressure). All this leads to the conclusion that large-scale implementation of PEMWE systems is hindered by high investment costs as well as the dependence on noble metal catalysts.

For these reasons, in the continuation of this paper, a detailed review of the scientific achievements and advances in the field of development of modified membranes and cathode materials with high application potential for hydrogen production by PEM water electrolysis is given, with the aim of presenting suitable materials that could affect initial investment cost reduction, as well as making this technology market competitive.

3.1. Components of PEMWE

The PEMWE system consists of a membrane electrode assembly, which actually consists of an integrated proton exchange membrane (PEM), an anode and cathode electrocatalyst, and a porous transport layer (PTL), also referred to as a gas diffusion layer (GDL) in the literature. The catalysts can be deposited directly on the membrane or on the PTL. Flow field plates or bipolar plates (BPs) serve as separation plates and allow heat, charge, and mass transfer (Figure 2) [30].

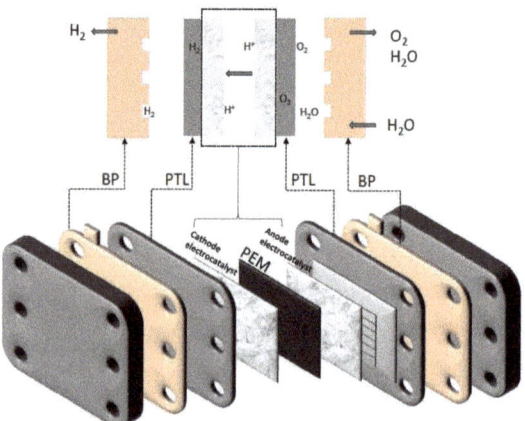

Figure 2. Key components of the PEMWE system. Modified from [30].

In the PEMWE system, water is supplied to the anode side of the cell, flows through the channels of BP and PTL, and reaches the anode catalyst layer. At the anode, the water molecules dissociate into O_2, protons, and electrons. O_2 evolves from the system, and electrons and protons pass through the membrane to the cathode catalyst layer, where H_2 is produced. The hydrogen formation reaction can proceed via two mechanisms: the Volmer–Heyrovsky mechanism, where protons are first adsorbed on the surface of the metal electrode, where M–H* is formed (Volmer step). In the Heyrovsky step, M–H* reacts with H$^+$ ions or electrons to form H$_2$. In contrast to the Volmer–Heyrovsky mechanism, in the Volmer–Tafel mechanism, M–H* can react with another M–H*, also forming H$_2$ [31].

There are various hydrogen storage systems that can be divided into physically based and material-based technologies. Physically based technologies include compressed gas, liquid, and cryocompressed forms [32]. The hydrogen produced by the reduction in protons on the cathode side must be stored in special vessels. To facilitate storage, it is recommended to produce H$_2$ at an elevated pressure and to use high-pressure electrolysis. High-pressure electrolysis has been successfully applied in practice, and several commercial systems operate at pressures up to 20 bar [33]. Further increasing the pressure to several hundred bar allows direct storage of the generated hydrogen in pressure vessels, but also requires overcoming certain difficulties. The biggest problem caused by working under high pressure is the so-called phenomenon of gas crossover, which occurs during electrolysis in the entire area of the proton exchange membrane, i.e., hydrogen and oxygen generated on both sides of the membrane could enter the membrane space and mix with each other, increasing the risk of gas explosion [34]. Besides all the advantages of physically based technologies for hydrogen storage, there are other disadvantages related to storage capacity, energy consumption, requirements for the vessels to withstand high pressures, and high costs, as well as safety issues related to leakage, bursting, and fire hazards. Due to the numerous drawbacks of physically based technologies, material-based (chemical) hydrogen storage technologies have been developed that offer a number of advantages, including higher hydrogen storage capacity at room temperature, lower hydrogen storage pressure, and slower hydrogen release rate. Metal hydrides, for example, have the potential to store both hydrogen and thermal energy. However, complex thermal management systems, expensive catalysts, and stability issues remain drawbacks, so most industries continue to rely on physically based hydrogen storage technologies [32,35].

3.1.1. Porous Transport Layers (PTLs)

PTLs are a porous medium located between PEM and BPs. They are used for liquid/gas transport, i.e., transport of water to the catalyst layer, where generation of O$_2$, H$^+$, and electrons takes place, and transfer of O$_2$/H$_2$ to the separator plates. Considering the functions of such a porous medium, the PTL material must meet certain requirements such as corrosion resistance, good electron conductivity, and good mechanical strength. Due to the highly acidic conditions, high overpotential, and the presence of O$_2$, the range of materials for producing such porous media is limited to metals or carbon materials [30].

Titanium and stainless steel are most commonly used for PTLs fabrication due to their high corrosion resistance [5,36]. Moreover, carbon materials can also be used for the fabrication of PTLs, but only for the fabrication of the cathode PTL, as the high oxidation potential of the anode would severely affect its mechanical strength. On the other hand, titanium is susceptible to passivation, which primarily leads to deterioration of durability and the formation of an oxide layer, which then affects the contact between PTLs and the fluid collector and also affects the conductivity of the medium. In addition to passivation, hydrogen embrittlement can also occur [30]. In the study by Rakousky et al., uncoated Ti–PTL was examined to determine its rate of degradation. It was found that the highest rate of degradation (194 µV h^{-1}) occurred at constant operation at 2 A cm^{-2} and was attributed to the voltage increase [37]. Hydrogen embrittlement and passivation can be prevented by coating titanium with noble metals, but this greatly affects the cost of the entire system. In their study, Kang et al. showed that novel gold (Au)-sputtered titanium thin/tunable liquid/gas diffusion layers (TT-LGDL) improved interfacial contacts and reduced ohmic and activation losses through their advantages of flat surface and thin structures [38]. Furthermore, Liu et al. showed that no degradation of the electrolyzer occurred due to the formation of a <10 nm thick IrOx layer on the iridium-coated PTL. It was shown that the Ir coating was oxidized, and the TiOx layer under the iridium was not further passivated, unlike the unprotected PTL, which may improve durability and reduce the cost of today's expensive PTLs, i.e., a balance should be found between capital cost and durability [39].

In general, Ti mesh, Ti felts, Ti foams, and sintered Ti powders are commonly used in PEMWE systems for the production of anode PTL [40,41]. On the other hand, the use of materials such as mesh, foam, etc., results in random pore size and pore size distribution. For this reason, researchers have focused on optimizing the structure, i.e., pore size and distribution, as well as the thickness of the porous medium. It was shown that the optimal pore size of PTL is between 10 and 13 μm [34], while the optimal porosity is about 30%. Therefore, it is important that the pores in the medium are not too large, as this reduces the efficiency of electron transport; the pores also cannot be too small, as the removal of the generated gas in the system is hindered, i.e., the transfer resistance increases.

3.1.2. Bipolar Plates (BPs)

BPs are multifunctional components whose main functions are to ensure the flow of reactants and products, the electrical connection of adjacent cells in the so-called stack, and mass and heat transport. Considering the harsh environment in PEMWE, the materials used for the production of BPs must have high mechanical strength, corrosion resistance, high conductivity, impermeability, and low cost [42].

The most commonly used materials for manufacturing BP are graphite, stainless steel, and titanium. Although graphite has high electrical conductivity, it has not proven to be a good option for PEMWE systems due to its low mechanical strength, susceptibility to corrosion, high cost, and the fact that it cannot be used for BP anodes but only for cathodes due to oxidation of the carbon surface. Considering that BP accounts for almost 50% of the total cost, 80% of the weight, and 50% of the volume of the entire PEMWE system, intensive work is being done to find suitable substitute materials and optimize the geometry [31,43,44].

Materials such as stainless steel, titanium, and their alloys have been shown to be good substitutes for graphite. However, as in the case of PTLs, titanium has excellent properties such as high mechanical strength and corrosion resistance, but is subject to passivation, i.e., the formation of an oxide layer. Therefore, coatings and alloys have been used as a solution for the protection of titanium and stainless steel plates. Titanium plates can be coated with noble metals to overcome deficiencies and achieve satisfactory durability and performance. For example, gold (Au) coatings can be applied to protect the titanium surface from oxidation under severe corrosion conditions. In the study by Jung et al., the stress degradation rate of the Au-coated titanium-based bipolar plate was shown to be five times lower than that of the conventional carbon bipolar plate [45]. However, coating with noble metals is very expensive from the perspective of large-scale electrolytic cells. Therefore, to reduce the cost, it is necessary to focus on the production of low-cost materials with satisfactory properties. Wakayama et al. investigated titanium (Ti) bipolar plates coated with titanium suboxide (Ti_4O_7) as a substitute for Pt-coated titanium bipolar plates as a cost-effective way to fabricate the PEM water electrolysis system. In their study, Ti_4O_7-sputtered Ti was shown to have a very low contact resistance (4–5 mΩ cm^{-2}) before and after voltage application, which is equivalent to that of gold or platinum coatings [46]. Moreover, in the study by Rojas et al., the multilayer Ti/TiN coating on SS 321 showed the best performance with 0.02% weight loss, current at 2 V_{SHE} to 436 μA cm^{-2}, and interfacial contact resistance after corrosion test of up to 9.9 mΩ cm^{-2}, indicating good protective properties of the coating [47].

In addition to optimizing the composition, optimizing the geometry of the flow channel also proved effective in increasing corrosion resistance and ensuring uniform flow distribution of reactants and products on the surface area of the field plate. Therefore, it is very important that the flow field plate ensures uniform distribution of reactants across the catalytic reaction surface in order to provide a way for collection of products and a conductive path to the reaction site. In the study by Toghyani et al., five flow field patterns were compared: parallel, one-way serpentine, two-way serpentine, three-way serpentine, and four-way serpentine to determine the best performance in terms of distribution of the molar fraction of hydrogen produced, current density, temperature, and pressure drop. It was found that the two-way serpentine provides the best performance for PEM electrolyzers due to relatively higher hydrogen production rate, more uniform temperature distribution, and reasonable pressure drop [48].

PFSA–PEM Membranes

A thin perfluorosulfonic acid (PFSA) membrane, known by the trade name Nafion®, is most commonly used as a solid electrolyte in PEMWE systems. Nafion® is a fluoropolymer of sulfonated polytetrafluoroethylene first commercialized by DuPont (Wilmington, DE, USA) in the mid-1970s. It is commercially available in thicknesses of approximately 100 µm [49]. Nafion® consists of a neutral semi-crystalline polymer backbone (polytetrafluoroethylene-PTFE) and a randomly tethered side chain of polysulfonyl fluoride vinyl ether with the ionic group SO_3^- linked to a specific counterion. Considering the length of its side chain, Nafion® can be described as "long-side chain" (LSC) polymer. The nature of the covalently bonded side chain and backbone result in phase separation, which is further enhanced by the addition of solvent molecules (water). Due to its phase-separated morphology, Nafion® has demonstrated excellent ion and solvent transport capabilities. Moreover, the main advantages of Nafion® are also its high proton conductivity and water permeability, as well as its high chemical and mechanical resistance. Despite its exceptional properties, there are certain disadvantages, which are mainly reflected in its high price, high disposal costs due to the presence of fluorine in the material skeleton, and reduced proton conductivity under high temperature conditions (>100 °C) [50,51].

Currently, state-of-the-art commercially available membranes, in addition to Nafion®, are Flemion® (Asahi Glass, Tokyo, Japan) and Aciplex® (Asahi Kasei, Tokyo, Japan), which are polymers with identical structures to Nafion®, but with shorter side chains. Although Flemion® and Aciplex® have good properties, they are chemically less stable than Nafion®. In the 1980s, Dow Chemical Company (Midland, MI, USA) introduced a perfluorinated ionomer with a short side chain (SSC-Dow Membrane) without a fluoroether group in the side chain, containing only two -CF$_2$ groups. In addition, Solvay Specialty Polymers (Brussels, Belgium) launched HyflonR Ion (known as Aquivion®). During the same period, the 3M™ Corporation (Maplewood, MN, USA) developed an ionomer (the 3M™ ionomer) with a fluoroether-free side chain containing four -CF$_2$ groups (the structures are given in Table 4). In addition, GORE—SELECT, a reinforced composite membrane, was introduced by W. L. Gore & Associates (Newark, DE, USA) by incorporating a strong hydrophobic reinforcement layer into a PFSA ionomer that improves dimensional stability in response to hydration [50,52].

Table 4. Commercially available PFSA membranes.

	Manufacturer	Structure	Parameters	Ref.
Nafion®	DuPont (Wilmington, DE, USA)		m = 1; n = 2; x = 5–13.5; y = 1000 e.g., EW * = 1000; x = ~5.5	[50,53]
Flemion®	Asahi Glass (Tokyo, Japan)		m = 0 or 1; n = 1–5	[53–55]
3M®	3M™ Corporation (Maplewood, MN, USA)		m = 0; n = 4; x = ~3–5 for EW = 660–825; e.g., for EW = 1000; x = ~6.5 or EW = 700; x = ~3	[50,53,56]
Aciplex®	Asahi Kasei (Tokyo, Japan)		M = 0–3; n = 2–5; x = 1.5–14 e.g., for EW = 1130; x = ~7	[53–55,57]
Aquivion®/Dow SSC®	Solvay Specialty Polymers (Brussels, Belgium)		m = 0; n = 2; x = 3.6–10 e.g., for EW = 1000; x = ~7	[50,53]

* EW stands for equivalent weight, grams of dry polymer per ionic group [g mol^{-1}].

There are several criteria for high efficiency of PEM membranes, such as high proton conductivity, low gas crossover capability, good thermal properties, low swelling ratio, sufficient water absorption, and high mechanical and chemical resistance [58,59]. In order to overcome the shortcomings of PFSA membranes and thus improve membrane performance, various modification strategies have been developed to address the decomposition and performance issues. To improve structure, stability, and performance, the side chain chemistry of PFSA can be modified by doping or chain modification itself, or the membrane can serve as a host for reinforcements and additives. Ionomers can be impregnated or doped with radical scavengers (to reduce radical formation), with inert hydrophobic mechanical support layers (to improve mechanical properties and dimensional stability under humidity fluctuations), with inorganic particles (to improve the stability of the membrane itself), or with hygroscopic fillers (to improve water retention inside the membranes at high temperatures).

In general, modifications can be divided into two categories: (1) reinforcement with another polymer such as expanded polytetrafluroethylene (ePTFA) or blending with other polymers by electrospinning; and (2) impregnation of the ionomer with additives, metal salts or inorganic dopants [50].

The incorporation of a reinforcing layer, e.g., a mesh made of ePTFA, has proven effective. Besides ePTFA, there are other porous reinforced materials such as poly (vinylidene fluoride) (PVDF) electrospinning microporous membrane, polyvinyl alcohol (PVA) microporous membrane, etc. The porosity of this hydrophobic mesh allows the use of membranes with smaller thicknesses and smaller EW ionomers that would otherwise severely compromise mechanical stability. The microporous support layer is incorporated in the center of the ionomer, where it is filled with ionomer, which contributes to the formation of a continuous proton transport channel and water transport throughout the thickness of the membrane. The consequences of membrane reinforcement can affect dimensional anisotropy. Reinforced membranes exhibit better in-plane stability, which means less in-plane swelling. The reinforced structure increases yield strength and elastic modulus and decreases in-plane swelling. Lower in-plane swelling is therefore critical for reducing swelling-induced mechanical stress and increases resistance to defect propagation. In addition, reinforced membranes have been shown to be less sensitive to humidity fluctuations. Examples of commercially available membranes include Gore Select membranes, Nafion XL, and HP, which are PFSA-reinforced membranes with ePTFE. Along with porous reinforcement materials, impregnation of PFSA ionomers with electrospun nanofibers such as poly(phenylsulfone) (PPSU), poly(acrylic acid) (PAA), or poly(ethylene oxide) (PEO) and reinforcing particles such as carbon nanotubes can be used to improve mechanical properties and water retention [50,60–62].

Impregnation of the ionomer with additives, metal salts, or inorganic dopants is used to improve membrane performance, e.g., dimensional stability, thermomechanical stability, or conductivity at lower humidity and higher temperatures. The most common fillers used to improve membrane performance are metal salts, hygroscopic inorganic fillers such as SiO_2 [63,64], ZrO_2 [65,66], TiO_2 [66,67], functionalized inorganic fillers, and particles such as carbon nanotubes [68]. Composite membranes with metal oxides (SiO_2, TiO_2, or WO_2) showed promising properties for use at high temperatures. In their research, Baglio et al. and Antonucci et al. studied the operation of Nafion–TiO_2 and Nafion–SiO_2 composite membranes at elevated temperatures (>100 °C) [69,70]. Both studies showed that due to the presence of inorganic hygroscopic fillers in the polymer mass, the performance at elevated temperatures exhibited better water retention and more uniform distribution of water in the composite membrane, reducing ohmic resistance and improving electrolytic performance [63,66].

Although great progress has been made in reducing the cost and eliminating the deficiencies of PFSA membranes, there is still a great need to improve their chemical and mechanical properties and to reduce their degradation, aging, or contamination while improving important properties such as proton conductivity.

Hydrocarbon-Based Membranes

Hydrocarbon-based membranes show potential for application in PEMWE systems. They showed good thermal stability, similar proton conductivity, and the production cost is lower compared to PFSA [52]. The application potential was particularly evident for hydrocarbon-based membranes fabricated from sulfonated derivatives such as sulfonated poly(phenylene sulfone)—sPPS, sulfonated poly(arylene ether sulfone)—sPAES, sulfonated polysulfone—sPSf, sulfonated poly(ether ether ketone)—sPEEK, and sulfonated polybenzimidazole—sPBI. Although the use of alternative membranes offers many advantages, a careful evaluation of their mechanical, chemical, and thermal stability and durability is very important before they are finally installed in PEMWE systems, where the systems must be in operation for more than 50,000 h [31].

In sPPS membranes, the aromatic ring bearing the sulfonic acid group is connected to two sulfone bonds ($-SO_2^-$) that are strongly electron-withdrawing (electron acceptor), i.e., the sulfonic acid groups are directly bonded to a strongly electron-withdrawing poly(phenylene sulfone) backbone [71]. This structural feature is the reason for the significantly higher acidity of sPPS compared to other sulfonated hydrocarbons, resulting in improved proton conductivity, especially at low relative humidity [72]. Furthermore, sPPS membranes showed good thermal stability and low gas crossover capability [31]. In the sPEEK membrane, the phenyl ring is linked to ether bonds and carbonyl groups, while the sulfonic acid group is linked to the phenyl ring [73]. The properties of the membranes largely depend on the degree of sulfonation (DS), whereby higher DS always favors excellent proton conductivity [74]. The potential for application is also evident in its low-cost fabrication and thermal, chemical, and mechanical stability [31]. Moreover, sPAES is widely used for PEM applications due to its simple and inexpensive production and modification as well as its excellent membrane properties. In addition, due to its excellent mechanical properties, thinner membranes can be produced, which helps to reduce ohmic resistance. However, the properties of hydrocarbon-based membranes, including SPAES, are also affected by the DS of the polymers [75]. The overview of recently reported PEM membranes (PFSA and hydrocarbon-based) in PEMWE systems is given in Table 5.

Hydrogen produced by electrolysis has the lowest emission rate, but its production cost is nevertheless the highest due to high capital and electricity costs. The Energy Transitions Commission (ETC) assumes that the cost of hydrogen in Europe today is 5.10 EUR/kg (assuming USD 780/kW capital cost) [76]. The price of hydrogen produced by the PEMWE system depends not only on the price of electricity, but also on the efficiency of the cell used and its lifetime, i.e., materials, catalysts, electrolytes (membranes), as well as working pressures and temperatures [77]. For example, Areva H_2Gen has shown that the price of hydrogen produced with PEMWE is USD 3.90/kg for 1 MW PEMWE (8000 operating hours per year) at an electricity price of about USD 55/MWh. International Energy Agency's (IEA) predicts that green hydrogen will cost about USD 1–2.50/kg by 2050 [78]. In order to reduce the price of hydrogen produced, it is necessary to invest mainly in PEMWE systems powered exclusively by renewable energy sources. In addition, the efficiency of electrolyzers must be improved by introducing new materials, focusing on low-cost, durable, high-performance materials such as durable and more active catalysts, thinner membranes, and less critical raw materials [77]. Table 6 contains PEMWE's key performance indicators related to advances in membrane materials and catalysts. Comparing Table 6 with Table 5, Table 7, and Table 8, it is clear that with respect to 2022, some of the targeted parameters are being improved by the introduction of new materials. However, the new materials implemented in PEMWE have not yet reached commercial levels, and further research is needed in order to achieve the 2050 targets. Furthermore, a review of PEMWE performance data reported in the literature shows that performance results vary so widely that it is difficult to draw conclusions about technological improvements and research and development directions. It is crucial that testing protocols be standardized to allow comparison of performance evaluations between different studies [79].

Materials **2023**, *16*, 6319

Table 5. Brief overview of recent reports on PEMWE with PFSA- and hydrocarbon-based membranes.

Membrane	Material	Thickness/μm	Cathode Loading /Cathode Catalyst	IEC	Conductivity/ mS cm⁻¹	T/°C	Current Density	Stability Test	Ref.
				PFSA					
Fumapem®/graphene	Per-Fluorinated Sulfonic Acid (PFSA)/PTFE copolymer; 0.38 w/v—graphene loading	112	-	0.82 mmol g⁻¹	115	80	-	-	[80]
(S-TiO₂)/Nafion	sulfated titania (S-TiO₂)-dopped Nafion	100–110	0.5 ± 0.1 mg cm⁻² of Pt Pt/Vulcan XC-72—30%	0.82 ± 0.01 meq g⁻¹	≈70	100	4 A cm⁻² at 2 V	-	[81]
biaxially stretched Nafion 117	PFSA, Nafion series 117	28.2 ± 1.7	0.4 mg cm⁻² of Pt Pt/C—0.5	0.92 meq g⁻¹	$\sigma_i = 73$ $\sigma_t = 54$	80	3 A cm⁻² at 1.9 V.	0.4 A cm⁻² for 50 h	[82]
hBN/Nafion	monolayer hexagonal boron nitride/Nafion	-	0.4 mg cm⁻² of Pt	-	18.7 ± 0.9	70	-	0.4 A cm⁻² for 100 h (50°C)	[83]
Aq830-PSU(5 wt %)	electrospun polysulfone fiber web/Aquivion®	45 ± 2	0.5 mg cm⁻² of Pt and 33 wt% Nafion® ionomer (5 wt% solution) Pt/C—40 wt%	-	220	80	2 A cm⁻² at 1.76 V	-	[84]
3M 729/ePTFE (annealed at 180°) AQ 720/ePTFE (annealed at 180°)	ePTFE porous support was impregnated with 3M 729 and AQ 720 and annealed at different temperatures	55–60	0.25–0.30 mg cm⁻² of Pt Pt/C—40 wt%	1.30 meq g⁻¹ 1.31 meq g⁻¹	106 112	80	-	-	[62]

Table 5. *Cont.*

Membrane	Material	Thickness/μm	Cathode Loading /Cathode Catalyst	IEC	Conductivity/ mS cm^{-1}	T/°C	Current Density	Stability Test	Ref.
NPP-95	Nafion/poly(acrylic acid)/poly(vinyl alcohol) 95:2.5:2.5	50–60	0.1 mg cm^{-2} of Pt	0.84 meq g^{-1}	189.2 ± 12.1	80	4.310 A cm^{-2} at 2.0 V	-	[59]
					Hydrocarbon membranes				
BPSH50 (random)	hydrocarbon-based sulfonated poly(arylene ether sulfone)	40–50	0.5 mg cm^{-2} of Pt Pt/C—0.4	1.86 meq g^{-1}	178	80	5.3 A cm^{-2} at 1.9 V	3 A cm^{-2} for 90 h	[85]
CSPPSU	crosslinked sulfonated polyphenylsulfone	70–130	0.3 mg cm^{-2} of Pt Pt/C—20 wt%	1.71 meq g^{-1}	30	150	0.456 A cm^{-2} at 1.8 V	-	[86]
sPPS	sulfonated poly(phenylene sulfone)	115 ± 12	0.5 mg cm^{-2} of Pt Pt/C—1.6 wt%	2.78 meq g^{-1}	-	80	3.48 ± 0.03 A cm^{-2} at 1.8 V	1 A cm^{-2} for 80 h	[87]
SPAES50	Sulfonated poly(arylene ether sulfone)	20	0.4 mg cm^{-2} of Pt with a 10 wt% P50 content Pt/C—40 wt%	1.89 meq g^{-1}	330.1 ± 6.0	90	1.069 A cm^{-2} at 1.6 V	-	[75]
12%MKT-NW/C-sPEEK	MXene/potassium titanate nanowire cross-linked sulfonated polyether ether ketone	-	-	1.88 meq g^{-1}	9.7	room temperature	-	-	[88]
4%MXene-Cu$_2$O/sPEEK	Titanium carbide-copper oxide cross-linked sulfonated poly ether ether ketone	-	-	1.66 meq g^{-1}	10.5	30	-	-	[89]

Table 5. *Cont.*

Membrane	Material	Thickness/μm	Cathode Loading /Cathode Catalyst	IEC	Conductivity/ mS cm^{-1}	T/°C	Current Density	Stability Test	Ref.
SPPNBP_3 SPPNBP_5	multi-block copolymer membranes consisting of sulfonated poly(p-phenylene) and naphthalene containing poly(arylene ether ketone)	42–57	0.5 mg cm^{-2} of Pt Pt/C—0.4	2.05 2.49 meq g^{-1}	200 152	80	4.8 5.5 A cm^{-2} at 1.9 V	-	[90]
G-sPSS-1.95	grafting a highly sulfonated poly-(phenylene sulfide sulfone) side chain onto a poly(arylene ether sulfone) main chain	50–60	0.4 mg cm^{-2} of Pt Pt/C—40 wt%	1.95 meq g^{-1}	290	90	6 A cm^{-2} at 1.9 V	1 A cm^{-2} for 50 h	[91]

Table 6. Target performance indicators of PEMWE related to advances in membrane materials and catalysts. Modified from [77].

	2022	Target 2050	Research and Development
Nominal current density	1–3 A cm^{-2}	4–6 A cm^{-2}	Membranes
Voltage	1.4–2.3 V	<1.7 V	Catalysts, membranes
Operating temperature	50–80 °C	80 °C	Durability of the membranes
Cell pressure	≤50 bar	>70 bar	Membranes, catalysts
Load Range	5–130%	5–300%	Membranes
H$_2$ purity	99.9–99.9999%	99.9–99.9999%	Membranes
Voltage efficiency (LHV)	50–68%	>80%	Catalysts
Electrical efficiency (stack)	44–66 kWh/kg H$_2$	<42 kWh/kg H$_2$	Catalysts, membranes
Lifetime (stack)	50,000–80,000 h	100,000–120,000 h	Catalysts, membranes

4. Electrocatalysts for Hydrogen Production

The question of using expensive materials for the fabrication of the electrodes in PEMWE technology dates back to the 1960s when the first PEMWE devices were developed. Regarding the empirical aspects, these early systems were considerably efficient, presenting performances of 1.88 V @ 1 A cm^{-2} or 2.24 V @ 2 A cm^{-2}, with a cell life of over 15,000 h without substantial performance degradation. Back then, the question concerning the high cost of used catalysts was also raised, and for these systems, catalyst layers were based on Ir and Pt black with high metal loading [5]. This work will put aside the analysis of anode electrocatalysts, and the focus will be on cathode ones.

Harsh electrochemical environments (high anodic overpotential, low pH, the presence of strong oxidants, possibility of operating at higher temperatures) require the use of precious metal compounds as electrocatalysts in PEMWE [92]. For HER, highly dispersed carbon-supported Pt-based materials, with low overpotential close to zero and a Tafel slope around 30 mV/decade, are currently benchmark catalysts, but the practical developments to satisfy the growing demands require the use of cheaper electrocatalysts [7]. The cathode catalyst represents a considerable portion of the total system cost, especially if degradation or corrosion of the carbon support occurs [5]. Nowadays, cathode side metal loading is maintained at approximately 0.5–1 mg cm^{-2}, and further decreases will be needed for values reaching below 0.2 mg cm^{-2} [2]. The PEMWE with the non-noble cathodes exhibited the current density of 0.35–0.73 A cm^{-2} at 2.0 V in the operating temperature range of 80–90 °C, which were still lower than that with noble Pt/C cathodes (1.46–2.71 A cm^{-2}) [93].

Typically, electrocatalysts are chosen based on their specific characteristics such as particle size, pore structure, good electrical conductivity with high surface area, and corrosion stability under oxidizing conditions [94]. For better electrochemical performance and maximum consumption of catalyst surface, the electrocatalysts are usually supported on the carbon because of its superior electrical conductivity, mechanical and thermal stability, large surface area, environmental friendliness, and relatively low cost [94]. Different types of carbons are studied for the application in PEMWE, such as carbon black (CB), carbon nanomaterials (CNMs), graphene, fullerenes, carbon nanotubes (CNTs), and heteroatom-doped CNMs(N-CNTs) [94].

There are different proposed ways of decreasing Pt loading in PEMWE, but most of the solutions require a finding of cheaper non-noble catalyst, the use of specific supports to achieve better dispersion and a higher catalytic surface, and the formation of noble metals alloys by addition of new elements to the main compound [95].

4.1. HER Electrocatalysts with Substituted Noble Metals Content

According to the volcano-plot theory, the electrocatalytic activity is controlled by the H adsorption free energy for HER, and only noble metals can efficiently electrolyze the HER [7]. But, during the operation of PEMWE, conventional HER catalysts suffer fewer kinetic and stability problems due to relatively facile reduction in protons in acidic media and the negative potential window for operation [96]. This opens the possibility of using non-noble metals as HER catalysts for PEMWE. But, pure transition metals, such as Ni (-0.280 V_{SHE}), Co (-0.277 V_{SHE}), Fe (-0.440 V_{SHE}), and Mo (-0.200 V_{SHE}) are also less stable in PEMWE electrolysis because they can undergo dissolution during HER because of the values of their standard reduction potentials [4,96]. The use of these transition metals could be improved by composite formation with other promising materials since the main requirements within this field are the improvement of the surface area of the electrodes and the optimization of their ability to reduce protons to molecular hydrogen [95]. Therefore, transition metal compounds including carbides, nitrides, phosphides, and sulfides have been actively investigated for acidic HER. Their higher HER activities than those of pure transition metals can be attributed to the suitable energies for hydrogen adsorption on the catalyst surface [93]. On the other side, the insufficient intrinsic activity of non-noble metals could be improved by tuning properties such as the composition and morphology of the used materials [97]. Non-metallic elements (C, N, P, S, and Se) that are being used in the composites together with metallic elements have high electronegativity and therefore draw electrons from the metal components, as well as attract protons due to the partial negative charge [98].

Although the performances of such materials are still lower than those with noble cathodes, due to the relatively low price of non-noble metals, there is still great scientific interest for research within these alternatives. Incorporation of high-performance low-cost HER electrocatalysts into PEM electrolyzers is an emerging area of research with still limited reports of PEM electrolyzers that utilize non-precious catalysts at the H_2 electrode side being published [99]. Most of the published works still only deal with the investigation of electrocatalytic activity of novel materials in acidic media by the application of three-electrode systems without further application in PEM water electrolyzers. Both of the mentioned applications will be presented in further sections.

Since the works of Hinnemann et al. [100] and Thomas et al. [101], who reported promising results of the HER electrolysis utilizing MoS$_2$, transition metal chalcogenides are among the most promising non-noble metal cathode electrocatalysts. The main characteristic of this family of materials is that they have a 2D lattice structure and, like graphene, the reactive sites are located along the edges while the basal plane is catalytically inactive [7]. Therefore, the aim of most of the works within this area is to increase the number of active sites by increasing the ratio of edges to the basal plane and to increase the electrical conductivity simultaneously while minimizing the overpotential required for HER [7]. For this purpose, many distinct molybdenum sulfide structures such as nanoparticles, nanowires, films, and mesopores were synthesized with the aim of maximizing the number of exposed edge sites [99].

Besides the aspect of active sites, electric conductivity is another crucial factor related to electrocatalytic activity because a high conductivity ensures fast electron transport during the catalytic process [102]. Because of that, MoS$_2$, which is more economical and 10^4 times more abundant than Pt, can be chemically bonded to RGO via a facile solvothermal approach [103]. Such a composite that contains highly exposed edges can exhibit HER activity with a small overpotential of ~0.1 V, large cathodic currents, and a Tafel slope of 41 mV/decade. Another research work conducted by Corrales-Sánchez et al. [104] explored the electrocatalytic activity of MoS$_2$/RGO hybrids, as well as of pristine MoS$_2$ and MoS$_2$ physically mixed with an electrically conducting carbon material (Vulcan® XC72) towards PEM electrolysis. As a reference, the performance of Pt black was also shown. Among tested MoS_2-based materials, 47 wt% MoS_2/Vulcan® gave the best performance in terms of current density at an encouraging level for practical application, while the MoS_2/RGO hybrids showed higher HER activity than pristine MoS_2. The poor performance of pristine MoS_2 can be contributed to poor electrical conductivity. Although improvement in electrocatalytic activity for the mentioned composites can be noticed, results still did not exceed those achieved with the use of Pt black.

Research conducted by Kumar et al. [102] confirmed that by controlling the reaction temperature and sulfur precursor employed, different MoS_2 nanostructures like nanosheets, nanocapsules, and nanoflakes could be obtained. Among all indicated materials, MoS_2 in the form of nanocapsules exhibits superior activity towards HER in 0.5 M H_2SO_4 with an overpotential of 120 mV vs. RHE. The following Mo-based catalyst was further incorporated into the PEM electrolyzer where the fabricated MEA consisted of a Nafion PEM sandwiched between iridium (IV) oxide and MoS_2-nanocapsules used as the anode and cathode catalyst. The designed cell was operated for 200 h at 2 V without any degradation of electrocatalytic activity.

Recently, another efficient and stable electrocatalyst composed of earth-abundant TiO_2 nanorods decorated with MoS_2 thin nanosheets was recorded [105]. This composite possesses hydrogen evolution activity in acidic media at an overpotential of 0.35 V and a Tafel slope of 48 mV/decade. It is very important to measure the Tafel slope because it is a primary and inherent property of the catalyst which indicates the rate-determining step involved in the HER [105]. In this case, the measured Tafel slope is very close to the one of benchmarking Pt/C (32 mV/decade) and indicates that electrochemical desorption is the rate-limiting step for the HER in acidic media.

Higher HER activity than MoS_2 can be obtained by the use of alternative Mo-based catalysts such as MoS_x, $[MoS_3S_{13}]^{2-}$ nanoclusters, and sulfur-doped molybdenum phosphide (MoP|S), loaded onto CB support. These carbon-supported catalysts, synthesized by Ng et al. [99], are electrochemically tested in a standard three-electrode electrochemical and subsequently integrated into PEM electrolyzer systems and operated continuously for 24 h. Related to PEM electrolyzer testing, the performance of each electrolyzer was examined by stepping the potential from 1.2 to 1.0 V at 50 mV intervals at a cell temperature of 80 °C. The MoS_x-CB-based electrolyzer required 1.86 ± 0.03 V to reach 0.5 A cm^{-2}, while the MoS_3S_{13}-CB and MoP|S-CB-based electrolyzers both required 1.81 ± 0.03 V to reach 0.5 A cm^{-2}. The best overall performance, also including three-electrode electrochemical data, was achieved with the (MoP|S) electrolyzer. Such results suggest that Mo-based catalysts hold promise for commercial applications with the possibility of replacing the Pt-based cathodes currently being used in PEM electrolyzers.

Other interesting transition metal chalcogenides applied as a cathode side within the field of PEMWE are iron sulfide materials, which have the great advantage of being widespread in nature. Pyrite (FeS_2) is the most abundant sulfide mineral, while pyrrhotite is an unusual iron sulfide mineral with variable iron content [$Fe_{(1-x)}S_{(x=0-0.2)}$] that often accompanies base metal sulfides in ore deposits [106,107]. The synthesis, characterization, and activity towards the HER of different stoichiometries of iron sulfide materials including the above-mentioned pyrite and pyrrhotite, as well as greigite (Fe_3S_4), were investigated by Di Giovanni et al. [108]. Finally, their performances were also investigated in situ in a PEM electrolyzer single cell under 80 °C. The MEAs were prepared by using pyrite, pyrrhotite, or greigite as the cathode catalyst and tested in an electrolysis single cell. The catalysts were not supported but were mixed with 20% of CB. Nafion 115 (125 μm) was used as the membrane and IrO_2 as the anode. According to the SEM results presented within the research, the thickness of the IrO_2 catalyst layer is ~6 μm, while the thickness of the FeS_2/CB catalyst layer is ~30–40 μm. Also, the experimental results have shown that all three catalysts allow~2100 mV at 1 A cm^{-2} to be reached, but both ex situ and in situ electrochemical experiments have revealed that pyrite (FeS_2) is more active than greigite

(Fe_3S_4), which is more active than pyrrhotite (Fe_9S_{10}). Generally, all three catalysts allow ~2100 mV at 1 A cm^{-2} to be reached.

The electronic structure of metal sulfide materials can be modified by doping metal atoms, which can optimize hydrogen adsorption energy and enhance HER catalytic activity. Such an example can be seen in the work of Wang et al. [109], where Co-doped iron pyrite FeS_2 nanosheets were hybridized with carbon nanotubes ($Fe_{1-x}Co_xS_2$/CNT). HER was tested in 0.5 M H_2SO_4 acidic solution in a three-electrode system without further application in PEMWE. Electrochemical measurements showed a low overpotential of ~0.12 V at 20 mA cm^{-2}, a Tafel slope of ~46 mV/decade, and long-term durability over 40 h of operation using bulk quantities of $Fe_{0.9}Co_{0.1}S_2$/CNT hybrid catalysts at high loadings (~7 mg cm^{-2}). Density functional theory (DFT) revealed that an increase in the catalytic activity comes from a large reduction in the kinetic energy barrier of H atom adsorption on FeS_2 surface upon Co doping in the FeS_2 structure.

Transition metal phosphides, such as CoP, NiP, FeP, and MoP, are viewed as a promising replacement of Pt because of their good stability and high activity in acidic media, and further improvement of intrinsic HER activity can be realized by employing more than one transition metal [110]. Therefore, FeCoP shows a near optimal hydrogen adsorption free energy (ΔG_H) that is similar to that of Pt and is significantly affected by the Fe/Co ratio [110]. The development of NiP catalysts is also on the ascending path, with the composites being developed by different research groups. NiP catalysts electrodeposited on carbon support were developed by Kim et al. [98] and applied as a cathode for a PEMW electrolyzer. The performance of the water electrolyzer was evaluated in a galvanostatic mode in the range of 0.02–4 A cm^{-2} after the activation process, and cell voltages of 1.96, 2.07, and 2.16 were required to obtain a current density of 1.2 and 3 A cm^{-2}. NiP nanoparticles, but with a different stoichiometric ration (Ni_2P) and support (multiwall carbon nanotubes), were designed by in situ thermal decomposition of nickel acetylacetonate as the nickel source and trioctylphosphine as the phosphorus source in an oleylamine solution of carbon nanotubes [111]. Electrocatalytic activity of this nanohybrid was evaluated in 0.5 M H_2SO_4 with an onset overpotential of 88 mV, a Tafel slope of 53 mV/decade, and an exchange current density of 0.0537 mA cm^{-2}.

Besides nickel–phosphide materials, nickel–carbon-based catalysts were developed by Fan et al. [112]. This work reveals the new area of tuning structure and functionality of metal–carbon-based catalysts at an atomic scale that may help accelerate the large-scale application of PEM electrolyzers. By the use of electrochemical methods, the indicated composite can be activated to obtain isolated nickel atoms anchored on graphitized carbon, consequently displaying high activity and durability for HER. Owing to their low-coordination and unsaturated atoms, isolated metal atoms have demonstrated more catalytic active than nanometer-sized metal particles. Other attempts at the improvement of the "noble-metal-free" electrocatalysts can be noticed at the use of group VI transition metal carbides that exhibit catalytic properties analogous to platinum group materials (PGMs) because of their unique d-band electronic structures [113]. Catalytic properties of carbide materials strongly depend on their surface structure and composition, which are closely associated with their method of synthesis [113]. Therefore, by the use of a simple and environmentally friendly carburization process, Chen et al. [113] synthesized Mo_2C covalently anchored to carbon supports (carbon nanotubes and XC-72R carbon black). The electrochemical impedance spectroscopy (EIS) results demonstrated that the incorporation of Mo_2C onto carbon supports enhanced the exchange current density (measured overpotential of 63 mV applied for driving 1 mA cm^{-2} of exchange current density), reduced charge-transfer resistance, and a change in the HER mechanism.

Excellent chemical stability has opened the possibility of the application of transition metal oxides, such as WO_2 in the field of clean hydrogen energy production. Metallic WO_2–C mesoporous nanowires with a high concentration of oxygen vacancies (OVs) were synthesized by Wu et al. [114]. All tests were carried out in 0.5 M H_2-saturated H_2SO_4, and the products exhibited promising performance for hydrogen generation with a Tafel slope of 46 mV per decade. For comparison, as already noted, corresponding to the literature [7], the value of the Tafel slope for commercial Pt/C was about 30 mV per decade. Other interesting Pt-free alternatives with the corresponding features are listed in Table 7.

Table 7. Non-noble cathode materials for the use in electrolytic acidic hydrogen generation and PEMWE systems.

Cathode Catalyst	Electrochemical Characterization	Membrane	T	Performance	Ref.
MoO_3 nanowires	Three-electrode cell with 1 M H_2SO_4 electrolyte	-	-	11.3 and 56.8 mA cm^{-2} at potential of 0.0 and 0.1 V with a Tafel slope of 116 mV/decade	[115]
Co-Cu alloys	PEMWE single cell with Co-Cu deposited on a carbon paper (CP) as a cathode and IrO_2 electrodeposited on a CP as an anode	N212 (DuPont)	90 °C	1.2 A cm^{-2} at 2.0 V$_{cell}$	[116]
CoP	Three-electrode cell with 0.5 M H_2SO_4 electrolyte	-	-	Current density of 20 mA cm^{-2} at an overpotential of 85 mV	[117]
CoP/CC	Three-electrode cell with 0.5 M H_2SO_4 electrolyte	-	-	Onset overpotential of 38 mV with a Tafel slope of 51 mV/decade	[118]
WC@NC	PEMWE single cell with WC@NC as the cathode and IrO_2 (Sunlaite) as an anode	N212 (DuPont)	80 °C	0.78 A cm^{-2} at 2.0 V$_{cell}$	[119]
OsP_2@NPC	Three-electrode cell with 0.5 M H_2SO_4 electrolyte	-	-	10 mA cm^{-2} at onset overpotential of 46 mV	[120]
NiMo/CF/CP	PEMWE single cell with NiMo/CF/CP as the cathode and IrO_2/CP as an anode	N212 (DuPont)	90 °C	~2.0 A cm^{-2} at 2.0 V$_{cell}$	[121]
Ni–Mo–N	Three-electrode cell with 0.5 M H_2SO_4 electrolyte	-	-	Overpotential of 53 mV at 20 mA cm^{-2}	[122]
NiS_2 $NiSe_2$ $NiTe_2$	Three-electrode cell with 0.5 M H_2SO_4 electrolyte	-	-	Overpotential of 213 mV at 10 mA cm^{-2} Overpotential of 156 mV at 10 mA cm^{-2} Overpotential of 276 mV at 10 mA cm^{-2}	[123]
MoP/C (NaCl)	Home-made electrolyzer using MoP/C (NaCl) as cathode and IrO_2 (Sunlaite) as an anode	N211 (DuPont)	80 °C	0.71 A cm^{-2} at 2.0 V$_{cell}$	[124]
MoP@PC	Three-electrode cell with 0.5 M H_2SO_4 electrolyte	-	-	Overpotential of 258 mV at 10 mA cm^{-2}, with a Tafel slope of 59.3 mV/decade	[125]
MoP@PC	Three-electrode cell with 0.5 M H_2SO_4 electrolyte	-	-	Overpotential of 51 mV at 10 mA cm^{-2} with a Tafel slope of 45 mV/decade	[126]

Table 7. *Cont.*

Cathode Catalyst	Electrochemical Characterization	Membrane	T	Performance	Ref.
MoP@PC	Three-electrode cell with 0.5 M H_2SO_4 electrolyte	-	-	Onset overpotential of 77 mV, overpotential of 153 mV at 10 mA cm^{-2}, with a Tafel slope of 66 mV/decade	[127]
MoP/NG	Three-electrode cell with 0.5 M H_2SO_4 electrolyte	-	-	Overpotential of 94 mV at 10 mA cm^{-2} with a Tafel slope of 50.1 mV/decade	[128]
MoP/NC	Three-electrode cell with 0.5 M H_2SO_4 electrolyte	-	-	Overpotential of 120 mV at 10 mA cm^{-2}	[129]
MoP|S	Three-electrode cell with 0.5 M H_2SO_4 electrolyte	-	-	Overpotential of 86 mV at 10 mA cm^{-2}	[130]
N–Mo_2C	Three-electrode cell with 0.5 M H_2SO_4 electrolyte	-	-	Onset overpotential of 78.1 mV for HER and a Tafel slope of 59.6 mV/decade	[131]
Mo_2C/C	Three-electrode cell with 0.5 M H_2SO_4 electrolyte	-	-	Tafel slope of 56 mV/decade	[132]
Mo_2C/C	Three-electrode cell with 0.5 M H_2SO_4 electrolyte	-	-	Overpotential of 180 mV at 10 mA cm^{-2}	[133]
Cu_xMo_{100-x}/CP	PEMWE single cell with $Cu_{93.7}Mo_{6.3}$/CP as the cathode and IrO_2/CP as an anode	N212 (DuPont)	90 °C	0.50 A cm^{-2} at 1.9 V_{cell}	[96]
$Cu_{1-x}Ni_xWO_4$	Three-electrode cell with 1 M H_2SO_4 electrolyte	-	-	4.3 mA cm^{-2} at the anodic peak potential of 0.09 V	[134]
Ni–P supported by copper foam (CF) on CP	PEMWE single cell with Ni–P/CF/CP as the cathode and IrO_2/CP as an anode	N212 (DuPont)	90 °C	0.67 A cm^{-2} at 2.0 V_{cell}	[135]
NiMo/CF/CP	PEMWE single cell with Ni–Mo/CF/CP as the cathode and IrO_2/CP as an anode	N212 (DuPont)	90 °C	2.0 A cm^{-2} at 2.0 V_{cell}	[121]
FeCo/N–G	Three-electrode cell with 1 M H_2SO_4 electrolyte	-	-	Onset overpotential of 88 mV and overpotential of 262 mV at 10 mA cm^{-2}	[136]
P-Ag@NC	Three-electrode cell with 1 M H_2SO_4 electrolyte	-	-	Overpotential of 78 mV at 10 mA cm^{-2}	[137]
Co@N–CNTs@RGO	Three-electrode cell with 0.5 M H_2SO_4 electrolyte	-	-	Overpotential of 87 mV at 10 mA cm^{-2}	[138]

4.2. HER Electrocatalysts with Reduced Noble Metals Content

The last section of this review paper will contribute to electrocatalyst cathode materials which are designed using a smaller proportion of noble metals compared to conventionally used ones. To achieve an environmentally sustainable society, the reduction in the consumption of noble metals is a topic of great importance. In the section above, substituted noble metals alternatives are shown; even though much scientific effort has been put to develop new and prosperous materials, their activities are rarely comparable to that of the benchmark catalyst Pt/C, and their application is still limited in real energy devices. Also, a lot of developed materials are only being tested in acidic media, without further tests in PEMWE systems, which should be another step forward to fully understand their potential use. For these reasons, strategies to synthesize catalysts have mostly been focused on the reduction in the content of noble metals, which is considered a more practical strategy for accelerating their industrial application [139]. Considering commercial PEMWE systems and their industrial applications, the typical catalyst loading is 1–2 mg cm^{-2}; therefore, it is responsible for 25% of the PEMWE stack cost [140,141].

To reduce Pt content, different approaches are listed. Some of the solutions imply the development of novel thin and tunable gas diffusion electrodes with a Pt catalyst thickness of 15 nm and a total thickness of about 25 μm, which can enhance catalyst mass activity up to 58 times higher than conventional catalyst-coated membrane (CCM) at 1.6 V under the operating conditions of 80 °C and 1 atm. [142] On the other hand, core-shell structures with Pt on the surface and Ru forming the core of the particles were developed by Ayers et al. [140]. This composition enables appropriate electrocatalytic activity to be achieved; by utilizing Pt spontaneous deposition on metallic Ru nanoparticles, an ultralow Pt-content catalyst was made with a 20:1 Ru:Pt atomic ratio. Furthermore, reactive spray deposition technology (RSDT) enabled one-step fabrication of two MEAs (86 cm^2) containing platinum group metal (PGM) loadings in amounts of only 0.2 and 0.3 mg$_{PGM}$ cm^{-2} loading in the cathode and anode electrodes, respectively. This assembly, involved in electrolysis operation conducted at 50 °C and 400 psi differential pressure with 1.8 A cm^{-2} current, demonstrated durability for over 3000 h of operation at industrially relevant operating conditions [143].

Consumption of energy to produce hydrogen strongly depends on the current and voltage applied. The cell voltage is composed of the anode and cathode potentials and the *IR* drop in the electrolyte. The reduction in the cathode potential can be achieved by modifying the composition of the catalyst or by increasing the surface area of the catalyst by reducing the particle size [144]. Such an example can be seen in the work of Ravichandran et al. [144] where, during the impregnation reduction synthesis of Pt catalyst, the addition of nonionic surfactant reduced the particle size. The MEAs of 4 cm^2 coating area, with a loading of 0.4 mg cm^{-2} Pt and 1.2 mg cm^{-2} IrO$_2$, was tested for HER and operated as a single cell at 2 V and 80 °C, achieving the highest current density of 1.5 A cm^{-2}. Similarly, the stack of the hydrogen generation capacity of about 1 N m^3 h^{-1} capacity was assembled and tested by the integration of five single MEAs of 500 cm^2 area. The stack displayed a current density of about 1.18 A cm^{-2} at 10.0 V and 80 °C; performance lasted up to 3000 h of operation with not much change in the current density noticed. The use of the impregnation reduction method of synthesis and surfactant are beneficial in reducing the particle size, which also confirmed earlier conducted studies [145,146]. Wang et al. [145] designed surfactant-stabilized Pt and Pt alloy electrocatalysts on carbon supports for the application in PEMFC and revealed the improved electrocatalytic activity due to the well dispersed and smaller catalytic particles; while Rajalakshmi et al. [146], through the impregnation reduction method, synthesized Pt-deposited Nafion® membrane as cathode without using any surfactant. The obtained material showed improved fuel cell performance in comparison to other methods of synthesis.

To create cathode formulations with cheaper characteristics than Pt, efforts are being addressed to develop Pd-, Rh-, and Ru-based catalysts. Pd is three times less expensive than Pt and was used in the work of Kumar et al. [94] to prepare a phosphorus-doped carbon-nanoparticles-supported palladium (Pd/P-CNPs) electrocatalyst. The structural modification of the carbon by phosphorus doping will be more effective than nitrogen doping since phosphorus has a much larger covalent radius (107 ± 3 pm) than carbon (73 ± 1 pm) compared with nitrogen (71 ± 1 pm). The synthesized electrocatalyst was used as the HER electrode for the fabrication of MEAs, and its performance was evaluated in house-fabricated PEMWE 25 cm^2 single-cell assemblies. The obtained results showed that the synthesized Pd/P-CNPs have shown similar electrochemical activity and stability compared to commercial Pt/C.

Rh–P catalysts exhibit very good HER performances due to the introduction of P into the Rh catalyst material, which induces a ΔG_{H^*} shift to more neutral values; this indicates greater active catalytic activity for a lower amount of Rh loading [97]. Facile fabrication of Rh and Rh–P electrodes on a carbon paper as substrate via electrodeposition at room temperature and ambient pressure was performed by Kim et al. [97] and evaluated for the acidic HER in terms of intrinsic and mass activity. Under the optimized deposition parameters, such as potential and time, a certain facet widespread at the surface of Rh electrodes (Rh (111) facet) demonstrated high intrinsic activity for HER in acidic medial, while the further enhancement of the catalyst performance was achieved by a modified electronic structure of Rh–P electrodes with intrinsic and mass activity greater than ones of Pt electrodes. Except for Rh–P, composites with different stoichiometric ratios of Rh and P are also highly represented within this field. Many different morphologies of Rh_2P are synthesized and tested for hydrogen production in acidic media with the potential of use in PEMWE. Yang et al. [147] performed a colloidal synthesis of monodisperse Rh_2P nanoparticles with an average size of 2.8 nm and with an overpotential of 140 mV achieved a current density of 10 mA cm^{-2} in 0.5 M H_2SO_4, while Duan et al. [148] synthesized rhodium phosphide nanocubes supported on high surface area carbon (Rh_2P/NCs). In the case of Rh_2P/C, the overpotential at the current density of 5 mA cm^{-2} is 5.4 mV, which is lower than Pt/C (8.0 mV) and Rh/C (68.4 mV). Carbon support was also used in the synthesis of wrinkled, ultrathin Rh_2P nanosheets (w-Rh_2P NS/C) for enhancing HER in 0.1 M HClO$_4$ [149]. To reach a current density of 10 mA cm^{-2}, the overpotential of 15.8 mV is required, which is 6.3 and 25.8 mV lower than those of commercial Pt/C (22.1 mV) and carbon-supported Rh nanosheets (Rh NS/C) (41.6 mV). The Tafel slope is 29.9 mV s^{-1}, which is comparable for commercial Pt/C and lower than that of Rh NS/C (37.4 mV/decade).

The core-shell structure of obtaining composites used in hydrogen production is well known, and such morphology is also presented in the work of Pu et al. [150], who synthesized Rh_2P nanoparticles encapsulated in an N-doped carbon (NC) core-shell structure ($Rh_2P@NC$) achieving overpotential of 9 mV at 10 mA cm^{-2} in 0.5 M H_2SO_4. Comparison in the electrocatalytic activity between $Rh_2P@NC$, Rh/NC, and Pt/C and belonging HER polarization curves shows that both $Rh_2P@NC$ and Pt/C exhibit high HER catalytic activities with 0 mV onset overpotential (η_{onset}), which is much smaller than that of Rh/NC. $Rh_2P@NC$ needs an overpotential (η_{10}) of 9 mV at the current density of 10 mA cm^{-2} with the corresponding Tafel slope of 26 mV/decade. The last part of the electrochemical measurements examined stability test where, after 1000 cyclic voltametric (CV) cycles at a scan rate of 100 mV/s in 0.5 M H_2SO_4 solution, the polarization curve retains an almost similar performance to the initial test.

Besides N-doping, carbon-supported materials can also be *double-codoped* with nitrogen and phosphorus. Two electrocatalysts composed of N and P codoped carbon (NPC) modified with noble metal phosphides (Rh_xP/NPC and RuP/NPC) with a low loading of Rh ($\approx$0.4 wt%) and Ru ($\approx$0.5 wt%) achieved promising electrocatalytic activities [139]. Rh_xP/NPC delivers Pt-like HER activity with an ultralow overpotential at 10 mA cm^{-2} (19 mV) and a small Tafel slope (36 mV/decade), while the RuP/NPC requires overpotential of 125 mV to achieve 10 mA cm^{-2} and a Tafel slope of 107 mV/decade. Besides metal phosphides, conductive oxides can also be considered as potential catalysts for the HER. The advantage of such materials is that, unlike their metallic counterparts and most prominently Pt, they are not prone to poisoning by underpotential deposition of less active metals that are always presented in the form of impurities in technological electrolytes [151]. Ru and Ir thin films, as well as their corresponding thermally oxidized RuO_2 and IrO_2 thin films, were developed by Cherevko et al. [151] and evaluated for HER in 0.1 M H_2SO_4. Metals, exhibit more extensive dissolution are found to be more active in catalyzing the

Materials **2023**, 16, 6319

hydrogen production, while metal oxides are easily blocked by hydrogen bubbles and show no dissolution during HER. Based on the results, it can be concluded that oxides may be considered to catalyze HER in case Pt contamination is an issue; even though metals are more active, their application as cathode materials is not feasible due to low stability. The dissolution of metals in acidic solutions is 2–3 magnitudes higher compared to their respective oxides.

Table 8 contains selected cathode materials recently synthesized and tested under acidic conditions with a high potential for later application in PEM water electrolysis, according to listed electrochemical parameters (overpotential at current density, Tafel slope, and stability) that are very close to the benchmark 20% Pt/C cathode catalyst material.

Table 8. Cathode materials with reduced noble metals content for use in acidic electrolytic hydrogen generation.

Cathode Catalyst	Electrolyte	Overpotential @Current Density	Tafel Slope	Stability	Ref.
Au@AuIr$_2$ (core-shell structure nanoparticles (NPs) with Au core and AuIr$_2$ alloy shell)	0.5 M H$_2$SO$_4$	29 mV@ of 10 mA cm^{-2}	15.6 mV/decade	40 h	[152]
PdCu/Ir core shell nanocrystals	0.5 M H$_2$SO$_4$	20 mV@ of 10 mA cm^{-2} • same overpotential as a commercial Pt/C	-	15 h@ 20 mA cm^{-2}	[153]
IrPdPtRhRu high entropy alloy (HEA) NPs	0.05 M H$_2$SO$_4$	33 mV@ of 10 mA cm^{-2} • much lower overpotentials to achieve a 10 mA cm^{-2} than the monometallic Ru (77.1 mV), Rh (58.6 mV), Pd (78.4 mV), Ir (47.8 mV) and Pt (48.9 mV)	-	CV for 3000 cycles	[154]
PtRu@RFCs (Pt is alloyed with Ru and embedded in porous resorcinol-formaldehyde carbon spheres) Pt loading 99.9% less than commercial Pt-based catalyst	0.5 M H$_2$SO$_4$	19.7 mV@10 mA cm^{-2} 43.1 mV @ 100 mA cm^{-2}	27.2 mV/decade for comparison: Pt/C (commercial) = 29.9 mV/decade	CV for 5000 cycles	[155]
RuP synthesized by dry chemistry method	0.1 M HClO$_4$	36 mV@10 mA cm^{-2} • benchmark Pt/C catalyst (20 wt%, Johnson Matthey) = 21 mA cm^{-2}	39.8 ± 0.5 mV/decade	CV for 8000 cycles	[156]
Pd$_4$S-SNC (palladium sulfide supported by S, N-doped carbon NPs)	0.5 M H$_2$SO$_4$	32 mV@ of 10 mA cm^{-2}	52 mV/decade	CV for 1000 cycles	[157]
PtN$_x$ cluster loaded on a TiO$_2$ support (PtN$_x$/TiO$_2$)	0.5 M H$_2$SO$_4$	67 mV@ of 10 mA cm^{-2}	52 mV/decade	CV for 5000 cycles	[158]

Table 8. *Cont.*

Cathode Catalyst	Electrolyte	Overpotential @Current Density	Tafel Slope	Stability	Ref.
Pt nanoclusters (NCs) anchored on porous TiO_2 nanosheets with rich oxygen vacancies (V_o-rich Pt/TiO_2)	0.5 M H_2SO_4	-	34 mV/decade • much smaller than the Tafel slope of commercial 20% Pt/C (116 mV/decade)	CV for 1000 cycles	[159]
Pt/OLC (onion-like nanospheres on carbon (OLC) with atomically dispersed Pt) • 0.27 wt% of Pt	0.5 M H_2SO	38 mV@10 mA cm^{-2}	36 mV/decade • for comparison: Pt/C (commercial, 20 wt% of Pt) = 35 mV/decade	100 h@ 10 mA cm^{-2}	[160]

5. Challenges and Insights for Future Clean Hydrogen Production Using PEM Water Electrolyzers

Hydrogen production by water electrolysis using electricity from renewable energy sources (solar energy, wind energy, etc.), although still insufficiently represented compared to the production of hydrogen produced from carbon-based sources, has experienced great progress in recent years, especially in the area of research and development.

This is due to the accelerated development of the hydrogen economy, aimed at achieving the targets set in various action plans and strategies at the EU level. In order to achieve the very ambitious EU targets, the future production of electrolyzers must be increased, which must be accompanied by cost reductions and efficiency improvements. Some of the conditions that must be fulfilled are (i) longevity of hydrogen electrolyzers, taking into account all the individual components from which they are composed: in particular, the durability of the electrode material; (ii) the ability to perform well-performing safety measurements during operation and monitoring; and (iii) the ability to quickly detect a potential performance problem and its solution.

In the short to medium term, electrolytically produced hydrogen is expected to find use in certain industries, such as semiconductors and food, where small quantities of high-purity hydrogen are needed to perform basic processes.

Focusing only on hydrogen production by PEM water electrolysis, it is of extreme importance to reduce the noble metal content, especially the platinum content, by at least an order of magnitude and replace it with non-noble metal alternatives in the future. The most promising alternatives are based on transition metals such as tungsten and molybdenum in the form of carbides, phosphides, and sulfides. Another important parameter is the need to increase the electrode area. This could be achieved by implementation of various HER catalysts with different morphological characteristics. A review of the literature shows that changes in the morphological structure of HER electrocatalysts, e.g., core-shell structures, have a positive effect on the increase in the active surface area of the catalytic material itself, but are not sufficient for a wide industrial use. Extensive scientific investigations are urgently needed to address these shortcomings. Furthermore, the role of the cathode material substrate surface is extremely important, and it must assist in the flow of electrons to the current collectors. Doping of carbon supports, which are mostly used as supports for the cathode electrocatalysts, can make them electrochemically active, enable them to have better interaction with the material they are supporting, and also participate in the charge transfer reactions.

All indications are that scientific collaboration is essential if hydrogen is to become the most important energy vector of the present. The planned energy goals for the widespread use of hydrogen will hardly be achieved with an individual approach. The solution with the highest probability of realization is a complementary approach, combining the proposals of leading experts in the field of electrocatalysis, physics, polymer chemistry, environmental chemistry, etc., and therefore, find the best solution with high practical application. Only in this way will PEM electrolyzers become a competitive technology that can be more easily deployed on a large scale.

6. Conclusions

Water electrolysis is a promising, renewable technique that enables applications that require small volumes of high purity hydrogen. Among the different types of this technique, of particular importance is PEM water electrolysis due to the dynamic range, reliability, and lack of corrosive electrolyte in comparison with alkaline electrolysis. Wider use at the megawatt scale is limited with the cost of catalyst-coated membrane. Pt supported on carbon black is a conventionally used cathode material that exhibits the best catalytic performance within this field. Since Pt is an extremely expensive noble metal, the increase in PEMWE application requires the development of cheap and long-lived hydrogen evolution reaction electrocatalysts that could substitute the use of Pt-based ones. The most promising non-noble metal alternatives are compounds based on transition metals in the form of sulfides, phosphides, and carbides. Even though much scientific effort is being put in the development of novel materials that do not contain noble metals in their structure, most of the obtained cathode materials still do not exceed the performances of Pt-based electrocatalysts. For these reasons, another part of the scientific community is committed to the development of composites consisting of a lower amount of platinum group materials in combination with other non-noble metal alternatives. Such Pd-, Ru-, Rh-based electrodes are highly efficient in hydrogen production with the potential of large-scale application. But still, further work is needed to improve the activity and stability of the mentioned catalyst, specifically within the context of industrial application. Integration of synthesized catalysts into commercial devices represents an important step forward towards sustainable hydrogen production through proton exchange membrane water electrolysis.

Author Contributions: Conceptualization, K.P., K.K., S.M. and A.J.; methodology, K.P. and S.M.; formal analysis, K.P. and S.M.; investigation, K.P. and S.M.; writing—original draft preparation, K.P. and S.M.; writing—review and editing, K.P., S.M., K.K. and A.J.; visualization, K.P. and S.M.; supervision, K.K. All authors have read and agreed to the published version of the manuscript.

Funding: This research received no external funding.

Institutional Review Board Statement: Not applicable.

Informed Consent Statement: Not applicable.

Data Availability Statement: Not applicable.

Conflicts of Interest: The authors declare no conflict of interest.

References

1. Khan, K.; Tareen, A.K.; Aslam, M.; Zhang, Y.; Wang, R.; Ouyang, Z.; Gou, Z.; Zhang, H. Recent Advances in Two-Dimensional Materials and Their Nanocomposites in Sustainable Energy Conversion Applications. *Nanoscale* **2019**, *11*, 21622–21678. [CrossRef] [PubMed]
2. Shiva Kumar, S.; Himabindu, V. Hydrogen Production by PEM Water Electrolysis—A Review. *Mater. Sci. Energy Technol.* **2019**, *2*, 442–454. [CrossRef]
3. Acar, C.; Dincer, I. Comparative Assessment of Hydrogen Production Methods from Renewable and Non-Renewable Sources. *Int. J. Hydrogen Energy* **2014**, *39*, 1–12. [CrossRef]
4. Aricò, A.S.; Siracusano, S.; Briguglio, N.; Baglio, V.; Di Blasi, A.; Antonucci, V. Polymer Electrolyte Membrane Water Electrolysis: Status of Technologies and Potential Applications in Combination with Renewable Power Sources. *J. Appl. Electrochem.* **2013**, *43*, 107–118. [CrossRef]
5. Carmo, M.; Fritz, D.L.; Mergel, J.; Stolten, D. A Comprehensive Review on PEM Water Electrolysis. *Int. J. Hydrogen Energy* **2013**, *38*, 4901–4934. [CrossRef]
6. Directives Directive (Eu) 2018/2001 of The European Parliament and of The Council of 11 December 2018 on the Promotion of the Use of Energy from Renewable Sources (Recast) (Text with EEA Relevance). *Official Journal of the European Union*, 21 December 2018.
7. Eftekhari, A. Electrocatalysts for Hydrogen Evolution Reaction. *Int. J. Hydrogen Energy* **2017**, *42*, 11053–11077. [CrossRef]

8. European Commission. *A Hydrogen Strategy for a Climate-Neutral Europe*; European Commission: Brussels, Belgium, 2020; pp. 1–24.
9. European Parliament. *European Council European Climate Law*; European Commission: Brussels, Belgium, 2021; Volume 2021, pp. 1–17.
10. European Commission. *Fit for 55*; European Commission: Brussels, Belgium, 2021.
11. European Commission. *REPowerEU*; European Commission: Brussels, Belgium, 2022.
12. European Clean Hydrogen Alliance. In Proceedings of the European Electrolyser Summit Joint Declaration, Brussels, Belgium, 5 May 2022; pp. 1–8.
13. Europen Investment Bank. Available online: https://www.eib.org/en/index (accessed on 5 May 2023).
14. Clean Hydrogen Partnership. Strategic Research and Innovation Agenda 2021–2027. *Unpublished* 2022; p. 179. Available online: https://www.clean-hydrogen.europa.eu/system/files/2022-02/Clean%20Hydrogen%20JU%20SRIA%20-%20approved%20by%20GB%20-%20clean%20for%20publication%20%28ID%2013246486%29.pdf (accessed on 8 May 2023).
15. Hydrogen Public Funding Compass. Available online: https://single-market-economy.ec.europa.eu/industry/strategy/hydrogen/funding-guide_en (accessed on 8 May 2023).
16. European Commission. *Clean Hydrogen Alliance Learbook on Hydrogen Supply Corridors*; European Commission: Brussels, Belgium, 2023; pp. 1–54.
17. European Clean Hydrogen Alliance. Available online: https://single-market-economy.ec.europa.eu/industry/strategy/industrial-alliances/european-clean-hydrogen-alliance_en (accessed on 9 May 2023).
18. Hydrogen Europe. Available online: https://hydrogeneurope.eu/ (accessed on 9 May 2023).
19. Clean Hydrogen Partnership. Available online: https://www.clean-hydrogen.europa.eu/index_en (accessed on 9 May 2023).
20. Hydrogen Valleys. Available online: https://www.clean-hydrogen.europa.eu/hydrogen-valleys-0_en (accessed on 10 May 2023).
21. Gospodinova, S.; Miccoli, F. *Hydrogen: Commission Supports Industry Commitment to Boost by Tenfold Electrolyser Manufacturing Capacities in the EU*; European Commission-Press Release: Brussels, Belgium, 2022; pp. 1–2.
22. IRENA. *Green Hydrogen Cost Reduction. Scaling up Electrolysers to Meet the 1.5 °C Climate Goal (IRENA)*; International Renewable Energy Agency (IRENA): Abu Dhabi, United Arab Emirates, 2020; pp. 1–106, ISBN 978-92-9260-295-6.
23. Russell, J.H.; Nuttall, L.J.; Fickett, A.P. Hydrogen Generation by Solid Polymer Electrolyte Water Electrolysis. *Am. Chem. Soc. Div. Fuel Chem. Prepr.* **1973**, *18*, 24–40.
24. Lettenmeier, P.; Wang, R.; Abouatallah, R.; Saruhan, B.; Freitag, O.; Gazdzicki, P.; Morawietz, T.; Hiesgen, R.; Gago, A.S.; Friedrich, K.A. Low-Cost and Durable Bipolar Plates for Proton Exchange Membrane Electrolyzers. *Sci. Rep.* **2017**, *7*, srep44035. [CrossRef]
25. Buttler, A.; Spliethoff, H. Current Status of Water Electrolysis for Energy Storage, Grid Balancing and Sector Coupling via Power-to-Gas and Power-to-Liquids: A Review. *Renew. Sustain. Energy Rev.* **2018**, *82*, 2440–2454. [CrossRef]
26. Marshall, A.; Børresen, B.; Hagen, G.; Tsypkin, M.; Tunold, R. Hydrogen Production by Advanced Proton Exchange Membrane (PEM) Water Electrolysers-Reduced Energy Consumption by Improved Electrocatalysis. *Energy* **2007**, *32*, 431–436. [CrossRef]
27. Badea, G.E.; Hora, C.; Maior, I.; Cojocaru, A.; Secui, C.; Filip, S.M.; Dan, F.C. Sustainable Hydrogen Production from Seawater Electrolysis: Through Fundamental Electrochemical Principles to the Most Recent Development. *Energies* **2022**, *15*, 8560. [CrossRef]
28. Mohammed-Ibrahim, J.; Moussab, H. Recent Advances on Hydrogen Production through Seawater Electrolysis. *Mater. Sci. Energy Technol.* **2020**, *3*, 780–807. [CrossRef]
29. Sun, X.; Xu, K.; Fleischer, C.; Liu, X.; Grandcolas, M.; Strandbakke, R.; Bjørheim, T.S.; Norby, T.; Chatzitakis, A. Earth-Abundant Electrocatalysts in Proton Exchange Membrane Electrolyzers. *Catalysts* **2018**, *8*, 657. [CrossRef]
30. Zhang, K.; Liang, X.; Wang, L.; Sun, K.; Wang, Y.; Xie, Z.; Wu, Q.; Bai, X.; Hamdy, M.S.; Chen, H.; et al. Status and Perspectives of Key Materials for PEM Electrolyzer. *Nano Res. Energy* **2022**, *1*, e9120032. [CrossRef]
31. Chen, Y.; Liu, C.; Xu, J.; Xia, C.; Wang, P.; Xia, B.Y.; Yan, Y.; Wang, X. Key Components and Design Strategy for a Proton Exchange Membrane Water Electrolyzer. *Small Struct.* **2023**, *4*, 2200130. [CrossRef]
32. Hassan, I.A.; Ramadan, H.S.; Saleh, M.A.; Hissel, D. Hydrogen Storage Technologies for Stationary and Mobile Applications: Review, Analysis and Perspectives. *Renew. Sustain. Energy Rev.* **2021**, *149*, 111311. [CrossRef]
33. Onda, K.; Kyakuno, T.; Hattori, K.; Ito, K. Prediction of Production Power for High-Pressure Hydrogen by High-Pressure Water Electrolysis. *J. Power Sources* **2004**, *132*, 64–70. [CrossRef]
34. Grigoriev, S.A.; Millet, P.; Korobtsev, S.V.; Porembskiy, V.I.; Pepic, M.; Etievant, C.; Puyenchet, C.; Fateev, V.N. Hydrogen Safety Aspects Related to High-Pressure Polymer Electrolyte Membrane Water Electrolysis. *Int. J. Hydrogen Energy* **2009**, *34*, 5986–5991. [CrossRef]
35. Rivard, E.; Trudeau, M.; Zaghib, K. Hydrogen Storage for Mobility: A Review. *Materials* **2019**, *12*, 1973. [CrossRef]
36. Mo, J.; Kang, Z.; Yang, G.; Retterer, S.T.; Cullen, D.A.; Toops, T.J.; Green, J.B.; Zhang, F.Y. Thin Liquid/Gas Diffusion Layers for High-Efficiency Hydrogen Production from Water Splitting. *Appl. Energy* **2016**, *177*, 817–822. [CrossRef]
37. Rakousky, C.; Reimer, U.; Wippermann, K.; Kuhri, S.; Carmo, M.; Lueke, W.; Stolten, D. Polymer Electrolyte Membrane Water Electrolysis: Restraining Degradation in the Presence of Fluctuating Power. *J. Power Sources* **2017**, *342*, 38–47. [CrossRef]

38. Kang, Z.; Mo, J.; Yang, G.; Li, Y.; Talley, D.A.; Retterer, S.T.; Cullen, D.A.; Toops, T.J.; Brady, M.P.; Bender, G.; et al. Thin Film Surface Modifications of Thin/Tunable Liquid/Gas Diffusion Layers for High-Efficiency Proton Exchange Membrane Electrolyzer Cells. *Appl. Energy* **2017**, *206*, 983–990. [CrossRef]

39. Liu, C.; Shviro, M.; Gago, A.S.; Zaccarine, S.F.; Bender, G.; Gazdzicki, P.; Morawietz, T.; Biswas, I.; Rasinski, M.; Everwand, A.; et al. Exploring the Interface of Skin-Layered Titanium Fibers for Electrochemical Water Splitting. *Adv. Energy Mater.* **2021**, *11*, 2002926. [CrossRef]

40. Hackemüller, F.J.; Borgardt, E.; Panchenko, O.; Müller, M.; Bram, M. Manufacturing of Large-Scale Titanium-Based Porous Transport Layers for Polymer Electrolyte Membrane Electrolysis by Tape Casting. *Adv. Eng. Mater.* **2019**, *21*, 1801201. [CrossRef]

41. Stiber, S.; Balzer, H.; Wierhake, A.; Wirkert, F.J.; Roth, J.; Rost, U.; Brodmann, M.; Lee, J.K.; Bazylak, A.; Waiblinger, W.; et al. Porous Transport Layers for Proton Exchange Membrane Electrolysis under Extreme Conditions of Current Density, Temperature, and Pressure. *Adv. Energy Mater.* **2021**, *11*, 2100630. [CrossRef]

42. Langemann, M.; Fritz, D.L.; Müller, M.; Stolten, D. Validation and Characterization of Suitable Materials for Bipolar Plates in PEM Water Electrolysis. *Int. J. Hydrogen Energy* **2015**, *40*, 11385–11391. [CrossRef]

43. Wang, T.; Cao, X.; Jiao, L. PEM Water Electrolysis for Hydrogen Production: Fundamentals, Advances, and Prospects. *Carbon Neutrality* **2022**, *1*, 1–19. [CrossRef]

44. Shirvanian, P.; van Berkel, F. Novel Components in Proton Exchange Membrane Water Electrolyzers (PEMWE): Status, Challenges and Future Needs. *Electrochem. Commun* **2020**, *114*, 106704. [CrossRef]

45. Jung, H.Y.; Huang, S.Y.; Ganesan, P.; Popov, B.N. Performance of Gold-Coated Titanium Bipolar Plates in Unitized Regenerative Fuel Cell Operation. *J. Power Sources* **2009**, *194*, 972–975. [CrossRef]

46. Wakayama, H.; Yamazaki, K. Low-Cost Bipolar Plates of Ti_4O_7-Coated Ti for Water Electrolysis with Polymer Electrolyte Membranes. *ACS Omega* **2021**, *6*, 4161–4166. [CrossRef]

47. Rojas, N.; Sánchez-Molina, M.; Sevilla, G.; Amores, E.; Almandoz, E.; Esparza, J.; Cruz Vivas, M.R.; Colominas, C. Coated Stainless Steels Evaluation for Bipolar Plates in PEM Water Electrolysis Conditions. *Int. J. Hydrogen Energy* **2021**, *46*, 25929–25943. [CrossRef]

48. Toghyani, S.; Afshari, E.; Baniasadi, E.; Atyabi, S.A. Thermal and Electrochemical Analysis of Different Flow Field Patterns in a PEM Electrolyzer. *Electrochim. Acta* **2018**, *267*, 234–245. [CrossRef]

49. Goñi-Urtiaga, A.; Presvytes, D.; Scott, K. Solid Acids as Electrolyte Materials for Proton Exchange Membrane (PEM) Electrolysis: Review. *Int. J. Hydrogen Energy* **2012**, *37*, 3358–3372. [CrossRef]

50. Kusoglu, A.; Weber, A.Z. New Insights into Perfluorinated Sulfonic-Acid Ionomers. *Chem. Rev.* **2017**, *117*, 987–1104. [CrossRef] [PubMed]

51. Gagliardi, G.G.; Ibrahim, A.; Borello, D.; El-Kharouf, A. Composite Polymers Development and Application for Polymer Electrolyte Membrane Technologies—A Review. *Molecules* **2020**, *25*, 1712. [CrossRef] [PubMed]

52. Shin, D.W.; Guiver, M.D.; Lee, Y.M. Hydrocarbon-Based Polymer Electrolyte Membranes: Importance of Morphology on Ion Transport and Membrane Stability. *Chem. Rev.* **2017**, *117*, 4759–4805. [CrossRef] [PubMed]

53. Qiu, D.; Peng, L.; Lai, X.; Ni, M.; Lehnert, W. Mechanical Failure and Mitigation Strategies for the Membrane in a Proton Exchange Membrane Fuel Cell. *Renew. Sustain. Energy Rev.* **2019**, *113*, 109289. [CrossRef]

54. Walkowiak-Kulikowska, J.; Wolska, J.; Koroniak, H. Polymers Application in Proton Exchange Membranes for Fuel Cells (PEMFCs). *Phys. Sci. Rev.* **2017**, *2*, 20170018.

55. Higashihara, T.; Matsumoto, K.; Ueda, M. Sulfonated Aromatic Hydrocarbon Polymers as Proton Exchange Membranes for Fuel Cells. *Polymer* **2009**, *50*, 5341–5357. [CrossRef]

56. Kusoglu, A.; Dursch, T.J.; Weber, A.Z. Nanostructure/Swelling Relationships of Bulk and Thin-Film PFSA Ionomers. *Adv. Funct. Mater.* **2016**, *26*, 4961–4975. [CrossRef]

57. Sunda, A.P.; Venkatnathan, A. Atomistic Simulations of Structure and Dynamics of Hydrated Aciplex Polymer Electrolyte Membrane. *Soft Matter* **2012**, *8*, 10827–10836. [CrossRef]

58. Park, S.; Lee, W.; Na, Y. Performance Comparison of Proton Exchange Membrane Water Electrolysis Cell Using Channel and PTL Flow Fields through Three-Dimensional Two-Phase Flow Simulation. *Membranes* **2022**, *12*, 1260. [CrossRef] [PubMed]

59. Al Munsur, A.Z.; Goo, B.H.; Kim, Y.; Kwon, O.J.; Paek, S.Y.; Lee, S.Y.; Kim, H.J.; Kim, T.H. Nafion-Based Proton-Exchange Membranes Built on Cross-Linked Semi-Interpenetrating Polymer Networks between Poly(Acrylic Acid) and Poly(Vinyl Alcohol). *ACS Appl. Mater. Interfaces* **2021**, *13*, 28188–28200. [CrossRef] [PubMed]

60. Kusoglu, A.; Karlsson, A.M.; Santare, M.H.; Cleghorn, S.; Johnson, W.B. Mechanical Behavior of Fuel Cell Membranes under Humidity Cycles and Effect of Swelling Anisotropy on the Fatigue Stresses. *J. Power Sources* **2007**, *170*, 345–358. [CrossRef]

61. Tang, Y.; Kusoglu, A.; Karlsson, A.M.; Santare, M.H.; Cleghorn, S.; Johnson, W.B. Mechanical Properties of a Reinforced Composite Polymer Electrolyte Membrane and Its Simulated Performance in PEM Fuel Cells. *J. Power Sources* **2008**, *175*, 817–825. [CrossRef]

62. Shin, S.H.; Nur, P.J.; Kodir, A.; Kwak, D.H.; Lee, H.; Shin, D.; Bae, B. Improving the Mechanical Durability of Short-Side-Chain Perfluorinated Polymer Electrolyte Membranes by Annealing and Physical Reinforcement. *ACS Omega* **2019**, *4*, 19153–19163. [CrossRef]

63. Laberty-Robert, C.; Vallé, K.; Pereira, F.; Sanchez, C. Design and Properties of Functional Hybrid Organic–Inorganic Membranes for Fuel Cells. *Chem. Soc. Rev.* **2011**, *40*, 961–1005. [CrossRef]

64. Di Noto, V.; Boaretto, N.; Negro, E.; Stallworth, P.E.; Lavina, S.; Giffin, G.A.; Greenbaum, S.G. Inorganic-Organic Membranes Based on Nafion, [(ZrO$_2$)·(HfO$_2$) 0.25] and [(SiO$_2$)·(HfO$_2$) 0.28] Nanoparticles. Part II: Relaxations and Conductivity Mechanism. *Int. J. Hydrogen Energy* **2012**, *37*, 6215–6227. [CrossRef]
65. Casciola, M.; Bagnasco, G.; Donnadio, A.; Micoli, L.; Pica, M.; Sganappa, M.; Turco, M. Conductivity and Methanol Permeability of Nafion-Zirconiumphosphate Compositemembranes Containing High Aspect Ratio Filler Particles. *Fuel Cells* **2009**, *9*, 394–400. [CrossRef]
66. Slade, S.M.; Smith, J.R.; Campbell, S.A.; Ralph, T.R.; Ponce De León, C.; Walsh, F.C. Characterisation of a Re-Cast Composite Nafion® 1100 Series of Proton Exchange Membranes Incorporating Inert Inorganic Oxide Particles. *Electrochim. Acta* **2010**, *55*, 6818–6829. [CrossRef]
67. Di Noto, V.; Piga, M.; Lavina, S.; Negro, E.; Yoshida, K.; Ito, R.; Furukawa, T. Structure, Properties and Proton Conductivity of Nafion/[(TiO$_2$) ṡ (WO$_3$)0.148]ψ TiO$_2$ Nanocomposite Membranes. *Electrochim. Acta* **2010**, *55*, 1431–1444. [CrossRef]
68. Cele, N.P.; Ray, S.S. Effect of Multiwalled Carbon Nanotube Loading on the Properties of Nafion® Membranes. *J. Mater. Res.* **2014**, *34*, 66–78. [CrossRef]
69. Baglio, V.; Ornelas, R.; Matteucci, F.; Martina, F.; Ciccarella, G.; Zama, I.; Arriaga, L.G.; Antonucci, V.; Aricò, A.S. Solid Polymer Electrolyte Water Electrolyser Based on Nafion-TiO$_2$ Composite Membrane for High Temperature Operation. *Fuel Cells* **2009**, *9*, 247–252. [CrossRef]
70. Antonucci, V.; Di Blasi, A.; Baglio, V.; Ornelas, R.; Matteucci, F.; Ledesma-Garcia, J.; Arriaga, L.G.; Aricò, A.S. High Temperature Operation of a Composite Membrane-Based Solid Polymer Electrolyte Water Electrolyser. *Electrochim. Acta* **2008**, *53*, 7350–7356. [CrossRef]
71. Schuster, M.; Kreuer, K.D.; Andersen, H.T.; Maier, J. Sulfonated Poly(Phenylene Sulfone) Polymers as Hydrolytically and Thermooxidatively Stable Proton Conducting Ionomers. *Macromolecules* **2007**, *40*, 598–607. [CrossRef]
72. Yazili, D.; Marini, E.; Saatkamp, T.; Münchinger, A.; de Wild, T.; Gubler, L.; Titvinidze, G.; Schuster, M.; Schare, C.; Jörissen, L.; et al. Sulfonated Poly(Phenylene Sulfone) Blend Membranes Finding Their Way into Proton Exchange Membrane Fuel Cells. *J. Power Sources* **2023**, *563*, 232791. [CrossRef]
73. Wu, G.; Lin, S.J.; Hsu, I.C.; Su, J.Y.; Chen, D.W. Study of High Performance Sulfonated Polyether Ether Ketone Composite Electrolyte Membranes. *Polymers* **2019**, *11*, 1177. [CrossRef]
74. Maria Mahimai, B.; Sivasubramanian, G.; Sekar, K.; Kannaiyan, D.; Deivanayagam, P. Sulfonated Poly(Ether Ether Ketone): Efficient Ion-Exchange Polymer Electrolytes for Fuel Cell Applications-a Versatile Review. *Mater. Adv.* **2022**, *3*, 6085–6095. [CrossRef]
75. Park, J.E.; Kim, J.; Han, J.; Kim, K.; Park, S.B.; Kim, S.; Park, H.S.; Cho, Y.H.; Lee, J.C.; Sung, Y.E. High-Performance Proton-Exchange Membrane Water Electrolysis Using a Sulfonated Poly(Arylene Ether Sulfone) Membrane and Ionomer. *J. Memb. Sci.* **2021**, *620*, 118871. [CrossRef]
76. Energy Transitions Commission. *Making the Hydrogen Economy Possible: Accelerating Clean Hydrogen in an Electrified Economy The Making Mission Possible Series*; Energy Transitions Commission: London, UK, 2021.
77. Chatenet, M.; Pollet, B.G.; Dekel, D.; Dionigi, F.; Deseure, J.; Millet, P.; Braatz, R.; Bazant, M.Z.; Eikerling, M.; Staffell, I. Water Electrolysis: From Textbook Knowledge to the Latest Scientific Strategies and Industrial Developments. *Chem. Soc. Rev.* **2022**, *51*, 4583–4762. [CrossRef]
78. IEA. *Net Zero by 2050—A Roadmap for the Global Energy Sector*; International Energy Agency: Paris, France, 2017.
79. Bender, G.; Carmo, M.; Smolinka, T.; Gago, A.; Danilovic, N.; Mueller, M.; Ganci, F.; Fallisch, A.; Lettenmeier, P.; Friedrich, K.A.; et al. Initial Approaches in Benchmarking and Round Robin Testing for Proton Exchange Membrane Water Electrolyzers. *Int. J. Hydrogen Energy* **2019**, *44*, 9174–9187. [CrossRef]
80. Ion-Ebrasu, D.; Pollet, B.G.; Spinu-Zaulet, A.; Soare, A.; Carcadea, E.; Varlam, M.; Caprarescu, S. Graphene Modified Fluorinated Cation-Exchange Membranes for Proton Exchange Membrane Water Electrolysis. *Int. J. Hydrogen Energy* **2019**, *44*, 10190–10196. [CrossRef]
81. Siracusano, S.; Baglio, V.; Nicotera, I.; Mazzapioda, L.; Aricò, A.S.; Panero, S.; Navarra, M.A. Sulfated Titania as Additive in Nafion Membranes for Water Electrolysis Applications. *Int. J. Hydrogen Energy* **2017**, *42*, 27851–27858. [CrossRef]
82. Lee, C.J.; Song, J.; Yoon, K.S.; Rho, Y.; Yu, D.M.; Oh, K.H.; Lee, J.Y.; Kim, T.H.; Hong, Y.T.; Kim, H.J.; et al. Controlling Hydrophilic Channel Alignment of Perfluorinated Sulfonic Acid Membranes via Biaxial Drawing for High Performance and Durable Polymer Electrolyte Membrane Water Electrolysis. *J. Power Sources* **2022**, *518*, 230772. [CrossRef]
83. Kim, T.; Sihn, Y.; Yoon, I.H.; Yoon, S.J.; Lee, K.; Yang, J.H.; So, S.; Park, C.W. Monolayer Hexagonal Boron Nitride Nanosheets as Proton-Conductive Gas Barriers for Polymer Electrolyte Membrane Water Electrolysis. *ACS Appl. Nano Mater.* **2021**, *4*, 9104–9112. [CrossRef]
84. Giancola, S.; Zatoń, M.; Reyes-Carmona, Á.; Dupont, M.; Donnadio, A.; Cavaliere, S.; Rozière, J.; Jones, D.J. Composite Short Side Chain PFSA Membranes for PEM Water Electrolysis. *J. Memb. Sci.* **2019**, *570–571*, 69–76. [CrossRef]
85. Han, S.Y.; Yu, J.H.; Mo, Y.H.; Ahn, S.M.; Lee, J.Y.; Kim, T.H.; Yoon, S.J.; Hong, S.; Hong, Y.T.; So, S. Ion Exchange Capacity Controlled Biphenol-Based Sulfonated Poly(Arylene Ether Sulfone) for Polymer Electrolyte Membrane Water Electrolyzers: Comparison of Random and Multi-Block Copolymers. *J. Memb. Sci.* **2021**, *634*, 119370. [CrossRef]
86. Kim, J.; Ohira, A. Crosslinked Sulfonated Polyphenylsulfone (Csppsu) Membranes for Elevated-Temperature Pem Water Electrolysis. *Membranes* **2021**, *11*, 861. [CrossRef]

87. Klose, C.; Saatkamp, T.; Münchinger, A.; Bohn, L.; Titvinidze, G.; Breitwieser, M.; Kreuer, K.D.; Vierrath, S. All-Hydrocarbon MEA for PEM Water Electrolysis Combining Low Hydrogen Crossover and High Efficiency. *Adv. Energy Mater.* **2020**, *10*, 1903995. [CrossRef]
88. Waribam, P.; Jaiyen, K.; Samart, C.; Ogawa, M.; Guan, G.; Kongparakul, S. MXene Potassium Titanate Nanowire/Sulfonated Polyether Ether Ketone (SPEEK) Hybrid Composite Proton Exchange Membrane for Photocatalytic Water Splitting. *RSC Adv.* **2021**, *11*, 9327–9335. [CrossRef]
89. Waribam, P.; Jaiyen, K.; Samart, C.; Ogawa, M.; Guan, G.; Kongparakul, S. MXene-Copper Oxide/Sulfonated Polyether Ether Ketone as a Hybrid Composite Proton Exchange Membrane in Electrochemical Water Electrolysis. *Catal. Today* **2023**, *407*, 96–106. [CrossRef]
90. Ko, E.J.; Lee, E.; Lee, J.Y.; Yu, D.M.; Yoon, S.J.; Oh, K.H.; Hong, Y.T.; So, S. Multi-Block Copolymer Membranes Consisting of Sulfonated Poly(p-Phenylene) and Naphthalene Containing Poly(Arylene Ether Ketone) for Proton Exchange Membrane Water Electrolysis. *Polymers* **2023**, *15*, 1748. [CrossRef] [PubMed]
91. Ahn, S.M.; Park, J.E.; Jang, G.Y.; Jeong, H.Y.; Yu, D.M.; Jang, J.K.; Lee, J.C.; Cho, Y.H.; Kim, T.H. Highly Sulfonated Aromatic Graft Polymer with Very High Proton Conductivity and Low Hydrogen Permeability for Water Electrolysis. *ACS Energy Lett.* **2022**, *7*, 4427–4435. [CrossRef]
92. Feng, Q.; Yuan, X.Z.; Liu, G.; Wei, B.; Zhang, Z.; Li, H.; Wang, H. A Review of Proton Exchange Membrane Water Electrolysis on Degradation Mechanisms and Mitigation Strategies. *J. Power Sources* **2017**, *366*, 33–55. [CrossRef]
93. Kim, H.; Kim, J.; Kim, S.K.; Ahn, S.H. A Transition Metal Oxysulfide Cathode for the Proton Exchange Membrane Water Electrolyzer. *Appl. Catal. B* **2018**, *232*, 93–100. [CrossRef]
94. Shiva Kumar, S.; Ramakrishna, S.U.B.; Rama Devi, B.; Himabindu, V. Phosphorus-Doped Carbon Nanoparticles Supported Palladium Electrocatalyst for the Hydrogen Evolution Reaction (HER) in PEM Water Electrolysis. *Ionics* **2018**, *24*, 3113–3121. [CrossRef]
95. Genova-Koleva, R.V. *Electrocatalyst Development for PEM Water Electrolysis and DMFC: Towards the Methanol Economy*; Universitat de Barcelona: Barcelona, Spain, 2017.
96. Kim, H.; Hwang, E.; Park, H.; Lee, B.S.; Jang, J.H.; Kim, H.J.; Ahn, S.H.; Kim, S.K. Non-Precious Metal Electrocatalysts for Hydrogen Production in Proton Exchange Membrane Water Electrolyzer. *Appl. Catal. B* **2017**, *206*, 608–616. [CrossRef]
97. Kim, J.; Kim, H.; Ahn, S.H. Electrodeposited Rhodium Phosphide with High Activity for Hydrogen Evolution Reaction in Acidic Medium. *ACS Sustain. Chem. Eng.* **2019**, *7*, 14041–14050. [CrossRef]
98. Kim, H.; Park, H.; Kim, D.K.; Choi, I.; Kim, S.K. Pulse-Electrodeposited Nickel Phosphide for High-Performance Proton Exchange Membrane Water Electrolysis. *J. Alloys Compd.* **2019**, *785*, 296–304. [CrossRef]
99. Ng, J.W.D.; Hellstern, T.R.; Kibsgaard, J.; Hinckley, A.C.; Benck, J.D.; Jaramillo, T.F. Polymer Electrolyte Membrane Electrolyzers Utilizing Non-Precious Mo-Based Hydrogen Evolution Catalysts. *ChemSusChem* **2015**, *8*, 3512–3519. [CrossRef]
100. Hinnemann, B.; Moses, P.G.; Bonde, J.; Jørgensen, K.P.; Nielsen, J.H.; Horch, S.; Chorkendorff, I.; Nørskov, J.K. Biomimetic Hydrogen Evolution: MoS₂ Nanoparticles as Catalyst for Hydrogen Evolution. *J. Am. Chem. Soc.* **2005**, *127*, 5308–5309. [CrossRef] [PubMed]
101. Jaramillo, T.F.; Jørgensen, K.P.; Bonde, J.; Nielsen, J.H.; Horch, S.; Chorkendorff, I. Identification of Active Edge Sites for Electrochemical H₂ Evolution from MoS₂ Nanocatalysts. *Science (1979)* **2007**, *317*, 100–102. [CrossRef]
102. Senthil Kumar, S.M.; Selvakumar, K.; Thangamuthu, R.; Karthigai Selvi, A.; Ravichandran, S.; Sozhan, G.; Rajasekar, K.; Navascues, N.; Irusta, S. Hydrothermal Assisted Morphology Designed MoS₂ Material as Alternative Cathode Catalyst for PEM Electrolyser Application. *Int. J. Hydrogen Energy* **2016**, *41*, 13331–13340. [CrossRef]
103. Li, Y.; Wang, H.; Xie, L.; Liang, Y.; Hong, G.; Dai, H. MoS₂ Nanoparticles Grown on Graphene: An Advanced Catalyst for the Hydrogen Evolution Reaction. *J. Am. Chem. Soc.* **2011**, *133*, 7296–7299. [CrossRef] [PubMed]
104. Corrales-Sánchez, T.; Ampurdanés, J.; Urakawa, A. MoS₂-Based Materials as Alternative Cathode Catalyst for PEM Electrolysis. *Int. J. Hydrogen Energy* **2014**, *39*, 20837–20843. [CrossRef]
105. Tahira, A.; Ibupoto, Z.H.; Mazzaro, R.; You, S.; Morandi, V.; Natile, M.M.; Vagin, M.; Vomiero, A. Advanced Electrocatalysts for Hydrogen Evolution Reaction Based on Core-Shell MoS₂/TiO₂ Nanostructures in Acidic and Alkaline Media. *ACS Appl. Energy Mater.* **2019**, *2*, 2053–2062. [CrossRef]
106. Chen, J.; Xu, Z.; Chen, Y. Chapter 2—Electronic Properties of Sulfide Minerals. In *Density Functional Theory and Applications*; Elsevier: Amsterdam, The Netherlands, 2020; pp. 13–81.
107. Vaughan, D.J. Sulfides. In *Encyclopedia of Geology*, 2nd ed; Academic Press: Cambridge, MA, USA, 2020; Volume 1–6, pp. 395–412, ISBN 9780081029091.
108. Di Giovanni, C.; Reyes-Carmona, Á.; Coursier, A.; Nowak, S.; Grenèche, J.M.; Lecoq, H.; Mouton, L.; Rozière, J.; Jones, D.; Peron, J.; et al. Low-Cost Nanostructured Iron Sulfide Electrocatalysts for PEM Water Electrolysis. *ACS Catal.* **2016**, *6*, 2626–2631. [CrossRef]
109. Wang, D.Y.; Gong, M.; Chou, H.L.; Pan, C.J.; Chen, H.A.; Wu, Y.; Lin, M.C.; Guan, M.; Yang, J.; Chen, C.W.; et al. Highly Active and Stable Hybrid Catalyst of Cobalt-Doped FeS₂ Nanosheets-Carbon Nanotubes for Hydrogen Evolution Reaction. *J. Am. Chem. Soc.* **2015**, *137*, 1587–1592. [CrossRef]

110. Kim, J.; Kim, H.; Kim, J.; Kim, J.H.; Ahn, S.H. Electrochemical Fabrication of Fe-Based Binary and Ternary Phosphide Cathodes for Proton Exchange Membrane Water Electrolyzer. *J. Alloys Compd.* **2019**, *807*, 148813. [CrossRef]
111. Pan, Y.; Hu, W.; Liu, D.; Liu, Y.; Liu, C. Carbon Nanotubes Decorated with Nickel Phosphide Nanoparticles as Efficient Nanohybrid Electrocatalysts for the Hydrogen Evolution Reaction. *J. Mater. Chem. A Mater.* **2015**, *3*, 13087–13094. [CrossRef]
112. Fan, L.; Liu, P.F.; Yan, X.; Gu, L.; Yang, Z.Z.; Yang, H.G.; Qiu, S.; Yao, X. Atomically Isolated Nickel Species Anchored on Graphitized Carbon for Efficient Hydrogen Evolution Electrocatalysis. *Nat. Commun.* **2016**, *7*, 10667. [CrossRef] [PubMed]
113. Chen, W.F.; Wang, C.H.; Sasaki, K.; Marinkovic, N.; Xu, W.; Muckerman, J.T.; Zhu, Y.; Adzic, R.R. Highly Active and Durable Nanostructured Molybdenum Carbide Electrocatalysts for Hydrogen Production. *Energy Environ. Sci.* **2013**, *6*, 943–951. [CrossRef]
114. Wu, R.; Zhang, J.; Shi, Y.; Liu, D.; Zhang, B. Metallic WO_2-Carbon Mesoporous Nanowires as Highly Efficient Electrocatalysts for Hydrogen Evolution Reaction. *J. Am. Chem. Soc.* **2015**, *137*, 6983–6986. [CrossRef] [PubMed]
115. Phuruangrat, A.; Ham, D.J.; Thongtem, S.; Lee, J.S. Electrochemical Hydrogen Evolution over MoO_3 Nanowires Produced by Microwave-Assisted Hydrothermal Reaction. *Electrochem. Commun* **2009**, *11*, 1740–1743. [CrossRef]
116. Kim, H.; Park, H.; Oh, S.; Kim, S.K. Facile Electrochemical Preparation of Nonprecious Co-Cu Alloy Catalysts for Hydrogen Production in Proton Exchange Membrane Water Electrolysis. *Int. J. Energy Res.* **2020**, *44*, 2833–2844. [CrossRef]
117. Popczun, E.J.; Read, C.G.; Roske, C.W.; Lewis, N.S.; Schaak, R.E. Highly Active Electrocatalysis of the Hydrogen Evolution Reaction by Cobalt Phosphide Nanoparticles. *Angew. Chem. Int. Ed.* **2014**, *53*, 5427–5430. [CrossRef] [PubMed]
118. Tian, J.; Liu, Q.; Asiri, A.M.; Sun, X. Self-Supported Nanoporous Cobalt Phosphide Nanowire Arrays: An efficient 3D hydrogen-evolving cathode over the wide range of pH 0–14. *J. Am. Chem. Soc.* **2014**, *136*, 7587–7590. [CrossRef]
119. Fen, Q.; Xiong, Y.; Xie, L.; Zhang, Z.; Lu, X.; Wang, Y.; Yuan, X.Z.; Fan, J.; Li, H.; Wang, H. Tungsten Carbide Encapsulated in Graphe-like N-Doped Carbon Nanospheres: One-Step Facile Synthesis for Low-Cost and Highly Active Electrocatalysts in Proton Exchange Membrane Water Electrolyzers. *ACS Appl. Mater. Interfaces.* **2019**, *11*, 25123–25132.
120. Chakrabartty, S.; Barman, B.K.; Raj, C.R. Nitrogen and Phosphorus Co-Doped Graphitic Carbon Encapsulated Ultrafine OsP_2 Nanoparticles: A pH Universal Highy Durable Catalyst for Hydrogen Evolution Reaction. *Chem. Commun.* **2013**, *55*, 4399–4402. [CrossRef]
121. Kim, J.H.; Kim, J.; Kim, H.; Kim, J.; Ahn, S.H. Facile Fabrication of Nanostructured NiMo Cathode for High-Performance Proton Exchange Membrane Water Electrolyzer. *J. Ind. Eng. Chem.* **2019**, *79*, 255–260. [CrossRef]
122. Wang, T.; Wang, X.; Liu, Y.; Zheng, J.; Li, X. A Highly Efficient and Stable Biphasic Nanocrystalline Ni-Mo-N Catalyst for Hydrogen Evolution in Both Acidic and Alkaline Electrolytes. *Nano Energy* **2016**, *22*, 111–119. [CrossRef]
123. Ge, Y.; Gao, S.P.; Dong, P.; Baines, R.; Ajayan, P.M.; Ye, M.; Shen, J. Insight into the Hydrogen Evolution Reaction of Nickel Dichalcogenide Nanosheets: Activities Related to Non-Metal Ligands. *Nanoscale* **2017**, *9*, 5538–5544. [CrossRef] [PubMed]
124. Feng, Q.; Zeng, L.; Xu, J.; Zhang, Z.; Zhao, Z.; Wang, Y.; Yuan, X.Z.; Li, H.; Wang, H. NaCl Template-Directed Approach to Ultrathin Lamellar Molybdenum Phosphide-Carbon Hybrids for Efficient Hydrogen Production. *J. Power Sources* **2019**, *438*. [CrossRef]
125. Li, J.S.; Zhang, S.; Sha, J.Q.; Wang, H.; Liu, M.Z.; Kong, L.X.; Liu, G.D. Confined Molybdenum Phosphide in P-Doped Porous Carbon as Efficient Electrocatalysts for Hydrogen Evolution. *ACS Appl. Mater. Interfaces* **2018**, *10*, 17140–17146. [CrossRef] [PubMed]
126. Han, S.; Feng, Y.; Zhang, F.; Yang, C.; Yao, Z.; Zhao, W.; Qiu, F.; Yang, L.; Yao, Y.; Zhuang, X.; et al. Metal-Phosphide-Containing Porous Carbons Derived from an Ionic-Polymer Framework and Applied as Highly Efficient Electrochemical Catalysts for Water Splitting. *Adv. Funct. Mater.* **2015**, *25*, 3899–3906. [CrossRef]
127. Yang, J.; Zhang, F.; Wang, X.; He, D.; Wu, G.; Yang, Q.; Hong, X.; Wu, Y.; Li, Y. Porous Molybdenum Phosphide Nano-Octahedrons Derived from Confined Phosphorization in UIO-66 for Efficient Hydrogen Evolution. *Angew. Chem. Int. Ed.* **2016**, *55*, 12854–12858. [CrossRef] [PubMed]
128. Huang, C.; Pi, C.; Zhang, X.; Ding, K.; Qin, P.; Fu, J.; Peng, X.; Gao, B.; Chu, P.K.; Huo, K. In Situ Synthesis of MoP Nanoflakes Intercalated N-Doped Graphene Nanobelts from MoO_3–Amine Hybrid for High-Efficient Hydrogen Evolution Reaction. *Small* **2018**, *14*, e1800667. [CrossRef]
129. Huang, Y.; Song, X.; Deng, J.; Zha, C.; Huang, W.; Wu, Y.; Li, Y. Ultra-Dispersed Molybdenum Phosphide and Phosphosulfide Nanoparticles on Hierarchical Carbonaceous Scaffolds for Hydrogen Evolution Electrocatalysis. *Appl. Catal. B* **2019**, *245*, 656–661. [CrossRef]
130. Kibsgaard, J.; Jaramillo, T.F. Molybdenum Phosphosulfide: An Active, Acid-Stable, Earth-Abundant Catalyst for the Hydrogen Evolution Reaction. *Angew. Chem. Int. Ed.* **2014**, *53*, 14433–14437. [CrossRef]
131. Jiang, R.; Fan, J.; Hu, L.; Dou, Y.; Mao, X.; Wang, D. Electrochemically Synthesized N-Doped Molybdenum Carbide Nanoparticles for Efficient Catalysis of Hydrogen Evolution Reaction. *Electrochim. Acta* **2018**, *261*, 578–587. [CrossRef]
132. Wang, D.; Guo, T.; Wu, Z. Hierarchical Mo_2C/C Scaffolds Organized by Nanosheets as Highly Efficient Electrocatalysts for Hydrogen Production. *ACS Sustain. Chem. Eng.* **2018**, *6*, 13995–14003. [CrossRef]
133. Wu, C.; Li, J. Unique Hierarchical Mo_2C/C Nanosheet Hybrids as Active Electrocatalyst for Hydrogen Evolution Reaction. *ACS Appl. Mater. Interfaces* **2017**, *9*, 41314–41322. [CrossRef]
134. Selvan, R.K.; Gedanken, A. The Sonochemical Synthesis and Characterization of Cu1-XNi XWO_4 Nanoparticles/Nanorods and Their Application in Electrocatalytic Hydrogen Evolution. *Nanotechnology* **2009**, *20*, 105602. [CrossRef] [PubMed]

135. Yeon, K.; Kim, J.; Kim, H.; Guo, W.; Han, G.H.; Hong, S.; Ahn, S.H. Electrodeposited Nickel Phosphide Supported by Copper Foam for Proton Exchange Membrane Water Electrolyzer. *Korean J. Chem. Eng.* **2020**, *37*, 1379–1386. [CrossRef]

136. Yang, Y.; Lun, Z.; Xia, G.; Zheng, F.; He, M.; Chen, Q. Non-Precious Alloy Encapsulated in Nitrogen-Doped Graphene Layers Derived from MOFs as an Active and Durable Hydrogen Evolution Reaction Catalyst. *Energy Environ. Sci.* **2015**, *8*, 3563–3571. [CrossRef]

137. Ji, X.; Liu, B.; Ren, X.; Shi, X.; Asiri, A.M.; Sun, X. P-Doped Ag Nanoparticles Embedded in N-Doped Carbon Nanoflake: An Efficient Electrocatalyst for the Hydrogen Evolution Reaction. *ACS Sustain. Chem. Eng.* **2018**, *6*, 4499–4503. [CrossRef]

138. Chen, Z.; Wu, R.; Liu, Y.; Ha, Y.; Guo, Y.; Sun, D.; Liu, M.; Fang, F. Ultrafine Co Nanoparticles Encapsulated in Carbon-Nanotubes-Grafted Graphene Sheets as Advanced Electrocatalysts for the Hydrogen Evolution Reaction. *Adv. Mater.* **2018**, *30*, e1802011. [CrossRef] [PubMed]

139. Qin, Q.; Jang, H.; Chen, L.; Nam, G.; Liu, X.; Cho, J. Low Loading of RhxP and RuP on N, P Codoped Carbon as Two Trifunctional Electrocatalysts for the Oxygen and Hydrogen Electrode Reactions. *Adv. Energy Mater.* **2018**, *8*, 1801478. [CrossRef]

140. Ayers, K.E.; Renner, J.N.; Danilovic, N.; Wang, J.X.; Zhang, Y.; Maric, R.; Yu, H. Pathways to Ultra-Low Platinum Group Metal Catalyst Loading in Proton Exchange Membrane Electrolyzers. *Catal. Today* **2016**, *262*, 121–132. [CrossRef]

141. Ayers, K.E.; Capuano, C.; Carter, B.; Dalton, L.; Hanlon, G.; Manco, J.; Niedzwiecki, M. Research Advances Towards Low Cost, High Efficiency PEM Electrolysis. *ECS Meet. Abstr.* **2010**, *33*, 3–15.

142. Kang, Z.; Yang, G.; Mo, J.; Li, Y.; Yu, S.; Cullen, D.A.; Retterer, S.T.; Toops, T.J.; Bender, G.; Pivovar, B.S.; et al. Novel Thin/Tunable Gas Diffusion Electrodes with Ultra-Low Catalyst Loading for Hydrogen Evolution Reactions in Proton Exchange Membrane Electrolyzer Cells. *Nano Energy* **2018**, *47*, 434–441. [CrossRef]

143. Mirshekari, G.; Ouimet, R.; Zeng, Z.; Yu, H.; Bliznakov, S.; Bonville, L.; Niedzwiecki, A.; Capuano, C.; Ayers, K.; Maric, R. High-Performance and Cost-Effective Membrane Electrode Assemblies for Advanced Proton Exchange Membrane Water Electrolyzers: Long-Term Durability Assessment. *Int. J. Hydrog. Energy* **2021**, *6*, 1526–1539. [CrossRef]

144. Ravichandran, S.; Venkatkarthick, R.; Sankari, A.; Vasudevan, S.; Jonas Davidson, D.; Sozhan, G. Platinum Deposition on the Nafion Membrane by Impregnation Reduction Using Nonionic Surfactant for Water Electrolysis—An Alternate Approach. *Energy* **2014**, *68*, 148–151. [CrossRef]

145. Wang, X.; Hsing, I.M. Surfactant Stabilized Pt and Pt Alloy Electrocatalyst for Polymer Electrolyte Fuel Cells. *Electrochim. Acta* **2002**, *47*, 2981–2987. [CrossRef]

146. Rajalakshmi, N.; Ryu, H.; Dhathathreyan, K.S. Platinum Catalysed Membranes for Proton Exchange Membrane Fuel Cells—Higher Performance. *Chem. Eng. J.* **2004**, *102*, 241–247. [CrossRef]

147. Yang, F.; Zhao, Y.; Du, Y.; Chen, Y.; Cheng, G.; Chen, S.; Luo, W. A Monodisperse Rh2P-Based Electrocatalyst for Highly Efficient and PH-Universal Hydrogen Evolution Reaction. *Adv. Energy Mater.* **2018**, *8*, 1703489. [CrossRef]

148. Duan, H.; Li, D.; Tang, Y.; He, Y.; Ji, S.; Wang, R.; Lv, H.; Lopes, P.P.; Paulikas, A.P.; Li, H.; et al. High-Performance Rh2P Electrocatalyst for Efficient Water Splitting. *J. Am. Chem. Soc.* **2017**, *139*, 5494–5502. [CrossRef] [PubMed]

149. Wang, K.; Huang, B.; Lin, F.; Lv, F.; Luo, M.; Zhou, P.; Liu, Q.; Zhang, W.; Yang, C.; Tang, Y.; et al. Wrinkled Rh2P Nanosheets as Superior PH-Universal Electrocatalysts for Hydrogen Evolution Catalysis. *Adv. Energy Mater.* **2018**, *8*, 1801891. [CrossRef]

150. Pu, Z.; Amiinu, I.S.; He, D.; Wang, M.; Li, G.; Mu, S. Activating Rhodium Phosphide-Based Catalysts for the PH-Universal Hydrogen Evolution Reaction. *Nanoscale* **2018**, *10*, 12407–12412. [CrossRef]

151. Cherevko, S.; Geiger, S.; Kasian, O.; Kulyk, N.; Grote, J.P.; Savan, A.; Shrestha, B.R.; Merzlikin, S.; Breitbach, B.; Ludwig, A.; et al. Oxygen and Hydrogen Evolution Reactions on Ru, RuO2, Ir, and IrO2 Thin Film Electrodes in Acidic and Alkaline Electrolytes: A Comparative Study on Activity and Stability. *Catal. Today* **2016**, *262*, 170–180. [CrossRef]

152. Wang, H.; Chen, Z.-N.; Wu, D.; Cao, M.; Sun, F.; Zhang, H.; You, H.; Zhuang, W.; Cao, R. Significantly Enhanced Overall Water Splitting Performance by Partial Oxidation of Ir through Au Modification in Core–Shell Alloy Structure. *J. Am. Chem. Soc.* **2021**, *143*, 4639–4645. [CrossRef] [PubMed]

153. Li, M.; Zhao, Z.; Xia, Z.; Luo, M.; Zhang, Q.; Qin, Y.; Tao, L.; Yin, K.; Chao, Y.; Gu, L.; et al. Exclusive Strain Effect Boosts Overall Water Splitting in PdCu/Ir Core/Shell Nanocrystals. *Angew. Chem. Int. Ed.* **2021**, *60*, 8243–8250. [CrossRef] [PubMed]

154. Wu, D.; Kusada, K.; Yamamoto, T.; Toriyama, T.; Matsumura, S.; Gueye, I.; Seo, O.; Kim, J.; Hiroi, S.; Sakata, O.; et al. On the Electronic Structure and Hydrogen Evolution Reaction Activity of Platinum Group Metal-Based High-Entropy-Alloy Nanoparticles. *Chem. Sci.* **2020**, *11*, 12731–12736. [CrossRef] [PubMed]

155. Li, K.; Li, Y.; Wang, Y.; Ge, J.; Liu, C.; Xing, W. Enhanced Electrocatalytic Performance for Hydrogen Evolution Reaction through Surface Enrichment of Platinum Nanocluster Alloying with Ruthenium In-Situ Embedded in Carbon. *Energy Environ. Sci.* **2018**, *11*, 1232–1239. [CrossRef]

156. Galyamin, D.; Torrero, J.; Elliott, J.D.; Rodríguez-García, I.; Sánchez, D.G.; Salam, M.A.; Gago, A.S.; Mokhtar, M.; de la Fuente, J.L.G.; Bueno, S.V.; et al. Insights into the High Activity of Ruthenium Phosphide for the Production of Hydrogen in Proton Exchange Membrane Water Electrolyzers. *Adv. Energy Sustain. Res.* **2023**, 2300059. [CrossRef]

157. Huang, Y.; Seo, K.-D.; Park, D.-S.; Park, H.; Shim, Y.-B. Hydrogen Evolution and Oxygen Reduction Reactions in Acidic Media Catalyzed by Pd4S Decorated N/S Doped Carbon Derived from Pd Coordination Polymer. *Small* **2021**, *17*, 104739. [CrossRef]

158. Cheng, X.; Lu, Y.; Zheng, L.; Cui, Y.; Niibe, M.; Tokushima, T.; Li, H.; Zhang, Y.; Chen, G.; Sun, S.; et al. Charge Redistribution within Platinum–Nitrogen Coordination Structure to Boost Hydrogen Evolution. *Nano Energy* **2020**, *73*, 104739. [CrossRef]

159. Wei, Z.W.; Wang, H.J.; Zhang, C.; Xu, K.; Lu, X.L.; Lu, T.B. Reversed Charge Transfer and Enhanced Hydrogen Spillover in Platinum Nanoclusters Anchored on Titanium Oxide with Rich Oxygen Vacancies Boost Hydrogen Evolution Reaction. *Angew. Chem.—Int. Ed.* **2021**, *60*, 16622–16627. [CrossRef]
160. Koca, A.; Ozkaya, A.R.; Akyuz, D. Atomically Dispersed Platinum Supported on Curved Carbon Supports for Efficient Electro-catalytic Hydrogen Evolution. *ECS Meet. Abstr.* **2016**, *MA2016-01*, 1728. [CrossRef]

Review

Materials Enabling Methane and Toluene Gas Treatment

Tong Lv and Rui Wang *

School of Environmental Science and Engineering, Shandong University, Qingdao 266237, China
* Correspondence: wangrui@sdu.edu.cn

Abstract: This paper summarizes the latest research results on materials for the treatment of methane, an important greenhouse gas, and toluene, a volatile organic compound gas, as well as the utilization of these resources over the past two years. These materials include adsorption materials, catalytic oxidation materials, hydrogen-reforming catalytic materials and non-oxidative coupling catalytic materials for methane, and adsorption materials, catalytic oxidation materials, chemical cycle reforming catalytic materials, and degradation catalytic materials for toluene. This paper provides a comprehensive review of these research results from a general point of view and provides an outlook on the treatment of these two gases and materials for resource utilization.

Keywords: adsorption; catalytic oxidation; hydrogen reforming; non-oxidative coupling; toluene; methane

Citation: Lv, T.; Wang, R. Materials Enabling Methane and Toluene Gas Treatment. *Materials* **2024**, *17*, 301. https://doi.org/10.3390/ma17020301

Academic Editor: Won San Choi

Received: 30 November 2023
Revised: 30 December 2023
Accepted: 3 January 2024
Published: 7 January 2024

1. Introduction

Greenhouse gases (GHGs) and volatile organic compounds (VOCs) are by far the two most damaging groups of substances to ecosystems, and methane and toluene are the typical representatives of these two groups, respectively. The treatment of methane and toluene represents, to some extent, the treatment of GHGs and VOCs. Methane is the hydrocarbon with the lowest carbon content and is the world's second-largest greenhouse gas after carbon dioxide. Although the atmospheric content of methane is much smaller than that of carbon dioxide, methane absorbs atmospheric thermal infrared radiation much more efficiently than carbon dioxide, and its global warming potential (GWP) per unit mass is 86 times higher than that of carbon dioxide over 20 years [1]. Therefore, the emission of methane into the air causes a serious greenhouse effect, which in turn threatens the human environment. Toluene is a colorless liquid with a specific aromatic odor and strong volatility. It is classified as a Class 3 automotive carcinogen according to the World Health Organization (WHO) list of carcinogens. Methane poses a serious threat to the environment as it pollutes the air, atmosphere, and water. Excessive emission of methane and toluene into the environment will therefore adversely affect human health and the ecological environment, and their emission must be restricted and properly handled. At the same time, these two gases have a certain value of resource utilization, which can be achieved if these two substances are treated by reasonable means. In the case of methane, in addition to adsorption and storage for reuse, we can use catalytic oxidation to convert it into methanol and other important raw materials for production and utilization, as well as decomposition and reforming to produce hydrogen. In addition, the aromatic coupling of methane in the absence of oxygen to produce aromatic compounds for industrial production is also an effective way to treat and create resources from methane. As for toluene, catalytic oxidation, catalytic decomposition, and chemical cycle reforming are common treatment methods. For the removal and resource utilization of these two substances, researchers have carried out many studies, in which the treatment of these two substances' auxiliary materials is crucial. This research area is currently one of intense interest. This paper categorically outlines and summarizes the latest research results on treatment aids for methane and toluene over the past two years and looks at future directions for these materials.

2. Materials for Methane Handling

2.1. Methane Adsorption Materials

Adsorption has been widely used for methane collection and treatment as a simple and economical methane gas treatment method, and methane adsorption materials constitute a popular research topic. Among them, metal-organic frameworks (MOFs) can be used as a good methane adsorption material, which can adsorb and store methane efficiently at moderate temperatures and pressures. Wu and Zhou found that MOFs with moderate interactions with methane and high densities have the greatest working capacity in natural gas storage, and that the flexible characteristics of some frameworks give these MOFs a high working capacity [2]. A multicomponent MOF[CuCeL(Cl$_4$-bdc)$_{0.5}$(H$_2$O)$_2$ (H$_2$O)$_6$]$_n$ (L = 1H-pyrazole-3,4,5-tricarboxylic acid, Cl$_4$-bdc = 2,3,5,6-tetrachloroterephthalate) with a column-layer structure was studied by Zhu et al. [3]. Figures 1 and 2 show schematic diagrams of the coordination patterns and connections of the catalytic material and the three-dimensional structure of the stereo.

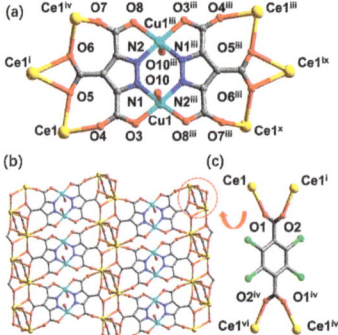

Figure 1. (**a**) Coordination pattern and connectivity of the Cu ion and L ligand. (**b**) 3d-4f Cu-Ce layer bridged by L ligand. (**c**) Coordination mode of the bridged Cl$_4$-bdc ligand. Symmetry codes: [i] $2 - X$, $1 - Y$, $1 - Z$; [ii] $1 + X$, $+Y$, $+Z$; [iii] $1 - X$, $1 - Y$, $-Z$; [iv] $-1 + X$, $+Y$, $+Z$; [v] $2 - X$, $2 - Y$, $1 - Z$; [vi] $+X$, $1 + Y$, $+Z$; [vii] $3 - X$, $1 - Y$, $1 - Z$; [viii] $2 + X$, $+Y$, $+Z$; [ix] $-1 + X$, $+Y$, $-1 + Z$; [x] $2 - X$, $1 - Y$, $-Z$ [3].

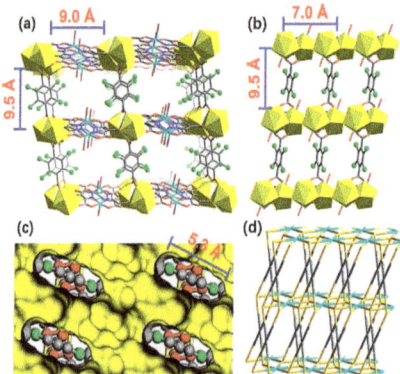

Figure 2. (**a,b**) Three-dimensional structural view of MOF[CuCeL(Cl$_4$-bdc)$_{0.5}$(H$_2$O)$_2$(H$_2$O)$_6$]$_n$ along the a- and c-axes (all free solvent molecules have been removed to clearly show the pore structure). (**c**) Internal structure of the pore along the b-axis with a yellow Connolly surface. (**d**) 1 Morphology along the c-axis. The atoms and structures represented by each color in the figure are the same as in Figure 1 [3].

The material has a high adsorption capacity of 28.41 cm^3/cm^3 for CH_4 and only 3.43 cm^3/cm^3 for N_2 (298 K, 1 bar), which provides a good methane–nitrogen separation.

Activated carbon is an important methane adsorption material, and its pore structure has an important influence on the methane adsorption effect. When biomass or biochar is activated by KOH, the necessary hierarchical porous network structure is formed, and the pore structure produced by different biomasses is different. Adlak et al. found that activated carbons made from coconut shells and pistachio shells had the proper size pore structure required for methane adsorption, but activated carbons prepared from other softer biomasses tended to have larger pore structures [4].

In order to examine the impact of activators on the pore opening of activated carbon (AC), Zaini et al. synthesized AC from palm kernel shell (PKS) using three different activators: steam, carbon dioxide, and carbon dioxide–steam. It was discovered that the PKS-made AC had a significant amount of methane adsorption potential. With a maximum adsorption capacity (MAC) of 4.5 mol/kg for methane gas (MG) and a specific surface area (SSA) of 869.82 m^2/g, the ACs produced from CO_2-steam exhibited varying rates of molecular diffusion and reabsorption. Additionally, their total pore volume (TPV) was measured at 0.47 cm^3/g. The Freundlich isotherm model fits all kinds of ACs extremely well, suggesting the emergence of multilayer adsorption (MLA). The formation of multilayer adsorption is indicated by this, and the adsorption kinetic data of the generated ACs conform to the pseudo-first-order model, in which the adsorption rate is primarily dictated by the physical adsorption between the pore surfaces and the methane gas, and the simultaneous influence of pore diffusion and outer diffusion on the adsorption of methane [5]. By thermochemically activating wood waste in H_3PO_4 at 1173 K, Pribylov et al. created the ES-103 microporous carbon adsorbent. At 303 K and 20 MPa, the adsorbent's methane adsorption capability was 22 weight percent. At 303 K and 20 MPa, the adsorbent's methane adsorption capability was 22 weight percent. Following the compaction of the ES-103 adsorbent with a binder, the bulk density of methane at a pressure of 10 MPa was 200 m^3NTP/m^3 [6].

Shale is a rock formed by dehydration and cementation of clay. It is dominated by clay minerals and has a unique thin-layered structure. Organic matter, also known as kerogen, is present in shale and is a microscopic structure with physical properties that are very different from the other components of shale. Despite their micron- and nanometer-scale composition, shales can adsorb and store large amounts of gases, thanks to the very large surface area of the cuticle. The use of shale for the adsorption and storage of methane and carbon dioxide is a promising technology for both reducing greenhouse gas emissions and improving the resource utilization of methane [7]. Adsorption isotherms of shale materials are important in understanding the mechanism of gas storage in shale. With the aid of isothermal adsorption experiments, Aji et al. examined the effects of total carbon (TOC), pore size distribution, and mineralogical properties on the adsorption capacity of shale. They also measured methane adsorption isotherms using gravimetric adsorption at a temperature of 120 °C and a maximum pressure of 10 MPa on four shale core samples from the Eagle Ford reservoir. The article's plots, like Figure 3, illustrate the relationship between the samples' measured TOC, adsorption capacity, and adsorption isotherms. Supercritical methane's calculated absolute adsorption capacity is higher than its excess adsorption capacity. At pressures greater than 9.6% of the critical methane pressure, the discrepancy between the absolute and excess adsorption capacities is more pronounced [8].

In order to investigate the methane absorption, desorption, and diffusion capacities in various coal samples, Zuo et al. employed both experimental and molecular simulation techniques. They then suggested an enhanced fracturing fluid formulation consisting of 0.8% CATB + 0.2% NaSal + 1% KCl + SiO_2 [9].

Compared with the coal sample treated with clean fracturing fluid, the methane adsorption capacity and desorption capacity of the coal sample treated with nano-fracturing fluid are improved to a certain extent. In addition, nano-particle-modified clean fracturing fluid can also reduce the damage caused by clean fracturing fluid to the desorption and diffusion capacity of coal seam. In addition, some studies on the effectiveness of some other materials for methane adsorption have been performed. David Ursueguia and colleagues produced composites consisting of HKUST-1 and Al_2O_3 particles, and evaluated their ability to adsorb methane. The samples were placed on a fixed bed and aged in the air at 100% relative humidity for 24 h in three successive cycles in order to test the materials' ability to adsorb methane both before and after wet treatment. It was demonstrated that the adsorption capacity of the composite with lower MOF loadings (<9%) rose by more than 38%, while the methane adsorption capacity of HKUST-1 decreased by around 37% [10].

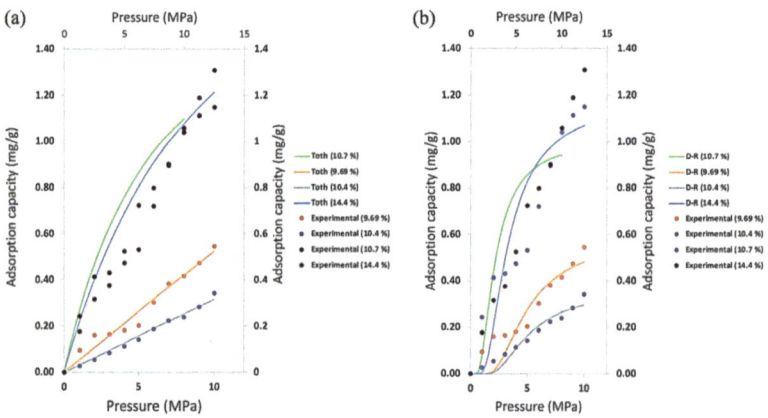

Figure 3. (**a**) Toth isotherm; (**b**) D-R isotherm [8].

2.2. Catalytic Oxidizing Materials

Methane poses environmental hazards and depletes resources; however, these problems can be effectively mitigated through the development of cost-effective catalysts and reactors for methane treatment and resource utilization. Zhao et al. created a unique non-precious metal catalyst for the catalytic oxidation of methane using oxalic acid etching of $La_{0.8}Sr_{0.2}MnO_3$ [11]. Figure 4 depicts a schematic design of the process of this catalyst's catalytic oxidation of methane.

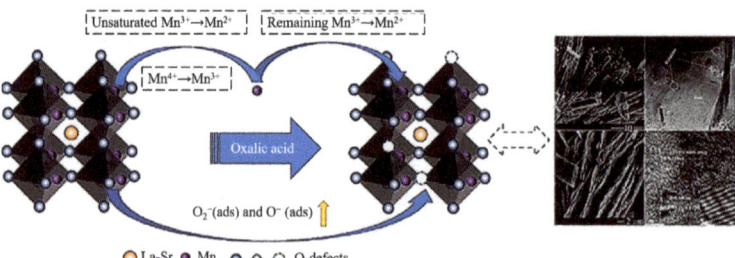

Figure 4. Catalyst structure and reaction diagram [11].

The catalyst etched with oxalic acid catalyzed the methane oxidation more efficiently than the unetched sample. Khatun et al. prepared a defect-rich Ni-Pt/CeO_2 catalyst by adding nickel and platinum to the ceramic lattice using a one-pot composite combustion process. The catalyst has excellent catalytic activity and stability, and can catalyze the reaction stably for up to 700 h. In this process, the conversion of CH_4 and CO_2 was close to 86% [12].

Through catalytic oxidation, methane is converted to methanol, which not only lessens methane's impact on the atmosphere but also yields methanol, a crucial raw material for both residential and industrial manufacturing. In their study on methane methanolization, Alvarez et al. discovered that the productivity of methanol is influenced by the adsorption period, and that it takes at least 60 min for the adsorption to reach equilibrium. Methane chemisorption's equilibrium and kinetics were modeled based on the synthesis of two adsorption precursors: carbon dioxide and methane. Additionally, they looked at the aerobic methane adsorption process to apply methane conversion to the methanol cycle in these situations [13]. Vitillo et al. compared the reaction profiles of four single iron-based catalysts for the direct oxidation of methane to methanol using two biomimetic models based on two enzymes (cytochrome P450 and taurine dioxygenase [Taub]) and two artificial reticulation frameworks (iron–BEA zeolites and a triple-ferric oxide-centered metal-organic frameworks) using the Kohn–Sham density functional method. When methane conversion was more than 1%, the biomimetic and inorganic catalysts' selectivity for methanol was nearly nonexistent at room temperature. The authors stressed that in the event of a lack of a methanol protection method, attaining high selectivity necessitates simulating the enzyme's reaction milieu beyond the initial iron coordination layer [14]. A metal-organic framework Fe/UiO-66 supports an iron catalyst, which allows methane to be selectively oxidized to methanol, as reported by Rungtaweevoranit et al. At 180 °C, methanol was consistently generated with strong selectivity for methanol at an excellent reaction rate of 5.9×10^{-2} $\mu mol_{MeOH} g_{Fe}^{-1} s^{-1}$ [15]. Zhu et al. doped bismuth oxychloride (BiOCl) with non-precious nickel sites and gave it a high oxygen vacancy content, which allowed them to directly oxidize CH_4 to CH_3OH in a single step. As methane oxidized to create methyl and adsorbed hydroxyl groups, the oxygen vacancies of the unsaturated Bi atoms adsorbed and activated CH_4, keeping the catalyst active. As much as 39.07 $\mu mol/(gcat\text{-}h)$ of O_2 and H_2O-based CH_3OH was converted at 420 °C and under flow conditions [16]. Pereira et al. developed a new process using copper-exchanged zeolite omega (Cu-MAZ) by discovering that carbon dioxide can be used as a substitute for oxygen and that CH_3OH yields are higher when carbon dioxide is involved in the reaction. The calculated energy diagram for carbon dioxide activation at the site is shown in Figure 5 [17].

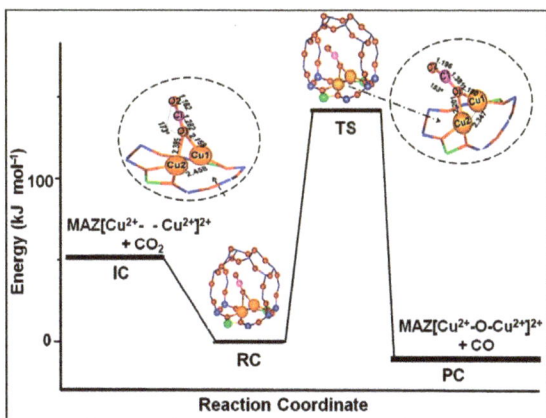

Figure 5. Calculated energy diagrams of carbon dioxide activation at sites. IC: initial complex, RC: reactant complex, TS: transition state, PC: product complex, where only the active site and zeolite pores are shown and other atoms of the zeolite are omitted. Blue (Si), red (O), green (Al), orange (Cu), and magenta (C) [17].

To further maximize the selectivity of methane oxidation to methanol, Zhou et al. generated 12 single-atom alloys (SAAs) for methane activation and screened those with good C-H bond dissociation catalytic activity. The methane dissociation activity is enhanced when Ir metal atoms are doped in inert covalent matrices (Ag, Au, and Cu). Ir1/Ag SAA is the most effective single-atom amalgam catalytic element for selectively oxidizing methane to methanol. This work offers crucial guidance for the future development of extremely active and effective methane methanolization catalysts [18]. Using several bismuth-based catalysts (BiPO$_4$, α;-Bi$_2$O$_3$, β;-Bi$_2$O$_3$) and molecular oxygen as the only oxidant, Matsuda et al. studied the direct oxidation of methane (CH$_4$) to formaldehyde (HCHO). When it came to the direct oxidation of CH$_4$, the monoclinic BiPO$_4$ nanoparticles (BiPO$_4$-DEG) that were created in diethylene glycol (DEG) and water-mixed solvent exhibited the maximum catalytic activity. BiPO$_4$-DEG outperforms FePO4 nanoparticles in the high-temperature range, where the reactive oxygen species on the surface of BiPO$_4$ combines with CH$_4$ to form HCHO, while FePO4 nanoparticles form HCHO for the phosphate units therein that react with CH$_4$ [19]. A proton-type zeolite catalyst containing Bronsted acid sites that are appropriate for CH$_4$ combustion was predicted and shown to work very well by Yasumura and associates. They investigated main-group elemental catalysts using silicon and aluminum for low-temperature CH$_4$ combustion with ozone, based on automated reaction route mapping. At 250 °C, it was discovered that catalysts containing strong Brønsted acid sites improved methane conversion. These catalysts show promise for methane combustion at low temperatures. At 190 °C, the benchmark catalyst (5 weight percent Pd-loaded Al$_2$O$_3$) catalyzed a reaction rate 442-fold lower than the main group catalyst (protonated β-zeolite). In light of autonomous reaction route mapping, this illustrates the sensible design of an earth-rich catalyst [20]. Using iron or copper oxides as essential intermediates, related researchers have created a variety of biomimetic molecular catalysts that are modeled after methane monooxygenases (MMOs). They are still far less effective than MMOs for catalyzing the oxidation of methane, though. In order to generate methane oxidation catalytic materials with better catalytic activity, Yamada et al. firmly stacked µN-bridged iron phthalocyanine dimers onto graphite surfaces. This graphite-supported µ-nitrogen-bridged iron phthalocyanine dimer could oxidize methane even at room temperature, and its catalytic activity was comparable to that of MMO, nearly 50 times higher than that of other powerful molecular methane oxidation catalysts in an aqueous solution containing H$_2$O$_2$ [21]. They also created a new mu-nitrogen-bridged heterodimer of iron porphyrin and iron phthalocyanine, and studied its catalytic CH$_4$ oxidation characteristics. When H$_2$O$_2$ was present in acidic aqueous solutions at 60 °C, the heterodimers demonstrated catalytic activity for CH$_4$ oxidation via high-valency iron–oxygen species. The µ-nitrogen-bridged iron porphyrin dimer's high-valency iron–oxygen species is extremely unstable under the same reaction circumstances [22]. Several CoxCeMgAlO mixed oxides with varied cobalt contents of 10% Ce and Mg/Al atomic ratios of 3 were obtained by Stoian et al. calcining layered double hydroxide (LDH) precursors at 750 °C. The complete oxidation of methane was used to assess their catalytic qualities. Because of their optimal redox characteristics and maximum Co/Ce surface atom ratio, the Co$_{40}$CeMgAlO mixed oxides exhibited the highest catalytic activity for methane combustion out of all of them. They also demonstrated the good stability of the catalytic activity by examining the impact of contact time [23]. Larger reaction temperatures are usually needed for the oxidative coupling of methane (OCM) in order to achieve larger C$_2$ yields; nevertheless, methane may be severely oxidized, leading to lower C$_2$ yields. Deep oxidation of methane causes CH$_3$ surface dimerization, which must be eliminated with a lower catalytic reaction temperature. In order to better understand OCM catalytic performance, Maulidanti et al. constructed M/Ce(Y) catalysts. The greatest catalytic performance was reported at 5% Fe$_3$O$_4$/CeO$_2$(A), with a 2:1 CH$_4$/O$_2$ ratio, 0.5 g of catalyst, and 27% CH$_4$ conversion and 1.4% C$_2$ production at 500 °C. Reducing the CH$_4$/O$_2$ ratio and raising the catalyst weight will increase the C$_2$ yield. Fe$_3$O$_4$ and CeO$_2$ together provide a large specific surface area and mesoporous pore size, which is highly useful for OCM at low temperatures [24].

2.3. Catalytic Materials for Dry Methane Conversion

In the hydrogen from the methane-reforming (DRM) process, methane and carbon dioxide can be simultaneously converted into syngas with H$_2$ and CO ratios close to 1, which provides a promising green pathway for large-scale utilization of methane in the chemical production of hydrogen. However, the industrialization of DRM has been seriously hindered by the sintering of nickel and carbon deposition in the process. Several scientists have made great efforts to solve this problem. By preparing a variety of uniformly distributed Fe-coated Ni/Al_2O_3 catalysts using atomic layer deposition (ALD), Zhao et al. found that adjusting the molar concentration of trace amounts of Fe (0.3–0.6%) could improve the low-temperature catalytic activity, accelerate the oxidation process of coke, and promote the dissolution of CH_4 on NiO. The activity of the 0.3% $Fe/Ni/Al_2O$ sample was essentially unchanged for 72 h at 650 °C. The activity of the 0.3% $Fe/Ni/Al_2O$ sample was found to be very low [25].

The modification of Fe achieved the dual effect of enhancing Ni/Al_2O_3-catalyzed CH_4 cracking and eliminating coke. Yadav et al. prepared monometallic (nickel, cobalt) substitutes and CeO_2 catalyst carriers using solution combustion synthesis and formaldehyde reduction methods. The DRM activity of the catalysts was significantly different. The co-substituted CeO_2 catalysts showed maximum stability in the DRM reaction at 800 °C. The catalysts showed significant stability. The catalysts showed significant stability. In addition, they found that the presence of graphitic and amorphous carbon was responsible for the deactivation, while the reactivity of surface lattice oxygen played an important role in the catalytic cracking stability, which in turn determined the steps involved in the process. The energy required to generate vacancies in cobalt-substituted CeO_2 is much lower compared to nickel-substituted CeO_2. It can therefore be confirmed that while nickel inhibits oxidation by decreasing the availability of surface oxygen required for the reaction, Co-substituted catalysts promote oxidation by increasing the availability of surface oxygen [26]. Density functional theory was utilized by Bandurist and Pichugina to simulate the methane dry-reforming process, which involves the breakdown of C-H bonds on Cu-rich Ni-Cu clusters. The nanoscale clusters $NiCu_{11}S_6(PH_3)_8$, $NiCu_{11}S_6$, $NiCu_{11}O_6(PH_3)_8$, and $NiCu_{11}O_6$ were used to simulate the catalysts. Based on the collected data, the most promising catalytic system for CH_4 activation was determined to be $NiCu_{11}O_6$, with an activation energy of 99 kJ/mol. This system also showed the best thermodynamic performance [27].

Zou et al. prepared a novel nickel–cobalt nanocomposite catalyst, and the synthesis process is schematically shown in Figure 6. Compared with the monometallic catalysts ($10Ni0Co/SiO_2$, $0Ni10Co/SiO_2$), the bimetallic $5Ni5Co/SiO_2$ nanocomposite catalysts showed the best catalytic activity, where the $10Ni0Co/SiO_2$ catalysts showed a 10% decrease in catalytic activity due to coking in the 50 h catalytic experiment. The $5Ni5Co/SiO_2$ bimetallic catalyst maintained its catalytic activity within 100 h. The catalytic activity of the $5Ni5Co/SiO_2$ bimetallic catalysts was also improved [28].

A series of Ni/MSS catalysts were prepared using four different methods (normal impregnation, glycine-assisted impregnation, glycol-assisted impregnation, and ammonia evaporation) such as that of Zhang. Among these methods, except for the catalysts prepared by normal impregnation, the other three methods, especially glycine-assisted impregnation, and ethylene glycol-assisted impregnation, could obtain smaller nickel particles, improve carrier–metal interactions, and effectively increase the coking and sintering resistance of the catalysts. In addition, the catalyst prepared by the ammonia evaporation method has the best stability [29]. On the other hand, Zhang et al. constructed NiMgAlOx/BN catalysts with closed interfaces (NiMAO/BN) using layered metal oxides (NiMgAlOx) and boron nitride (BN). They found that the triple interface between Ni, BN, and MgAlOx oxides in the catalyst enhanced the sintering resistance of the catalyst [30]. In addition, Zheng et al. investigated a material combining nickel, the oxygen carrier $Ca_2Fe_2O_5$, and CaO for a cyclic DMR process. The activated $NiO/Ca_2Fe_2O_5/CaO$ material produced large amounts of syngas and H_2 during cycling, and the material remained active for more than 30 cycles [31]. In the course of studying how supported precursors affected the catalytic activity of nickel catalysts (20% $Ni/5\%$ La_2O_3-95% Al_2O_3) used in the methane reforming process, Zakrzewski et al. discovered that catalysts made with precursors containing chloride had extremely low methane and carbon dioxide conversion rates [32]. For methane cracking on Ni(111) under steam-reforming circumstances, Yadavalli et al. created an ab initio kinetic

Monte Carlo (KMC) model. In a reasonable amount of computing time, the model provides insight into the coking state of graphene/coke by capturing the C-H activation kinetics and characterizing the production mechanism of graphene flakes at the thermodynamic level. They employed ever more fidelity clusters to compare the predictions of a KMC model that integrates these clusters into a mean field micromotion model in order to methodically assess the impact of effective cluster interactions between C and CH species on the coking state. The findings demonstrate that there is a considerable correlation between the coking condition and cluster faithfulness. Furthermore, at low temperatures, C-CH islands/rings are virtually disconnected, whereas at high temperatures, they wrap completely around the Ni(111) surface, according to high-fidelity simulations [33].

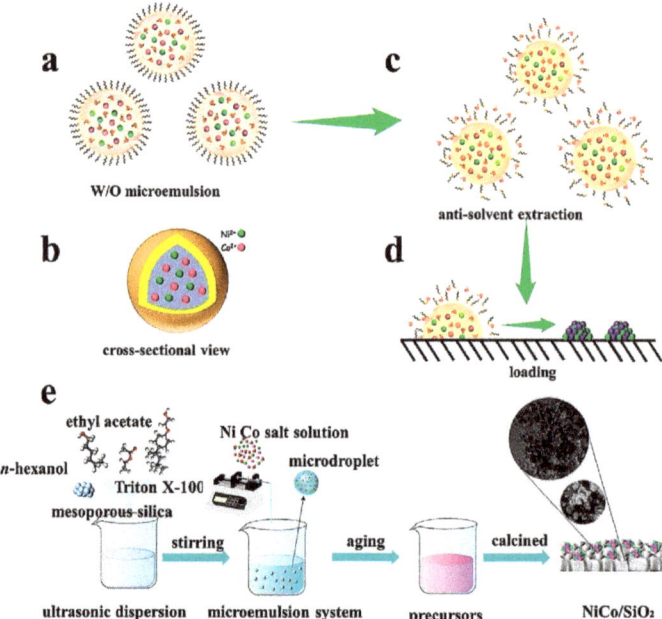

Figure 6. Schematic diagram of the catalyst synthesized by the microdroplet method [28]. (**a**) Water-in-Oil Microemulsion Droplet; (**b**) Cross-sectional view of water-in-oil microemulsion droplets; (**c**) Gradual diffusion of water from the droplets into the ethyl acetate phase; (**d**) Co-precipitation of precursor salts by microemulsion systems disrupted by water loss; (**e**) Catalyst preparation flow chart.

In addition to research on catalyst resistance to carbon deposition, there have been advances in the study of new materials for methane reforming catalysis. Hanifa et al. investigated the effects of solvent and pretreatment processes on the Ni catalytic performance of palm oil fuel ash (POFA) catalysts loaded with 10% Ni. The POFA carriers prepared by the ultrasonic pretreatment process had smaller particle sizes compared to the conventional stirring process, and had 71% initial CH_4 conversion and 4.2% initial H_2 yield in methanolysis [34]. V_2CTx/V_2AlC catalysts modified with nickel were fabricated using impregnation/precipitation and high-frequency etching techniques by Wysocka et al. Their catalytic function was studied at 800 °C and 1 bar in the DRM process. It was shown that the V_2CTx/V_2AlC-based material undergoes a phase transition to produce aluminum oxide, vanadium oxide (V_2O_3), and vanadium carbide (V_8C_7). Nickel is necessary for the catalytic activity of the V_2CTx/V_2AlC catalysts. Catalytic activity fluctuations were related to changes in the nickel phase and sodium ion intercalation processes during the preparation process. It was demonstrated that the presence of alkali ions and their distribution of the copper layer affected the catalysts' stability and activity. The Ni-V_2CTx/V_2AlC_IMP

catalyst using just the technique of impregnation and no precipitant was found to have the maximum activity and stability in the DRM process. The nickel phase of the catalyst was uniformly distributed and produced a needle-like shape. During a 20 h test, the CO_2 and CH_4 conversions were 90–93% and 90–80%, respectively. After 20 h of DRM, the molar ratio of H_2 to CO in the outflow stream varied from 1.05 during the first hour of the process to 0.95. The unmodified V_2CTx/V_2AlC, unetched V_2AlC, and Ni-SiO_2 catalysts were less active and less stable than the Ni-V_2CTx/V_2AlC_IMP catalysts [35]. Pourghadiri and Sari created an inverted flow microchannel monolithic reactor to improve the enzyme-mediated partial oxygenation of hydrocarbons over Rh/Al_2O_3 catalysts for the production of syngas. They used a one-dimensional heterogeneous non-stationary model to simulate the reactor behavior. The full GRI 3.0 mechanism model was used for the gas-phase reactions, and the Langmuir–Hinshelwood surface mechanism was referenced for the catalytic reactions. Investigations were conducted into the impacts of feed preheating temperature, feed CH_4/O_2 ratio, reaction pressure, and flow switching time on yield, quality, and methane conversion. The results show that reverse flow operation significantly increases syngas yield and reduces the minimum preheating temperature required for reactor ignition. After the establishment of the cyclic steady state, syngas yields of up to 75% were achieved with feed CH_4/O_2 ratios close to 1.6, and H_2/CO ratios close to 2.7. Compared with unidirectional operation, reverse-flow operation increased methane conversion and syngas yields by at least 8% and 76%, respectively [36].

Apart from the catalyzed dry reforming of methane (DRM) reaction, another process that is more likely to achieve thermodynamic equilibrium and yield higher feedstock gas conversions without carbon displacement is heating plasma-activated replication (CRM) of CH_4-CO_2 to the syngas under non-catalytic conditions. Zhou et al. investigated the conversion process and optimum situation of a hot plasma-activated CRM reaction system. As the CO_2/CH_4 molar ratio rose, the CO_2 conversion and H_2 selectivity decreased, while the CO_4 transformation and CO discrimination rose. When the molar ratio of CO_2/CH_4 was 6/4, the selectivity of both carbon dioxide and water rose to 87.0% and 80.8%, respectively. This discovery provides a new perspective on the CRM reaction and advances our knowledge of how it converts when heated plasma is utilized to initiate it with no requirement for catalysis. [37].

2.4. Catalytic Materials for Hydrogen from Methane Decomposition

Nowadays, steam methane reforming is the primary method used to manufacture H_2. However, this process is not ecologically friendly, releasing 10 kg of carbon dioxide for every kilogram of H_2 produced. On the other hand, hydrogen from methane breakdown (TCD), where catalytic materials are essential, is an extremely green process that effectively creates H_2 and valuable solid carbon with almost minimal CO_2. Catalytic materials have been extensively studied for TCD.

Catalytic methane breakdown to produce high-purity H_2 and high-value-added carbon nanotubes (CNTs) is a promising technology for methane resource extraction, but significant challenges are encountered in controlling CNT manufacturing. To that end, Song et al. used a structural reconfiguration strategy based on a microporous confinement process to prepare LTA zeolite-derived pebble-shell catalysts, and for the first time, molecular dynamics simulations were used to realize the tip growth of multi-walled carbon nanotubes (MWCNTs) on nickel particles, and it was confirmed that nickel particle size was closely related to the methane decomposition rate/dissolved carbon penetration/carbon nanotube growth. Furthermore, they discovered that the evolution of the nickel metal active sites paralleled the synthesis of MWCNT, demonstrating the penetration process of dissolved carbon on the nickel metal clusters, and this discovery paved the way for the construction of functionalized CDM catalysts [38]. In order to generate smaller particles from 5 to 30 nm to 5 to 10 nm, Chudakova et al. observed that adding potassium (0.25 wt% K_2O) to the nickel catalysts altered the particle size distribution. This shift led to increased H_2 yields. The Ni-0.25%K_2O/Al_2O_3 and Ni-1%K_2O/Al_2O_3 catalysts had the highest hydrogen yields of 11.6 g/gcat and 17.0 g/gcat, respectively, and the Ni-1%K$_2$O/Al$_2$O$_3$ catalyst produced the best carbon material yield (51.0 g/gcat) [39]. In order to study the process of producing hydrogen in situ from the conversion of shale gas (methane) using microwave heating of a methane stream passing through a filled shale sample, Yan et al. carried out a number of tests. With conversion rates of 40.5% and 100% at reaction temperatures of 500 °C and 600 °C, respectively, they discovered that the methane conversion was greatly enhanced in the presence of Fe and Fe$_3$O$_4$ particles acting as catalysts. At lower reaction temperatures, the conversion of methane was made easier by the minerals in the shale's catalytic action [40]. Zaghloul, Nada, and colleagues studied the catalytic pyrolysis of methane in a molten metal bubble tower to produce hydrogen gas and separable carbon. Tin was used as the base metal, and catalytically active metals like nickel and copper were added. They took into account both traditional non-catalytic reactions in the gas and parallel catalytic surface reactions on the bubble surface. The impact of temperature, pore size, and kind of molten metal on the bubble surface–volume ratio was examined and analyzed, along with the consequences of these variables on the gas-phase reactions and catalytic surface reactions within the bubbles. Finally, the best performance was obtained with the 5-weight percent Ni-Sn mixture. Xavier Jr et al. investigated the viability of CMD under the reaction conditions of graphene nanoribbon zigzag (12-ZGNR) and armchair (AGRN) edges using dispersion-corrected density functional theory (DFT). They started by looking into the desorption of H and H$_2$ at 1200 K on passivated 12-ZGNR and 12-AGNR edges. On the passivated 12-ZGNR and 12-AGNR edges, the results demonstrated a very low level of H and H$_2$ desorption. The most beneficial H$_2$ desorption pathway is determined by the rate-determining step of hydrogen atom diffusion on the passivated edge, which has activation-free energies of 4.17 eV on 12-ZGNR and 3.45 eV on 12-AGNR. With a free-energy barrier of 1.56 eV, the 12-AGNR edge exhibits the most advantageous H$_2$ desorption. This indicates that the bare-carbon active site is accessible for catalytic applications. The chemisorption of CH$_4$ through direct dissociative means is not dissociative in nature. The optimal method for non-passivation of the 12-ZGNR edge is direct dissociative chemisorption, which has an activation-free energy of 0.56 eV. Furthermore, they delineated reaction pathways for the comprehensive catalytic dehydrogenation of methane on the 12-ZGNR and 12-AGNR boundaries, putting forth a mechanism via which the solid carbon that forms on the margins functions as a refreshed active site. There is a stronger tendency for regeneration at the active site on the 12-AGNR edge. Through the bare carbon edges of graphene nanoribbons, this study offers essential insights into the fabrication of carbon-based catalysts for CMD, with a methane decomposition performance comparable to that of widely used metal and bimetallic catalysts [41]. When it comes to pyrolyzing methane, molten manganese chloride (MnCl$_2$) is an appealing liquid catalyst at high temperatures. Researchers Bae et al. looked at the mechanism and kinetics of the reaction for the pyrolysis of CH$_4$ in mixtures of different mono- or divalent chlorides with molten MnCl$_2$. A significant difference was seen between the apparent activation energy of the uncatalyzed process and the pyrolysis of CH$_4$ in molten MnCl$_2$ mixtures. Only the MnCl$_2$-KCl mixture's apparent activation energies were less than those of pure manganese chloride among the manganese chloride mixes. Furthermore, MnCl$_2$-KCl yielded the highest crystallinity of the final solid carbon product and the greatest quantity of CH$_2$*, indicating a distinct pathway for CH$_4$ dehydrogenation and carbon formation in this system. This is supported by the fact that MnCl$_2$-KCl forms CH$_2$ at a more thermodynamically favorable rate than MnCl$_2$. The pathway for the formation of graphitic carbon layers in molten MnCl$_2$-KCl is more facile. Furthermore, an NVT ab initio molecular dynamics simulation shows that frequent reversible desorption and adsorption of CHx* intermediates in MnCl$_2$-KCl promotes gas-phase C-C coupling. These fundamental insights into the reasons for the enhanced high-temperature decomposition of CH$_4$ in MnCl$_2$-KCl could be useful for the development of more reactive molten salt catalysts [42]. Copper ferrite (CoFe$_2$O$_4$) was used as a catalyst to break down methane, according to Alharthi, Abdulrahman I. The reaction temperature was between 800 and 900 °C, and the catalytic reaction was conducted in a fixed-bed reactor operating at atmospheric pressure and a gas flow rate of 20 to 50 mL/min. As the reaction temperature rose, the amount of methane converted and hydrogen formed increased, while catalyst stability and induction time dropped. For methane cracking, $CoFe_2O_4$'s total catalytic activity dropped at gas flow rates higher than 20 mL/min. At 20 mL/min, 900 °C, and 50 mL/min, 800 °C, the greatest carbon deposited was 70.46%.

2.5. Non-Oxidatively Coupled Catalytic Materials

Although the non-oxidative coupling reaction (NOCM) of methane is highly intriguing, its applications are hindered by its low catalyst stability and harsh reaction conditions (T > 800 °C), which can quickly deactivate the catalyst and cause significant coking. Metal carbides are used in a variety of catalytic processes, from thermal coupling of methane to electrochemically driven reactions, due to their platinum-like properties. Zhang et al. looked at the role carbides play in the active creation of C_2 products during methane coupling at high temperatures. Whereas tungsten carbide (WC) loses selectivity because of the depletion of surface carbon by slow diffusion, molybdenum carbide (Mo_2C) exhibits persistent C_2 selectivity for an extended length of time in the gas stream due to its quick carbon diffusion kinetics. This result implies that the catalyst's carbons are important, with the metal carbons being in charge of the production of methyl radicals, and that the nonoxidative coupling of methane is facilitated by a carbon process similar to the Mars–Van Krevelen type [43]. As seen in Figure 7, a different Zhang et al. research group created siliceous [Fe] zeolites with MFI and CHA topologies, and suggested a polyaromatic hydrocarbon reaction network. In gas-phase products, this siliceous [Fe] zeolite exhibits remarkable selectivity for ethylene and ethane (MFI > 90%, CHA > 99%). Moreover, burning coke in the air can replenish deactivated [Fe]zeolites. Oxide semiconductors coated with metals are significant photocatalysts for the nonoxidative coupling of methane [44].

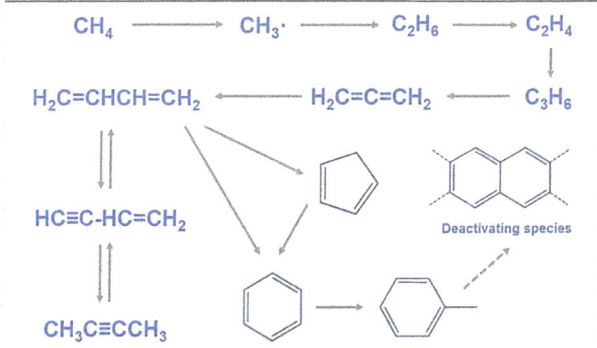

Figure 7. Potential pathways of reactions for methane conversion using [Fe] zeolite catalysts [44].

Zhang et al. looked into methane nonoxidative coupling reactions with a variety of noble metal-decorated oxides ($NaTaO_3$, $CaTiO_3$, $LiNbO_3$, and TiO_2) adorned with Ag, Au, Pt, Pd, Cu, and Ni [45]. It is generally accepted that the active areas for CH_4's H extraction and C-C coupling are spatially separated; that is, metal nanoparticles complete the final C-C coupling, while $NaTaO_3$ only completes the first H extraction during CH_4 activation. Because they greatly reduce the CH_4 dissociation energy barrier and enhance C-C interaction, precious metals predominate in the NOCM. With C_2H_6 yields as high as 194 mu mol g^{-1} h^{-1} in Ag/NaTaO$_3$-catalyzed NOCM, silver is the metal that prefers the weak adsorption of CH_3 intermediates at the centroid and subsequent metal-induced CC coupling among the other metals. This study provides a molecular understanding of the mechanism of CH_4 coupling on metal-decorated photocatalysts. In the work of a different Zhang et al. team, molybdenum-doped CeO_2 samples with isolated molybdenum sites were created using flame-jet pyrolysis. They underwent screening and evaluation for their effectiveness in the catalytic non-oxidative coupling of methane. It was discovered that the gas-phase products could have a selectivity of up to 98% for high-value-added C_2 hydrocarbons, such as ethylene and ethane. Molybdenum-oxygen species that were separated from the produced catalysts during the procedure were reduced, changed into molybdenum (oxy)carbide species, and used as methane activation sites. This work highlights the significance of lowering the reactor's free volume to restrict secondary gas-phase reactions, and offers insights into the development of effective catalysts for the non-oxidative coupling of methane [46]. In prior work, Ryu et al. demonstrated that Mo/HZSM-5's methane conversion and BTX selectivity could be greatly enhanced by the addition of nickel oxide through a straightforward physical mixing process. In a more recent study, the effective diameters of promoter NiO particles (4, 22, 36, 45, and 101 nm) were assessed, and it was discovered that 36 nm was the ideal size for raising Mo/HZSM-5's activity. Inactive $NiMoO_4$ was more likely to develop when NiO particle sizes lower than 22 nm caused significant agglomeration and limited dispersion of MoCx [47]. It has been possible for Andrey A. Stepanov and associates to improve Mo/ZSM-5 catalysts for the dehydroaromatization of methane. An investigation into the catalytic impact of Mo/ZSM-5 catalysts—which depends on high-purity silica zeolites of the ZSM-5 variety with microporous and micro mesoporous structures—was carried out. It was found that the addition of carbon black during the ZSM-5 zeolite synthesis stage did not cause structural changes, and the generated samples exhibited 100% crystallinity. Throughout the methane dehydrogenation process, the Mo/ZSM-5 catalysts' stability and activity are increased by their capacity to take on a microporous form. The 4.0 percent Mo/ZSM-5 catalyst developed utilizing zeolite synthesized from 1.0% carbon black displayed the greatest transformation of methane, achieving 13 percent after 20 min, and benzene synthesis, reaching 7.0% [48]. In order to create 4%Mo/ZSM-5 catalysts for the non-oxidative conversion of methane to aromatic hydrocarbons, Stepanov et al. examined the impact of secondary mesoporous structure formation in ZSM-5 zeolites on the catalytic characteristics of those catalysts. Carbon black was added to zeolites that were manufactured, zeolites treated with aqueous citric acid, and zeolites featuring a microporous structure without binders, in order to create 4%Mo/ZSM-5 catalysts. It was demonstrated that molybdenum modification of zeolites with microporous structure reduced the concentration and strength of the strong acid centers in the aromatization of methane, independent of the synthesis method; the most efficient treatment of the 4%Mo/ZSM-5 catalyst was with a 0.3 N citric acid solution [49].

3. Toluene Handling Materials

3.1. Toluene Adsorption Materials

Eliminating toluene is crucial for lowering air pollution, since it is a common example of an aromatic volatile organic compound (VOC). Adsorption of volatile organic compounds (VOCs) from gases, including toluene, is a proven method that is frequently employed in industrial production processes to eliminate VOCs while reusing them for future use. As shown in Figure 8, Zhang et al. synthesized a size-controlled, mallow-like copper-carbon tetrachloride-loaded biochar composite (Cu-BTC@biochar) using a simple one-pot method. The adsorption rate of Cu-BTC@biochar on toluene, a typical component of volatile organic compounds (VOCs), could reach 501.8 mg/g and 88.8 mg/g at medium and high temperatures (60 and 150 °C), respectively [50].

The pore distribution properties of porous adsorbent materials are among their most significant features, and many studies have looked at how to accurately and conveniently tailor usable porous carbons for the adsorption of volatile organic compounds (VOCs). Because of this, Wang et al. produced porous carbon for model experiments using precursors with various lignocellulose mass ratios. Additionally, they verified the pore shape and distribution properties of the porous carbon by using bacterial targeting of bagasse decomposition to validate the application of these mechanisms in real-world biomass materials. While the microvolume of the mesopores exhibited the opposite tendency, the microvolume of the ultramicro pores decreased as the cellulose concentration decreased. Both BACs-36 and BACs-48 showed excellent toluene adsorption capabilities at low concentrations of 635 mg/g and across ten cycles, respectively. Poor humidity stability in the presence of air humidity and the need for powder molding have greatly impeded the removal of volatile organic compounds (VOCs) from the metal-organic framework (MOF) MIL-101(Cr) powders [51]. Zhang et al. employed an in situ growth confinement technique to create the hydrophobic composite P-MIL-101(Cr)@PA-NH$_2$ in order to address this problem. On the macroporous surfaces of PA-NH$_2$ substrates, the MIL-101(Cr) nanocrystals' confinement growth method is schematically shown in Figure 9. The material's working capacity was 6.3 times larger than that of the original MIL-101(Cr) powder at 1000 mg/m^3 toluene and 50% relative humidity. Some 81.1% of the material's initial adsorption capacity remained after 30 days in a humid environment. The molecular sieve has problems with poor moisture resistance, high-temperature desorption, and a low-temperature desorption rate when it comes to purifying volatile organic compounds [52].

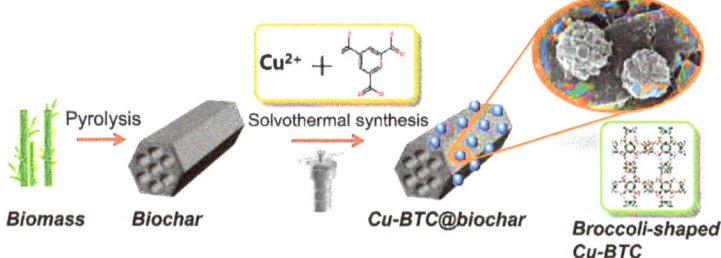

Figure 8. Diagram showing the steps involved in creating Cu-BTC@biochar [50].

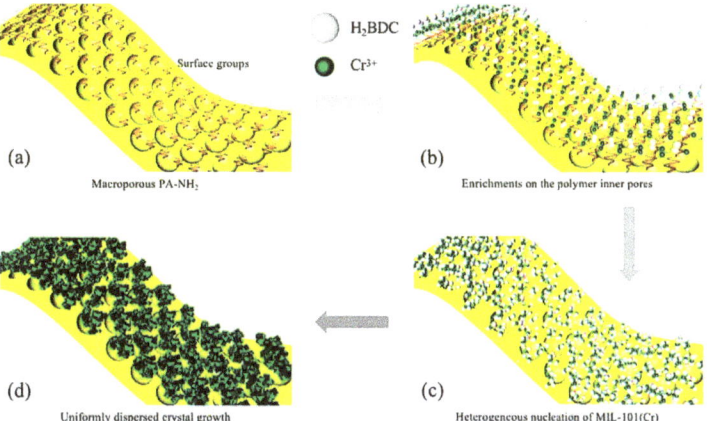

Figure 9. Diagram showing the limited development method of MIL-101(Cr) nanocrystals on the PA-NH$_2$ substrate's macroporous surface [52].

Zhou et al., on the other hand, developed a microporous bifunctional system for adsorption/catalysis by changing the mass proportion of Pt/microporous silica (Pt/MS) from 1.0 wt% to 3.5 wt% in order to achieve the best adsorption/catalytic performance for trace toluene. They additionally hydrophobically modified the created Pt/MS catalyst to make Pt/MS-H in order to boost the Pt/MS catalyst's resistance to water vapor. At a desorption temperature of 60 °C, 2.0 Pt/MS-H exhibited an optimal saturation adsorption capacity of 56.09 mg/g for toluene. Following five cycles of adsorption and desorption, the toluene adsorption capacity was kept at roughly 95%, with a good peak smoothing

Materials **2024**, 17, 301

effect. Trace toluene conversion capacity at T90 = 149 and DEG; C was optimal. For volatile organic molecules, it can be a powerful auxiliary material when used in conjunction with adsorption–catalysis [53]. The removal of toluene from laboratory to industrial scale utilizing a unique thermal mass exchanger was evaluated for the first time by Villarim et al., who also clarified the molecular process of absorption. They ascertained the absorption capacities and vapor–liquid partition coefficients (K) of three benzhydrols, namely propylene glycol, acetic acid, and their aqueous mixes. They discovered that in the solvents under study, the absorption of volatile organic molecules reduced as the concentration of water rose. Overall though, the water/benzyl alcohol mixes (60:40 wt%) show a high absorption capacity that is comparable to other organic solvents, and has considerable promise for the treatment of industrial air contaminated with toluene [54].

3.2. Catalytic Oxidizing Materials

Catalytic oxidation has been extensively studied as an efficient, resource-aware method of removing toluene from indoor air and industrial exhaust gases without additional treatment. Transition metal-functionalized γ-alumina carriers are widely used as industrial catalysts for the complete oxidation of volatile organic molecules at elevated temperatures, and Zumbar et al. achieved full oxidation activity of toluene at lower temperatures (200–380 °C) by rationally designing bimetallic CuFe-γ-alumina catalysts [55].

Fe–oxalate complexes under UV and visible light irradiation are the basis of a novel technique Zhao et al. devised for the effective recovery of platinum. Under UV radiation, the process can recover 98.9% of platinum in 30 min, and in sunshine, it can recover 95.6% of platinum in 8 h. This work reduces the cost of producing new catalysts by developing an environmentally responsible method of recycling platinum from discarded catalysts [56]. Conversely, Zhou et al. synthesized a variety of ceria-based high-entropy oxide catalysts using a solid-state reaction method. Among these, the Ce-HEO-T sample had a 100% conversion rate when it came to catalyzing the oxidation of toluene at 328 °C. After the addition of varying amounts of gold, Ce-HEO-500, which has the lowest toluene oxidation temperature, can perform even better. When compared to the Ce-HEO-500 support, the somewhat gold-containing Au/Ce-HEO-500 sample demonstrated about 70 °C lower toluene combustion, at 260 °C. It also showed good stability, converting 98% of the toluene over 60 h. Furthermore, demonstrating exceptional water resistance, the toluene conversion at 5% H_2O vapor stayed constant, and even significantly enhanced during the conversion in dry air [57].

To maximize lattice oxygen activation of pristine CoMnOx (CMO-E0) during toluene oxidation by acid treatment, Wang et al. synthesized nanostructured cobalt–manganese oxides (CoMnOx) with cationic defects (CMO-Ex, where x represents the acid concentration). The maximum toluene catalytic degradation activity was shown by the CMO-E$_{0.05}$ sample that had been adjusted with an ideal manganese and cobalt defect content. Furthermore, the CMO-E$_{0.05}$ sample exhibits superior water resistance and catalytic stability [58]. Alpha-MnO_2 doped with four metals (Cu, Ce, Co, and Fe) was synthesized by Jiang et al. using redox co-precipitation. MnCu exhibited the maximum activity, with 90% toluene conversion at a temperature of 224 °C and a weight-hourly space velocity of 30,000 $mL \cdot g^{-1} \cdot h^{-1}$. The most active was MnFe, which had a lot of surface defects and a relatively active surface lattice of oxygen, which sped up toluene adsorption and activation. The inadequate migration ability of lattice oxygen prevented the deep oxidation of toluene. Furthermore, the catalyst surface may be covered with adsorbed toluene and certain intermediates, which would prevent toluene from oxidizing continuously. Consequently, when designing mixed-oxide catalysts based on manganese dioxide, the profound oxidation of toluene is far more crucial than adsorption [59]. As adsorbents and catalysts, Zhou et al. synthesized a range of Mn/ZSM-5 catalysts with varying Mn concentrations (2, 4, and 6 wt%) to examine the regeneration performance of toluene adsorption and ozone-catalyzed oxidation at ambient temperature. Toluene was oxidized using a catalyst containing 4 weight percent Mn/ZSM-5, with a 90% conversion rate at 30 °C. Good toluene adsorption capability was still maintained by the 4 weight percent Mn/ZSM-5 after four successive adsorption–ozone cycles [60]. CexMn1-x O_2 catalysts were synthesized by Zhou et al. using the sol-gel method, and their catalytic activity for the oxidation of toluene was shown to be better than those of single oxides. The catalysts were prepared with different mixing ratios. The toluene oxidation reaction's ring-opening reaction was aided by the doped manganese's increased oxygen vacancy count and capacity to activate aromatic rings. Due to its readily available nature and low energy usage, photocatalytic technology has garnered significant attention for the removal of toluene. However, heat is produced by light, and when temperature rises, pollutants may desorb from the catalyst, making deterioration less likely [61]. Accordingly, Yu et al. conducted a detailed analysis of toluene adsorption on UiO-66 (Zr) and postulated a synergistic interaction between heat-induced adsorption and photocatalysis. According to the findings, UiO-66 (Zr) had a greater ability to adsorb toluene at higher temperatures. At the ideal temperature of 30 °C, the removal rate of toluene was 69.6% [62].

3.3. Catalytic Materials for Chemical Cycle Reforming

Syngas with a high hydrogen content is a valuable industrial feedstock and a sustainable energy source. The chemical equilibrium of the water–gas shift reaction limits the composition of the syngas produced in the traditional chemical looping steam reforming (CLSR) process. A two-step chemical cycle-reforming (TS-CLR) process can reduce the equilibrium limitation and generate syngas with larger CO and H_2 concentrations, according to research by Xu et al. The important components in this are oxygen carriers with outstanding cycle performance and good partial oxidation performance [63]. In order to enhance the cycling and partial oxidation ability of the oxygen carrier, Mao et al. proposed an embedding technique to modify the microreactor environment of the oxygen source carrier. As shown in Figure 10, they made Fe_2O_3@SBA-15 catalysts by embedding Fe_2O_3 into SBA-15 via wet impregnation, and investigated its catalytic role in the toluene chemical cycle-reforming reaction [64].

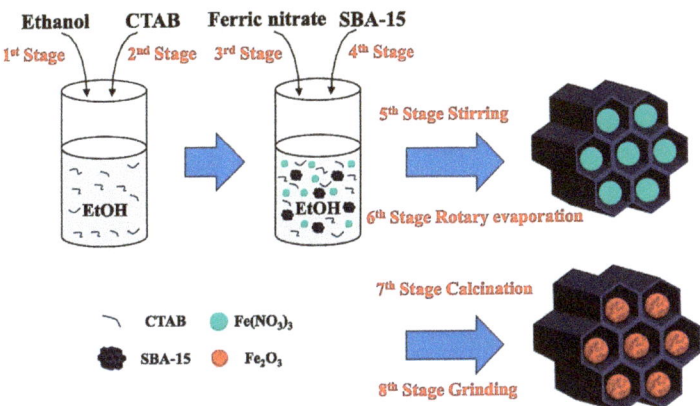

Figure 10. Fe_2O_3@SBA-15 schematic diagram created by wet impregnation [64].

It was found that catalytic cracking and partial oxidation are the two steps that make up the toluene chemical cycle reforming process, where the first stage mainly produces syngas and the second stage produces coke and H_2. Comparing Fe_2O_3@SBA-15 to pure Fe_2O_3, the CO selectivity rose from 25.9% to 96.2%. Significant improvements were made in the lattice oxygen usage and toluene conversion. The toluene conversion dropped by just 1.9% after ten cycles. By embedding metal oxides in molecular sieves, Liu et al. developed NiO@SBA-15, Fe_2O_3@SBA-15, and $NiFe_2O_4$@SBA-15 based on the decoupling method of the biomass chemical cycle gasification process. They investigated catalytic toluene chemical cycle regeneration. The results showed that $NiFe_2O_4$@SBA-15-catalyzed toluene could reach a highest conversion rate of 93.4% with a CO selectivity of 80.7%. By enhancing the dispersion and nanocrystallization of metal oxides in the oxygen carrier, the embedding technique can substantially decrease sintering. The ideal weight–time rate was 1.168 h^{-1}, and the ideal reaction temperature was 750 °C. Over 10 testing cycles, the average toluene conversion was 95.34%, and the moderate CO selectivity was 94.83% [65]. Zhang et al. also created $NiFe_2O_4$@SBA-15 by employing an impregnation-based embedding technique. At 750 °C, methane and toluene underwent a chemical cycle-reforming process. Figure 11 depicts the reaction's route diagram. Toluene conversion was 97.5%. The methane and toluene chemical cycle reaction was conducted using the impregnation method [66].

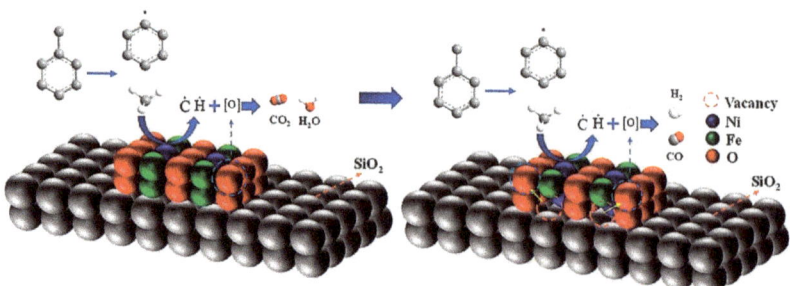

Figure 11. Chemical cycle reforming of $NiFe_2O_4$@SBA-15 with toluene: a reaction route diagram [66].

Sun and colleagues introduced a novel approach for biomass pyrolysis of volatiles: chemical loop reforming (CLR). The method is based on a decoupling technique, and its reaction kinetics and cycle performance were investigated for the toluene CLR process on $LaFe_{0.6}Co_{0.4}O_3$@SBA-15 OC. They discovered that the toluene conversion rose from 52.3% to 79.7%, the CO selectivity climbed from 57.0% to 87.4%, and the oxygen release (OR) increased by 100% when $LaFeO_3$ was encapsulated in SBA-15. Improved reaction performance, decreased sintering, and enhanced dispersion of the metal oxides were the outcomes of the encapsulation effect. Co-inclusion produced the best results, with a toluene conversion of 81.6% and a CO selectivity of 96.8%. During ten testing cycles, $LaFe_{0.6}Co_{0.4}O_3$@SBA-15's toluene conversion and CO selectivity stayed between 90.0% and 92.0% and 93.0% and 96.0%, respectively. This work offers recommendations for the use of organic compounds (OC) for the biomass pyrolysis volatile chemical cycle-reforming process [67]. Luo et al. used 6Ni, $4Ni_2Cu$, and $4Ni_2Fe$ as oxygen carriers and toluene as a tar model molecule. They used different temperatures and water vapor–carbon molar ratios (S/C) to study the chemical cycle reforming reaction of toluene. It was found that the metal oxygen carriers acted as both oxidizing and catalyzing agents in the chemical cycle reforming of toluene. After ten cycles, $4Ni_2Fe$ outperformed the other two oxygen carriers in terms of carbon deposition resistance, stability, and regeneration [68]. Rong et al. developed a two-step sol-gel process to manufacture Ca-Al-Fe, a hybrid adsorbent/catalyst for calcium cycle gasification (CLG) based on calcium oxide. With the least amount of coke deposited, the highest average H_2 production and an average hydrogen concentration of roughly 68.8% throughout five toluene-reforming cycles were demonstrated by Ca-Al-Fe. The percentage of hydrogen was roughly 68.8% on average. This is roughly 26.41% greater than the typical CaO conversion. Figure 12 displays the syngas concentration throughout the multi-cycle toluene conversion [69].

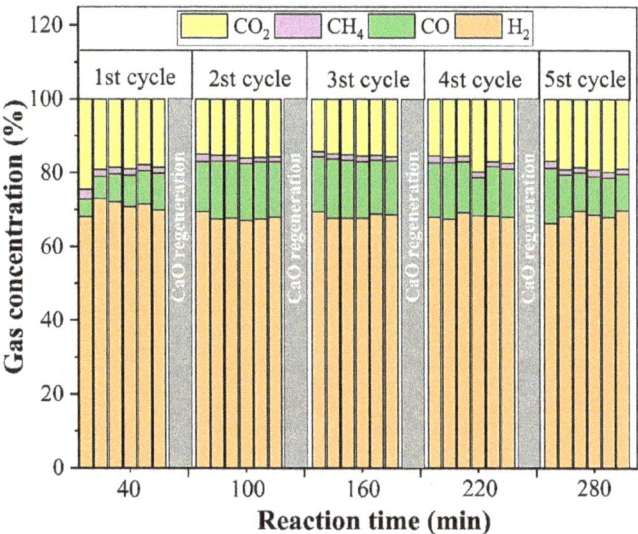

Figure 12. The concentration of syngas during multicycle toluene conversion with a Ca-Al-Fe absorber present (single column image) [69].

3.4. Degradation of Catalytic Materials

VOCs can be broken down using a very promising green technique called non-thermal plasma (NTP). In accordance with the increasing amount of research being carried out in this field, Yue et al. explored the mechanism of toluene degradation in air/H_2O dielectric barrier discharge (DBD) plasma, finding that at P = 115 W, Cin, toluene = 1000 ppm, the degradation efficiency of toluene was >82%. As illustrated in Figure 13, the breakdown mechanism of toluene is as follows: toluene → phenyl → benzaldehyde → benzene → phenoxy → cyclopentadiene → polycarbonate/alkyne → CO_2/H_2O. This advances the development of non-thermal plasma breakdown of volatile organic molecules and offers fresh insights into the plasma-catalyzed process [70].

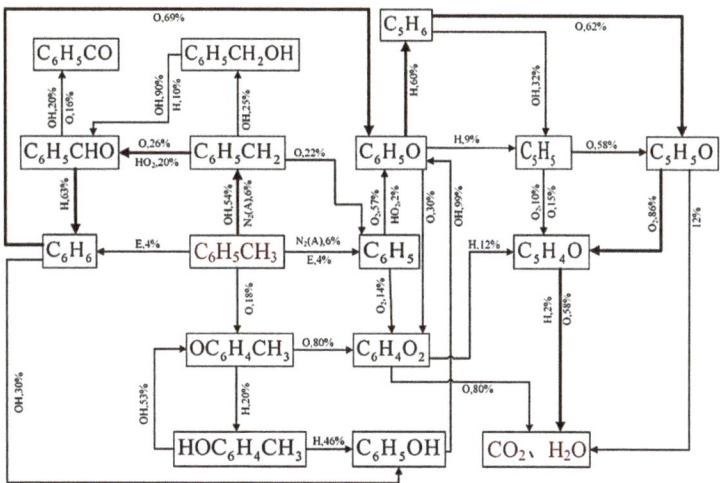

Figure 13. Degradation route of toluene and the ratio of products produced at each stage of the route [70].

One important use of photocatalysis is the breakdown of volatile organic molecules. Liu et al. used a straightforward sol-gel technique to adsorb vanadium–nitrogen co-doped TiO_2 onto a honeycomb ceramic (V/N-TiO_2@HC), creating a continuous-flow photocatalytic degradation reactor. The insertion of V/N dopants decreased the band gap and expanded the light absorption range of TiO_2, resulting in a continuous and nearly full photocatalytic degradation of toluene in this reactor. The reactor's abundant ordered pores in HC facilitated the mass transport of toluene. The photocatalyst's reusable application was made possible by HC's distinct and sturdy structure. The toluene gas degradation rate reached 97.8% and stayed at 96.7% over 24 cycles of photocatalytic degradation [71]. Using atmospheric surface photovoltaic (ASPV) spectroscopy, After heating TiO_2 in a hydrogen environment to produce more bridging hydroxyl groups (OHBs), Zhong et al. investigated the effects of surface site modifications on the transfer of charge carriers generated by photosynthesis to the reacting substances (O_2, H_2O, and toluene). The researchers found that treatment with hydrogen produced gap transmission channels for toluene and water and electron transmission routes for oxygen, and reduced H_2O interference with oxygen's ability to use electrons. The hydrogen found in the freshly formed OHB changed the distribution of electrons on the surface of TiO_2 by introducing electrons. The reaction was accelerated by the OHB and the surrounding Ti_{4-x} ions, which changed the adsorbed form of H_2O and helped transport electrons to oxygen (O_2) and gaps to toluene. Due to the great photothermal conversion ability and good thermocatalytic activity of transition metal oxides, it is possible to further enhance their photothermal catalytic capacity by logically triggering the photoelectric effect of semiconductors [72]. On this basis, Zhao et al. synthesized Mn_3O_4/Co_3O_4 composites with S-type heterojunctions for the photothermal catalytic degradation of toluene under ultraviolet–visible (UV-Vis) light. Under UV-Vis irradiation, the rapid electron transfer between the surfaces of the Mn_3O_4/Co_3O_4 composites promoted the generation of more reactive radicals, and the energy band bending and the intrinsic electric field at the Mn_3O_4/Co_3O_4 interface enhanced the photogenerated carrier transport pathway and maintained high redox potential. The elimination of toluene by Mn_3O_4/Co_3O_4 was improved from 53.3% and 47.5% to 74.7% compared to single metal oxides, respectively.

This work provides important guidance for the creation of effective narrow-band semiconductor heterojunction photothermal catalysts [73].

Toluene degradation by heterojunctions composed of rutile–rutile TiO_2 is thought to be a successful method; however, photogenerated electron usage is still inadequate. In order to enhance the photocatalyst's catalytic efficiency, Zhang et al. created a type II heterojunction using rutile-coated Lavoisier Institute material (MIL-101). It was discovered that in addition to the transfer of photogenerated electrons to anatase's oxygen vacancies, which encourages the production of oxygen-containing radicals, the enhancement of the photocatalytic performance also depends on the anatase's ability to encapsulate MIL-101, its capacity to absorb light, and the contact area between the two heterojunctions. The substance illustrates the complementary nature of heterojunction and heterojunction design, offering a theoretical foundation for their use in the breakdown of volatile organic molecules [74].

One efficient method to encourage the deep decomposition of volatile organic compounds (VOCs) is to modify the interactions between the metal and the support. In order to encourage the deep degradation of toluene, Bi et al. produced a Pd@ZrO_2 catalyst utilizing the Zr-based metal-organic framework (MOF) Pd@UiO-66, grown in situ, as a precursor. Figure 14 illustrates the degradation pathway [75].

When Pd@ZrO_2-Zr(OH)$_4$ was synthesized using Zr(OH)$_4$ as a precursor, MOF-derived Pd@ZrO_2 catalysts with varying calcination periods performed better in terms of toluene degradation, water resistance, and stability. Toluene's profound breakdown into CO_2 and H_2O happened more quickly. This work offers recommendations for optimizing interfacial interactions to enhance the deep degradation of volatile organic compounds (VOCs) using MOF-derived catalysts. Zhu and colleagues built a coaxial dielectric barrier discharge (DBD) reactor and loaded a sequence of Cu-MnO$_2$/gamma-Al$_2$O$_3$ catalysts into the plasma device via impregnation and redox processes to break down a toluene and o-xylene combination. The results showed that the Cu-doped MnO$_2$ catalysts drastically reduced the production of byproducts while also considerably improving CO$_2$ selectivity and pollutant removal. Toluene and o-xylene were both able to achieve a 100% removal rate and 92.73% CO$_2$ selectivity, with Cu$_{0.15}$Mn/gamma-Al$_2$O$_3$ exhibiting the greatest removal rate among them. This work offers theoretical direction and a useful foundation for the use of mixed benzene series volatile organic compounds (VOCs) catalyzed by DBD [76].

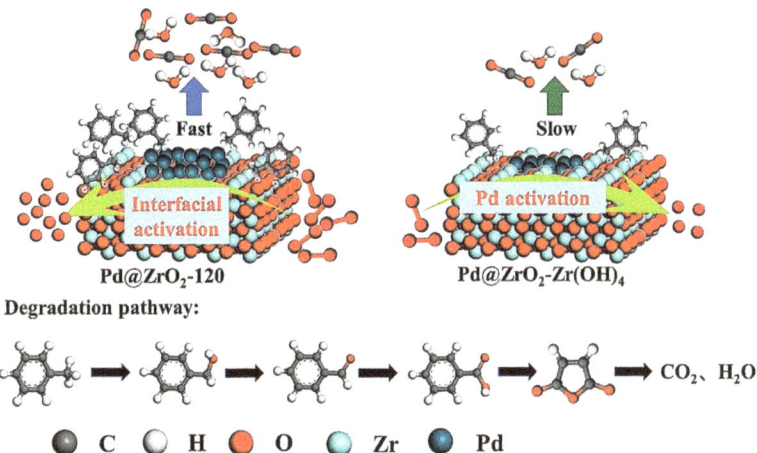

Figure 14. Mechanism of degradation of Pd@ZrO$_2$-120 and Pd@ZrO$_2$-Zr(OH)$_4$-oxidized toluene [75].

One efficient method to encourage the catalytic breakdown of volatile organic compounds (VOCs) is to modify the strength of the metal–oxygen connection to activate surface lattice oxygen (O-latt). Following the Mars–van Krevelen mechanism, Zhu et al. demonstrated that toluene could be oxidized by fast dehydrogenation of methyl groups with the help of highly active surface O-latt, followed by additional ring opening and deep mineralization to CO$_2$. This work offers a fresh approach to investigating interfacially enhanced transition metal catalysts for effective VOC abatement and surface O-latt activation [77].

Biofiltration is an efficient and economical treatment technology for the removal of volatile organic compounds (VOCs) from exhaust gas streams. Pineda et al. evaluated the effect of inoculum type on the removal of toluene, cyclohexane, and n-hexane mixtures in three biofilters (BF$_1$, BF$_2$, BF$_3$), and the reaction flow chart and related data are shown in Figure 15 [78].

The three biofilter inoculum types were BF$_1$: a compost and wood chip mixture and *Erythrobacter* spp.; BF$_2$: a compost and wood chip mixture and acclimatized activated sludge with *Erythrobacter* spp.; and BF$_3$: an expanded perlite and *Erythrobacter* spp. The three biofilters were operated for 374 days at different inlet loads and empty bed residence times. It was found that toluene was removed first, followed by cyclohexane, and hexane was removed last. At each stage of operation, BF$_2$ outperformed BF$_1$ and BF$_3$, with average maximum removal capacities as follows: toluene: 21 ± 3 g·m^{-3}·h^{-1}; cyclohexane: 11 ± 2 g·m^{-3}·h^{-1}; and hexane: 6.2 ± 0.9 g·m^{-3}·h^{-1}. Despite the differences in inoculum, the dominant microorganisms in all the biofilters were Rhodococcus, Mycobacterium, and Hexane. Rhodococcus, Mycobacterium, and Pseudonocardia genera; only the relative abundance was different. This study provides new ideas for the removal of VOCs.

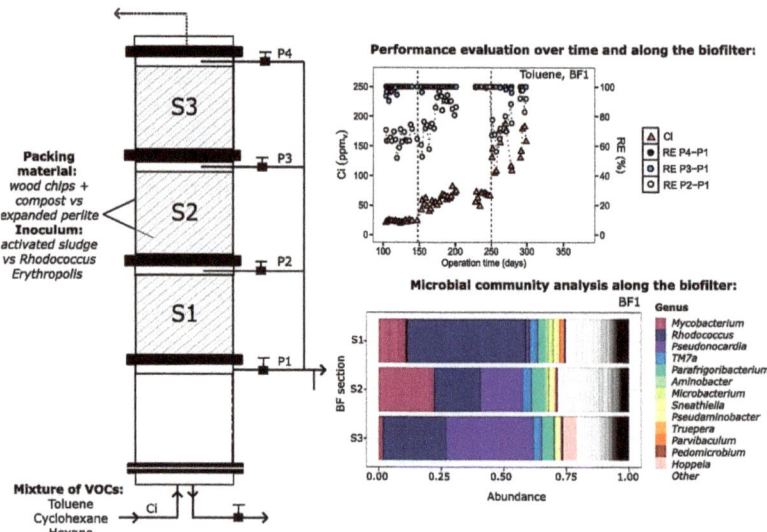

Figure 15. The experiment's flow chart, performance assessments along the biofilter and over time, and a microbiological community study along the biofilter [78].

From saline soil in Wadi An Natrun, Barghoth et al. isolated sixteen pure halophilic bacterial isolates capable of producing energy and carbon from toluene as the only source. Isolate M_7 had the best development and most similar characteristics out of all of these isolates. With 99% similarity to Exiguobacterium mexicanum, strain M7 belongs to the Exiguobacterium genus. With toluene serving as the only carbon source, strain M_7 exhibited good growth characteristics over a broad temperature range of 20–40 °C, pH values of 5–9, and salinity concentrations of 2.5–10%, w/v. The ideal growth conditions were found to be 35 °C and pH values of 8–9% and 5%, respectively. In 48 h, strain M_7 broke down 88.32% of toluene under ideal circumstances. The present results suggest that strain M_7 can be used as a biotechnological tool for the treatment of toluene waste [79].

4. Summary and Outlook

A variety of methods and materials are available here for the treatment of methane and toluene. Adsorption, catalytic oxidation, reforming, degradation, and other technologies all allow for the removal and resource utilization of methane and toluene. Many of these materials can also be very environmentally friendly and safe for the treatment and resource use of methane and toluene, and many can achieve conversion rates of 90% or more and remain active after reuse more than 10 times. However, there are still some areas for improvement. Firstly, certain auxiliary materials for the treatment of methane and toluene often need to function at high temperatures, high pressures, precious metals, etc., which undoubtedly requires the use of more energy and resources. Therefore, we need to further optimize these conditions and develop more inexpensive catalytic materials to reduce the use of energy and expensive resources.

Second, although there has been more study on materials for treating single gases, we must carry out more research into materials for treating mixed gases that cause pollution.

Thirdly, a lot of materials are only appropriate for use in laboratories, and cannot be used in industrial production or application. This is significant because if industrial production and application can be achieved, people will likely be able to deal with these two gases on a large scale, which will likely have a positive effect on both the environment and human health.

All things considered, future studies on related materials can begin with the aforementioned three principles and be refined regularly to provide more direction and support for the preservation of the environment worldwide, as well as the economical use of energy.

Author Contributions: Writing—original draft preparation, T.L.; writing—review and editing, T.L.; R.W.; supervision, R.W.; project administration, R.W.; funding acquisition, R.W. All authors have read and agreed to the published version of the manuscript.

Funding: Financial support from the PetroChina Innovation Foundation (2013D-5006-0507) is gratefully acknowledged.

Institutional Review Board Statement: This study did not require ethical approval.

Informed Consent Statement: This study did not involve humans.

Data Availability Statement: No new data were created in this study.

Conflicts of Interest: The authors declare no conflict of interest.

References

1. Jackson, R.B.; Saunois, M.; Bousquet, P.; Canadell, J.G.; Poulter, B.; Stavert, A.R.; Bergamaschi, P.; Niwa, Y.; Segers, A.; Tsuruta, A. Increasing anthropogenic methane emissions arise equally from agricultural and fossil fuel sources. *Environ. Res. Lett.* **2020**, *15*, 071002. [CrossRef]
2. Wu, Y.; Zhou, K. Evaluation of Metal-Organic Frameworks for Natural Gas Storage. In Proceedings of the International Conference on Optoelectronic Materials and Devices (ICOMD) 2022, Guangzhou, China, 10–12 December 2021.
3. Zhu, L.-M.; Li, W.-L.; Li, T.-R.; Shi, L.-P.; Li, L.-T.; Yao, Z.-Q.; Huang, H.-L.; Zhao, J.-P.; Liu, F.-C. A dense 3d–4f metal–organic framework with "gas pockets" for highly efficient CH_4/N_2 separation. *Inorg. Chem. Front.* **2023**, *10*, 2438–2443. [CrossRef]
4. Adlak, K.; Chandra, R.; Vijay, V.K.; Pant, K.K. Suitability analysis of sustainable nanoporous adsorbents for higher biomethane adsorption and storage applications. *Int. J. Energy Res.* **2022**, *46*, 14779–14793. [CrossRef]
5. Zaini, M.S.M.; Arshad, M.; Syed-Hassan, S.S.A. Adsorption Isotherm and Kinetic Study of Methane on Palm Kernel Shell-Derived Activated Carbon. *J. Bioresour. Bioprod.* **2023**, *8*, 66–77. [CrossRef]
6. Pribylov, A.A.; Fomkin, A.A.; Shkolin, A.V.; Men'shchikov, I.E. Adsorption of Methane onto Microporous Activated Carbon in a Volumetric Storage System. *Prot. Met. Phys. Chem. Surf.* **2023**, *59*, 14–18. [CrossRef]
7. Algazlan, M.; Pinetown, K.; Saghafi, A.; Grigore, M.; Roshan, H. Role of Organic Matter and Pore Structure on CO_2 Adsorption of Australian Organic-Rich Shales. *Energy Fuels* **2022**, *36*, 5695–5708. [CrossRef]
8. Aji, A.Q.M.; Mohshim, D.F.; Maulianda, B.; Elraeis, K.A. Supercritical methane adsorption measurement on shale using the isotherm modelling aspect. *RSC Adv.* **2022**, *12*, 20530–20543. [CrossRef]
9. Zuo, W.; Qi, M.; Liu, Y.; Li, H.; Han, H.; Wang, Y.; Long, L.; Wu, S. Study on Gas Adsorption-Desorption and Diffusion Behaviour in Coal Pores Modified by Nano Fracturing Fluid. *Acs Omega* **2023**, *8*, 29213–29224. [CrossRef]
10. Ursueguia, D.; Diaz, E.; Ordonez, S. MOF-alumina composites for improved methane adsorption under wet conditions. *Microporous Mesoporous Mater.* **2023**, *360*, 112712. [CrossRef]
11. Zhao, S.; Wang, Y.; Zhang, Y.; Bai, J.; Zhang, Y.; Wang, S.; Duan, E. Enhancement of the redox reactions of the $La_{0.8}Sr_{0.2}MnO_3$ catalyst by surface acid etching: A simple synthesis strategy to high-performance catalysts for methane combustion. *Fuel* **2023**, *345*, 128258. [CrossRef]
12. Khatun, R.; Pal, R.S.; Shoeb, M.A.; Khurana, D.; Singhl, S.; Siddiqui, N.; Poddar, M.K.; Khan, T.S.; Bal, R. Generation of active oxygen species by CO_2 dissociation over defect-rich Ni-Pt/CeO_2 catalyst for boosting methane activation in low-temperature dry reforming: Experimental and theoretical study. *Appl. Catal. B* **2024**, *340*, 123243. [CrossRef]
13. Alvarez, M.; Marin, P.; Ordonez, S. Upgrading of methane emissions via chemical looping over copper-zeolites: Experiments and modelling. *Chem. Eng. Sci.* **2022**, *259*, 117818. [CrossRef]
14. Vitillo, J.G.; Lu, C.C.; Bhan, A.; Gagliardi, L. Comparing the reaction profiles of single iron catalytic sites in enzymes and in reticular frameworks for methane-to-methanol oxidation. *Cell Rep. Phys. Sci.* **2023**, *4*, 101422. [CrossRef]
15. Rungtaweevoranit, B.; Abdel-Mageed, A.M.M.; Khemthong, P.; Eaimsumang, S.; Chakarawet, K.; Butburee, T.; Kunkel, B.; Wohlrab, S.; Chainok, K.; Phanthasri, J.; et al. Structural Evolution of Iron-Loaded Metal-Organic Framework Catalysts for Continuous Gas-Phase Oxidation of Methane to Methanol. *ACS Appl. Mater. Interfaces* **2023**, *15*, 26700–26709. [CrossRef]
16. Zhu, C.; Guo, G.; Li, W.; Wu, M.; Jiang, Y.; Wu, W.; Zhang, H. Direct Catalytic Oxidation of Low-Concentration Methane to Methanol in One on Ni-Promoted BiOCl. *Acs Omega* **2023**, *8*, 11220–11232. [CrossRef]
17. Pereira, T.C.P.; Vieira, J.V.R.; da Cunha, C.H.F.; Mizuno, S.C.M.; Carvalho, Y.O.; Faheina, T.; Picinini, M.; Blanco, A.L.; Tello, A.C.M.; Urquieta-Gonzalez, E.A.; et al. Conversion of methane to methanol over Cu-MAZ (zeolite omega): An oxygen-free process using H_2O and CO_2 as oxidants. *Appl. Catal. B Environ. Energy* **2024**, *342*, 123370. [CrossRef]
18. Zhou, L.; Su, Y.-Q.; Hu, T.-L. Theoretical insights into the selective oxidation of methane to methanol on single-atom alloy catalysts. *Sci. China Mater.* **2023**, *66*, 3189–3199. [CrossRef]

19. Matsuda, A.; Obara, K.; Ishikawa, A.; Tsai, M.-H.; Wang, C.-H.; Lin, Y.-C.; Hara, M.; Kamata, K. Bismuth phosphate nanoparticle catalyst for direct oxidation of methane into formaldehyde. *Catal. Sci. Technol.* **2023**, *13*, 5180–5189. [CrossRef]
20. Yasumura, S.; Saita, K.; Miyakage, T.; Nagai, K.; Kon, K.; Toyao, T.; Maeno, Z.; Taketsugu, T.; Shimizu, K.-i. Designing main-group catalysts for low-temperature methane combustion by ozone. *Nat. Commun.* **2023**, *14*, 3926. [CrossRef]
21. Yamada, Y.; Morita, K.; Sugiura, T.; Toyoda, Y.; Mihara, N.; Nagasaka, M.; Takaya, H.; Tanaka, K.; Koitaya, T.; Nakatani, N.; et al. Stacking of a Cofacially Stacked Iron Phthalocyanine Dimer on Graphite Achieved High Catalytic CH$_4$ Oxidation Activity Comparable to That of pMMO. *JACS Au* **2023**, *3*, 823–833. [CrossRef]
22. Yamada, Y.; Miwa, Y.; Toyoda, Y.; Phung, Q.M.; Oyama, K.-I.; Tanaka, K. Evaluation of CH$_4$ oxidation activity of high-valent iron-oxo species of a μ-nitrido-bridged heterodimer of iron porphycene and iron phthalocyanine. *Catal. Sci. Technol.* **2023**, *13*, 1725–1734. [CrossRef]
23. Stoian, M.C.; Romanitan, C.; Craciun, G.; Culita, D.C.; Papa, F.; Badea, M.; Negrila, C.; Popescu, I.; Marcu, I.-C. Multicationic LDH-derived Co(x)CeMgAlO mixed oxide catalysts for the total oxidation of methane. *Appl. Catal. A* **2023**, *650*, 119001. [CrossRef]
24. Maulidanti, E.G.; Awaji, M.; Asami, K. Low temperature oxidative coupling of methane over cerium oxide based catalyst. *Gas Sci. Eng.* **2023**, *116*, 205057. [CrossRef]
25. Zhao, R.; Du, X.; Cao, K.; Gong, M.; Li, Y.; Ai, J.; Ye, R.; Chen, R.; Shan, B. Highly dispersed Fe-decorated Ni nanoparticles prepared by atomic layer deposition for dry reforming of methane. *Int. J. Hydrogen Energy* **2023**, *48*, 28780–28791. [CrossRef]
26. Yadav, P.K.; Patrikar, K.; Mondal, A.; Sharma, S. Ni/Co in and on CeO$_2$: A comparative study on the dry reforming reaction. *Sustain. Energy Fuels* **2023**, *7*, 3853–3870. [CrossRef]
27. Bandurist, P.S.; Pichugina, D.A. Quantum-Chemical Study of C-H Bond Activation in Methane on Ni-Cu Oxide and Sulfide Clusters. *Kinet. Catal.* **2023**, *64*, 362–370. [CrossRef]
28. Zou, Z.; Zhang, T.; Lv, L.; Tang, W.; Zhang, G.; Gupta, R.K.; Wang, Y.; Tang, S. Preparation adjacent Ni-Co bimetallic nano catalyst for dry reforming of methane. *Fuel* **2023**, *343*, 128013. [CrossRef]
29. Zhang, Y.; Zhang, G.; Liu, J.; Li, T.; Zhang, X.; Wang, Y.; Zhao, Y.; Li, G.; Zhang, Y. Insight into the role of preparation method on the structure and size effect of Ni/MSS catalysts for dry reforming of methane. *Fuel Process. Technol.* **2023**, *250*, 107891. [CrossRef]
30. Zhang, X.; Shen, Y.; Liu, Y.; Zheng, J.; Deng, J.; Yan, T.; Cheng, D.; Zhang, D. Unraveling the Unique Promotion Effects of a Triple Interface in Ni Catalysts for Methane Dry Reforming. *Ind. Eng. Chem. Res.* **2023**, *62*, 4965–4975. [CrossRef]
31. Zheng, Y.; Sukma, M.S.; Scott, S.A. The exploration of NiO/Ca$_2$Fe$_2$O$_5$/CaO in chemical looping methane conversion for syngas and H$_2$ production. *Chem. Eng. J.* **2023**, *465*, 142779. [CrossRef]
32. Zakrzewski, M.; Shtyka, O.; Rogowski, J.; Ciesielski, R.; Kedziora, A.; Maniecki, T. Influence of Lanthanum Precursor on the Activity of Nickel Catalysts in the Mixed-Methane Reforming Process. *Int. J. Mol. Sci.* **2023**, *24*, 975. [CrossRef] [PubMed]
33. Yadavalli, S.S.; Jones, G.; Benson, R.L.; Stamatakis, M. Assessing the Impact of Adlayer Description Fidelity on Theoretical Predictions of Coking on Ni(111) at Steam Reforming Conditions. *J. Phys. Chem. C* **2023**, *127*, 8591–8606. [CrossRef] [PubMed]
34. Hanifa, N.H.E.; Ismail, M.; Ideris, A. Utilization of palm oil fuel ash (POFA) as catalyst support for methane decomposition. In Proceedings of the International Symposium of Reaction Engineering, Catalysis and Sustainable Energy (RECaSE), Electr Network, 2022, Virtual, 6 April 2021; pp. 1136–1141.
35. Wysocka, I.; Karczewski, J.; Golabiewska, A.; Lapinski, M.; Cieslik, B.M.; Maciejewski, M.; Koscielska, B.; Rogala, A. Nickel phase deposition on V$_2$CT$_x$/V$_2$AlC as catalyst precursors for a dry methane reforming: The effect of the deposition method on the morphology and catalytic activity. *Int. J. Hydrogen Energy* **2023**, *48*, 10922–10940. [CrossRef]
36. Pourghadiri, E.; Sari, A. Theoretical Investigation on Syngas Production by Catalytic Partial Oxidation of Methane in a Reverse-Flow Microchannel Monolithic Reactor. *Ind. Eng. Chem. Res.* **2023**, *62*, 9433–9452. [CrossRef]
37. Zhou, Y.; Chu, R.; Fan, L.; Zhao, J.; Li, W.; Jiang, X.; Meng, X.; Li, Y.; Yu, S.; Wan, Y. Conversion mechanism of thermal plasma-enhanced CH$_4$-CO$_2$ reforming system to syngas under the non-catalytic conditions. *Sci. Total Environ.* **2023**, *866*, 161453. [CrossRef]
38. Song, G.; Li, C.; Zhou, W.; Wu, L.; Lim, K.H.; Hu, F.; Wang, T.; Liu, S.; Ren, Z.; Kawi, S. Catalytic decomposition of methane for controllable production of carbon nanotubes and high purity H2 over LTA zeolite-derived Ni-based yolk-shell catalysts. *Chem. Eng. J.* **2023**, *474*, 145643. [CrossRef]
39. Chudakova, M.V.; Popov, M.V.; Korovchenko, P.A.; Pentsak, E.O.; Latypova, A.R.; Kurmashov, P.B.; Pimenov, A.A.; Tsilimbaeva, E.A.; Levin, I.S.; Bannov, A.G.; et al. Effect of potassium in catalysts obtained by the solution combustion synthesis for co-production of hydrogen and carbon nanofibers by catalytic decomposition of methane. *Chem. Eng. Sci.* **2024**, *284*, 119408. [CrossRef]
40. Yan, K.; Jie, X.; Li, X.; Horita, J.; Stephens, J.; Hu, J.; Yuan, Q. Microwave-enhanced methane cracking for clean hydrogen production in shale rocks. *Int. J. Hydrogen Energy* **2023**, *48*, 15421–15432. [CrossRef]
41. Xavier Jr, N.F.F.; Payne, A.J.R.; Bauerfeldt, G.F.; Sacchi, M. Theoretical insights into the methane catalytic decomposition on graphene nanoribbons edges. *Front. Chem.* **2023**, *11*, 1172687. [CrossRef]
42. Bae, D.; Kim, Y.; Ko, E.H.; Han, S.J.; Lee, J.W.; Kim, M.; Kang, D. Methane pyrolysis and carbon formation mechanisms in molten manganese chloride mixtures. *Appl. Energy* **2023**, *336*, 120810. [CrossRef]
43. Zhang, S.B.X.Y.; Pessemesse, Q.; Latsch, L.; Engel, K.M.; Stark, W.J.; van Bavel, A.P.; Horton, A.D.; Payard, P.-A.; Coperet, C. Role and dynamics of transition metal carbides in methane coupling. *Chem. Sci.* **2023**, *14*, 5899–5905. [CrossRef]

44. Zhang, H.; Bolshakov, A.; Meena, R.; Garcia, G.A.; Dugulan, A.I.; Parastaev, A.; Li, G.; Hensen, E.J.M.; Kosinov, N. Revealing Active Sites and Reaction Pathways in Methane Non-Oxidative Coupling over Iron-Containing Zeolites. *Angew. Chem. Int. Ed.* **2023**, *62*, e202306196. [CrossRef]

45. Zhang, J.; Shen, J.; Li, D.; Long, J.; Gao, X.; Feng, W.; Zhang, S.; Zhang, Z.; Wang, X.; Yang, W. Efficiently Light-Driven Nonoxidative Coupling of Methane on Ag/NaTaO$_3$: A Case for Molecular-Level Understanding of the Coupling Mechanism. *ACS Catal.* **2023**, *13*, 2094–2105. [CrossRef]

46. Zhang, H.; Su, Y.; Kosinov, N.; Hensen, E.J.M. Non-oxidative coupling of methane over Mo-doped CeO$_2$ catalysts: Understanding surface and gas-phase processes. *Chin. J. Catal.* **2023**, *49*, 68–80. [CrossRef]

47. Ryu, H.W.; Nam, K.; Lim, Y.H.; Kim, D.H. Effect of the NiO particle size on the activity of Mo/HZSM-5 catalyst physically mixed with NiO in methane dehydroaromatization. *Catal. Today* **2023**, *411*, 113875. [CrossRef]

48. Stepanov, A.A.; Korobitsyna, L.L.; Vosmerikov, A.V. Investigation of the Properties of Mo/ZSM-5 Catalysts Based on Zeolites with Microporous and Micro-Mesoporous Structures. *Chemistry* **2023**, *5*, 1256–1270. [CrossRef]

49. Stepanov, A.A.; Korobitsyna, L.L.; Budaev, Z.B.; Vosmerikov, A.V.; Gerasimov, E.Y.; Ishkildina, A.K. Influence of the Secondary Mesoporous Structure of Zeolite on the Properties of Mo/ZSM-5 Catalysts for Non-Oxidative Methane Conversion. *Izv. Vyss. Uchebnykh Zaved. Khimiya I Khimicheskaya Tekhnologiya* **2023**, *66*, 58–66. [CrossRef]

50. Zhang, J.; Shao, J.; Zhang, X.; Rao, G.; Li, G.; Yang, H.; Zhang, S.; Chen, H. Facile synthesis of Cu-BTC@biochar with controlled morphology for effective toluene adsorption at medium-high temperature. *Chem. Eng. J.* **2023**, *452*, 139003. [CrossRef]

51. Wang, Y.; Zhu, W.; Zhao, G.; Ye, G.; Jiao, Y.; Wang, X.; Yao, F.; Peng, W.; Huang, H.; Ye, D. Precise preparation of biomass-based porous carbon with pore structure-dependent VOCs adsorption/desorption performance by bacterial pretreatment and its forming process. *Environ. Pollut.* **2023**, *322*, 121134. [CrossRef]

52. Zhang, X.; Wang, Y.; Mi, J.; Jin, J.; Meng, H. Dual hydrophobic modification on MIL-101(Cr) with outstanding toluene removal under high relative humidity. *Chem. Eng. J.* **2023**, *451*, 139000. [CrossRef]

53. Zhou, M.; Li, S.; Cao, M.; Wang, T.; Nie, L.; Li, W.; Zhao, F.; Chen, Y. Enhanced hydrophobic microporous Pt/silica with high adsorption and catalytic oxidation for trace toluene removal. *J. Environ. Chem. Eng.* **2023**, *11*, 110821. [CrossRef]

54. Villarim, P.; Gui, C.; Genty, E.; Lei, Z.; Zemmouri, J.; Fourmentin, S. Toluene absorption from laboratory to industrial scale: An experimental and theoretical study. *Sep. Purif. Technol.* **2024**, *328*, 125070. [CrossRef]

55. Zumbar, T.; Arccon, I.; Djinovic, P.; Aquilanti, G.; Zerjav, G.; Pintar, A.; Risticc, A.; Drazic, G.; Volavsek, J.; Mali, G.; et al. Winning Combination of Cu and Fe Oxide Clusters with an Alumina Support for Low-Temperature Catalytic Oxidation of Volatile Organic Compounds. *ACS Appl. Mater. Interfaces* **2023**, *15*, 28747–28762. [CrossRef] [PubMed]

56. Zhao, X.; Liang, G.; Wang, H.; Qu, Z. New Insight toward Synergetic Effect for Platinum Recovery Coupling with Fe(III)-Oxalate Complexes Degradation through Photocatalysis. *ACS Sustain. Chem. Eng.* **2023**, *11*, 11580–11589. [CrossRef]

57. Zhou, J.; Zheng, Y.; Zhang, G.; Zeng, X.; Xu, G.; Cui, Y. Toluene catalytic oxidation over gold catalysts supported on cerium-based high-entropy oxides. *Environ. Technol.* **2023**. [CrossRef] [PubMed]

58. Wang, W.; Huang, Y.; Rao, Y.; Li, R.; Lee, S.; Wang, C.; Cao, J. The role of cationic defects in boosted lattice oxygen activation during toluene total oxidation over nano-structured CoMnO$_x$spinel. *Environ. Sci. Nano* **2023**, *10*, 812–823. [CrossRef]

59. Jiang, Y.; Jiang, Y.; Su, C.; Sun, X.; Xu, Y.; Cheng, S.; Liu, Y.; Dou, X.; Yang, Z. Clarify the effect of different metals doping on α-MnO$_2$ for toluene adsorption and deep oxidation. *Fuel* **2024**, *355*, 129402. [CrossRef]

60. Zhou, B.; Ke, Q.; Chen, K.; Wen, M.; Cui, G.; Zhou, Y.; Gu, Z.; Weng, X.; Lu, H. Adsorption and catalytic ozonation of toluene on Mn/ZSM-5 at low temperature. *Appl. Catal. A* **2023**, *657*, 119146. [CrossRef]

61. Zhou, W.; Li, H.; Song, B.; Ma, W.; Liu, Z.; Wang, Z.; Xu, Z.; Meng, L.; Wang, Y.; Qin, X.; et al. Catalytic Oxidation Mechanism of Toluene over Ce$_x$Mn$_{1-x}$O$_2$: The Role of Oxygen Vacancies in Adsorption and Activation of Toluene. *Langmuir* **2023**, *39*, 8503–8515. [CrossRef]

62. Yu, J.; Wang, X.; Wang, Y.; Xie, X.; Xie, H.; Vorayos, N.; Sun, J. Heating-induced adsorption promoting the efficient removal of toluene by the metal-organic framework UiO-66 (Zr) under visible light. *J. Colloid Interface Sci.* **2024**, *653*, 1478–1487. [CrossRef]

63. Xu, T.; Wang, X.; Xiao, B.; Zhao, H.; Liu, W. Optimisation of syngas production from a novel two-step chemical looping reforming process using Fe-dolomite as oxygen carriers. *Fuel Process. Technol.* **2022**, *228*, 107169. [CrossRef]

64. Mao, X.; Liu, G.; Yang, B.; Shang, J.; Zhang, B.; Wu, Z. Chemical looping reforming of toluene via Fe$_2$O$_3$@SBA-15 based on controlling reaction microenvironments. *Fuel* **2022**, *326*, 125024. [CrossRef]

65. Liu, G.; Sun, Z.; Zhao, H.; Mao, X.; Yang, B.; Shang, J.; Wu, Z. Chemical looping reforming of toluene as bio-oil model compound via NiFe$_2$O$_4$@SBA-15 for hydrogen-rich syngas production. *Biomass Bioenergy* **2023**, *174*, 106851. [CrossRef]

66. Zhang, B.; Sun, Z.; Li, Y.; Yang, B.; Shang, J.; Wu, Z. Chemical looping reforming characteristics of methane and toluene from biomass pyrolysis volatiles based on decoupling strategy: Embedding NiFe$_2$O$_4$ in SBA-15 as an oxygen carrier. *Chem. Eng. J.* **2023**, *466*, 143228. [CrossRef]

67. Sun, Z.; Zhang, B.; Zhang, R.; Mao, X.; Shang, J.; Yang, B.; Wu, Z. Chemical Looping Reforming of Toluene as Volatile Model Compound over LaFe$_x$M$_{1-x}$O$_3$@SBA via Encapsulation Strategy. *Ind. Eng. Chem. Res.* **2023**, *62*, 7731–7743. [CrossRef]

68. Luo, M.; Shen, R.; Qin, Y.; Liu, H.; He, Y.; Wang, Q.; Wang, J. Conversion and syngas production of toluene as a biomass tar model compound in chemical looping reforming. *Fuel* **2023**, *345*, 128203. [CrossRef]

69. Rong, N.; Han, L.; Ma, K.; Liu, Q.; Wu, Y.; Xin, C.; Zhao, J.; Hu, Y. Enhanced multi-cycle CO_2 capture and tar reforming via a hybrid CaO-based absorbent/ catalyst: Effects of preparation, reaction conditions and application for hydrogen production. *Int. J. Hydrogen Energy* **2023**, *48*, 9988–10001. [CrossRef]
70. Yue, W.; Lei, W.; Dong, Y.; Shi, C.; Lu, Q.; Cui, X.; Wang, X.; Chen, Y.; Zhang, J. Toluene degradation in air/H_2O DBD plasma: A reaction mechanism investigation based on detailed kinetic modeling and emission spectrum analysis. *J. Hazard. Mater.* **2023**, *448*, 130894. [CrossRef]
71. Liu, G.; Han, L.; Wang, J.; Yang, Y.; Chen, Z.; Liu, B.; An, X. Continuous near-complete photocatalytic degradation of toluene by V/N-dope d TiO_2 loade d on honeycomb ceramics under UV irradiation. *J. Mater. Sci. Technol.* **2024**, *174*, 188–194. [CrossRef]
72. Zhong, Z.; Shen, Z.; Zhang, Y.; Li, J.; Lyu, J. Effect of bridging hydroxyl on the interfacial charge transfer for photocatalytic degradation of toluene. *Appl. Surf. Sci.* **2023**, *628*, 157306. [CrossRef]
73. Zhao, J.; Li, C.; Yu, Q.; Zhu, Y.; Liu, X.; Li, S.; Liang, C.; Zhang, Y.; Huang, L.; Yang, K.; et al. Interface engineering of Mn_3O_4/Co_3O_4 S-scheme heterojunctions to enhance the photothermal catalytic degradation of toluene. *J. Hazard. Mater.* **2023**, *452*, 131249. [CrossRef] [PubMed]
74. Zhang, X.; Zhu, Z.; Rao, R.; Chen, J.; Han, X.; Jiang, S.; Yang, Y.; Wang, Y.; Wang, L. Highly efficient visible-light-driven photocatalytic degradation of gaseous toluene by rutile-anatase TiO_2@MIL-101 composite with two heterojunctions. *J. Environ. Sci.* **2023**, *134*, 21–33. [CrossRef]
75. Bi, F.; Ma, S.; Gao, B.; Liu, B.; Huang, Y.; Qiao, R.; Zhang, X. Boosting toluene deep oxidation by tuning metal-support interaction in MOF-derived Pd@ZrO_2 catalysts: The role of interfacial interaction between Pd and ZrO_2. *Fuel* **2024**, *357*, 129833. [CrossRef]
76. Zhu, Y.; Li, D.; Ji, C.; Si, P.; Liu, X.; Zhang, Y.; Liu, F.; Hua, L.; Han, F. Non-Thermal Plasma Incorporated with Cu-Mn/γ-Al_2O_3 for Mixed Benzene Series VOCs' Degradation. *Catalysts* **2023**, *13*, 695. [CrossRef]
77. Zhu, D.; Huang, Y.; Li, R.; Peng, S.; Wang, P.; Cao, J.-j. Constructing Active Cu^{2+}-O-Fe^{3+} Sites at the CuO-Fe_3O_4 Interface to Promote Activation of Surface Lattice Oxygen. *Environ. Sci. Technol.* **2023**, *57*, 17598–17609. [CrossRef]
78. Pineda, P.A.L.; Demeestere, K.; Gonzalez-Cortes, J.J.; Alvarado-Alvarado, A.A.; Boon, N.; Devlieghere, F.; Van Langenhove, H.; Walgraeve, C. Effect of inoculum type, packing material and operational conditions on the biofiltration of a mixture of hydrophobic volatile organic compounds in air. *Sci. Total Environ.* **2023**, *904*, 167326. [CrossRef]
79. Barghoth, M.G., Desouky, S.E.; Radwan, A.A.; Shah, M.P.; Salem, S.S. Characterizations of highly efficient moderately halophilic toluene degrading *Exiguobacterium mexicanum* M7 strain isolated from Egyptian saline sediments. *Biotechnol. Genet. Eng. Rev.* **2023**. [CrossRef]

MDPI AG
Grosspeteranlage 5
4052 Basel
Switzerland
Tel.: +41 61 683 77 34

Materials Editorial Office
E-mail: materials@mdpi.com
www.mdpi.com/journal/materials